AF361642

Methods in Dating and Other Applications using Luminescence

Methods in Dating and Other Applications using Luminescence

Special Issue Editor

James K. Feathers

MDPI • Basel • Beijing • Wuhan • Barcelona • Belgrade • Manchester • Tokyo • Cluj • Tianjin

Special Issue Editor
James K. Feathers
University of Washington
USA

Editorial Office
MDPI
St. Alban-Anlage 66
4052 Basel, Switzerland

This is a reprint of articles from the Special Issue published online in the open access journal *Methods and Protocols* (ISSN 2409-9279) (available at: https://www.mdpi.com/journal/mps/special_issues/MDOAL).

For citation purposes, cite each article independently as indicated on the article page online and as indicated below:

LastName, A.A.; LastName, B.B.; LastName, C.C. Article Title. *Journal Name* **Year**, *Article Number*, Page Range.

ISBN 978-3-03928-792-5 (Hbk)
ISBN 978-3-03928-793-2 (PDF)

Contents

About the Special Issue Editor

James K. Feathers has directed the Luminescence Dating Laboratory at the University of Washington in Seattle for 27 years. His research has focused on the application of luminescence in archaeology.

Editorial

Methods and Applications in Trapped Charge Dating

James K. Feathers

Department of Anthropology, University of Washington, Seattle, WA 98195, USA; jimf@uw.edu

Received: 13 March 2020; Accepted: 20 March 2020; Published: 24 March 2020

Abstract: Trapped charge dating is a commonly used chronological tool in Earth Sciences and Archaeology. The two principle methods are luminescence dating and electron spin resonance. Both are based on stored energy produced by the absorption of natural radioactivity in common minerals such as quartz and feldspars and in some biological materials such as tooth enamel. Methodological developments in the last 20 years have substantially increased accuracy and precision. This essay introduces a compilation of papers that offers a taste of recent research into both method and application.

Keywords: luminescence; electron spin resonance; chronology; earth sciences; archaeology

---

Trapped charge dating is a fast developing field that provides chronological and other information, principally in the geological and archaeological sciences. There are two main methods, luminescence dating and electron spin resonance (ESR) dating. Both are based on the storage of energy in certain materials as a function of natural radioactivity. When radiation impinges on such material, for example, quartz and feldspar minerals or tooth enamel, ionization produces detached electrons and electron vacancies, that can move about the crystal lattice. Most of them recombine and return to the ground state, but some become attracted to localized charge deficiencies associated with defects in the crystal. They are "trapped" at these defects until heat or sunlight provides sufficient energy to release them. The trapped charge builds up as a function of time, according to the rate of absorbed radiation. The amount of stored energy is thus proportional to the time when the material was last exposed to heat or sunlight (which cleans out the traps), or in the case of organic material such as teeth and shell, the time of crystalline formation. The accumulated energy in the traps can be related to radiation by calibration with artificial radiation in the laboratory, resulting in a quantity called equivalent dose, which is the amount of radiation dose necessary to produce the amount of trapped charge. Dividing the equivalent dose by the natural dose rate yields an age, or the time since the traps were last emptied. Thus, the time can be determined when ceramics or rocks were last heated, when sediments were last exposed to light (at time of burial) or when teeth or shells formed. This is possible because the long half-lives of the principle components of natural radioactivity mean that the dose rate is, in most cases, effectively constant through time.

Luminescence methods, which are mainly applied to quartz or feldspars, measure the trapped charge by stimulating with heat or light to release the charge. Recombination then produces light, called luminescence, whose intensity is proportional to the amount of stored energy. When stimulated by heat, the resulting signal is called thermoluminescence (TL). When stimulated by light (or more properly by photons), the resulting signal is called either optically stimulated luminescence (OSL)when the stimulation is with visible light or infrared stimulated luminescence (IRSL) when stimulation is with infrared light. ESR measures the trapped charge directly in the traps by inducing absorption resonance between two spin states by microwave radiation in a magnetic field. The amplitude of the resonance is proportion to the number of trapped electrons. ESR if often applied to tooth enamel and shells but also to quartz. Reviews of different aspects of both methods can be found in Rink and Thompson [1].

While these methods first developed during the last half of the 20th century, significant developments in instrumentation, method and application have occurred during the last 20 years, making luminescence, at least, the second most utilized chronological metric in Quaternary science, after radiocarbon dating. Applications have spread beyond dating to studies of provenience, exhumation rates and erosion rates, but have included novel extensions in dating, such as surface dating of rocks, complementing cosmogenic dating methods. Method improvement for both equivalent dose and dose rate have improved both accuracy and precision.

This compilation is an eclectic assortment of papers, which, while not fully representative of the wide scope of trapped charge methods, gives a taste of the range of methods and applications.

Two of the papers deal with techniques, the improvement of which has become necessary as the range of applications broaden. In archaeological and cultural heritage studies, minimum destruction of the record is imperative. Sample collection needs to be done in the least invasive way. Nelson et al. [2] explored the range of options in collecting sediment samples for luminescence measurements by coring, obviating the need for expensive and destructive excavation. Coring is also important for reaching otherwise hard-to-get targets, such as marine and ocean sediments. Obtaining equivalent dose values on single grains has become a major tool in luminescence dating for evaluating the integrity of deposits and dealing with mixed age sediments. This has also put a premium on getting additional information, such as composition (for dose rate determinations, among other things) and shape, from the individual grains measured. Doverbratt and Alexanderson [3] detail methods for transferring grains from the single-grain disks used to measure luminescence to other media, so that information from the same grains can be obtained.

While most research is directed toward determining an accurate measure of the equivalent dose, dose rate measurements are equally important, even if given less attention. Hood and Highcock [4] consider problems in determining the dose rate in complex environments, which are often encountered at archaeological sites. They demonstrate the use of DosiVox, a new computer program for reconstructing the radioactive environment, in this case for pottery vessels from Egyptian monuments. Blackwell et al. [5] discuss the need for intensive sampling to disentangle varied and high dose rates to tooth fragments, in the context of ESR dating, in cave sediments in the Caucasus of southern Russia.

A constant theme in trapped charge studies is the attempt to extend the possible dating range, both for very young samples and for very old samples. Spencer et al. [6], in a study of active fluvial processes in the Amazon River catchment, demonstrated the possibility of obtaining OSL ages as young as 13-14 years. This allows one to understand fluvial dynamics in the context of recent land use changes. The upper dating range of trapped charge method is ultimately defined by saturation of the traps. This occurs later for ESR than luminescence, and later for IRSL of feldspars than the OSL of quartz. The problem of saturation in quartz and the resulting underestimation of age is explored by Groza-Săcuciu et al. [7] in their study of Romanian loess, in the context of using different grain sizes of quartz. This problem is far from resolved in luminescence studies, and the authors present a systematic inquiry into various possible causes of the age underestimation of fine-grain compared to coarse-grain quartz for equivalent dose values of more than 50 Gy.

IRSL of feldspar saturates at a higher level than quartz and so is increasingly turned to for dating older sediments. Feldspar, however, suffers from an athermal loss of signal over time, called anomalous fading. While corrections for fading are possible, they work less well for older sediments. A non-fading signal has been documented for a protocol called post-IRSL IRSL (pIRSL), where a higher temperature stimulation follows a lower temperature one. The higher temperature stimulation taps traps less likely to fade. Zhang and Li [8] present a comprehensive review of the various pIRSL methods and also introduce the possibility of standard growth curves (luminescence versus dose) for feldspars.

Standard growth curves (SGC) are also discussed for quartz by Hu et al. [9]. They construct different SGCs for different groups of quartz grains, measured at the single-grain level. Dating older samples with quartz depends on isolating those quartz grains with high characteristic doses (which define the shape of saturating exponential functions).

The intensity of luminescence and ESR signals has also proved a fruitful subject of study. The intensity of quartz is known to increase with the number of cycles of exposure and burial experienced by the sediment. Thus, samples close to the bedrock source are less sensitive than those which have been transported a long way from the source. Sawakuchi et al. [10] have used differences in sensitivity to differentiate provenience of Amazon sediments. In this paper, they test methods for streamlining the measurement procedure, allowing expanded measurement probabilities, including in situ measurements with portable equipment. Tsukamoto et al. [11] show that the intensity of the ERS signal from oxidized iron increases with age and use this information to date gut strings from historic plucked instruments. For instruments of a known age, the method can determine if the strings are as old as the instrument.

Finally, Zhang et al. [12] discuss the problems of dating fluvial terraces in China. They show that one sample from each of the terraces is not sufficient. Rather, systematic sampling of each terrace is required to see how the terraces evolve, how they relate to each other, and to determine sedimentation rates.

References

1. Rink, W.J.; Thompson, J.W.; Jeroen, W. *Encyclopedia of Scientific Dating Methods*; Springer: Dordrecht, The Netherlands, 2015.

2. Nelson, M.; Rittenour, T.; Cornachione, H. Sampling methods for luminescence dating of subsurface deposits from cores. *Methods Protoc.* **2019**, *24*, 88. [CrossRef] [PubMed]

3. Doverbratt, I.; Alexanderson, H. Transferring grains from single-grain luminescence discs to SEM specimen stubs. *Methods Protoc.* **2019**, *2*, 87. [CrossRef] [PubMed]

4. Hood, A.G.E.; Highcock, E.G. Using DosiVox to reconstruct radiation transport through complex archaeological environments. *Methods Protoc.* **2019**, *2*, 91. [CrossRef] [PubMed]

5. Blackwell, B.A.B.; Kazi, M.F.; Huang, C.L.C.; Doronicheva, E.V.; Golovanova, L.V.; Doronichev, V.B.; Singh, I.K.C.; Blickstein, J.I.B. Sedimentary dosimetry for the Saradj-Chuko Grotto: A cave in a lava tube in the north-central Caucasus, Russia. *Methods Protoc.* **2020**, *3*, 20. [CrossRef] [PubMed]

6. Spencer, J.Q.G.; Huot, S.; Archer, A.W.; Caldas, M.M. Testing luminescence dating methods for small samples from very young fluvial deposits. *Methods Protoc.* **2019**, *2*, 90. [CrossRef] [PubMed]

7. Groza-Săcaciu, S.M.; Panaiotu, C.; Timar-Gabor, A. Single aliquot regeneration (SAR) optically stimulated luminescence dating protocols using different grains-sizes of quartz: Revisiting the chronology of Mircea Vodă loess-paleosol master section (Romania). *Methods Protoc.* **2020**, *3*, 19. [CrossRef] [PubMed]

8. Zhang, J.; Li, S.H. Review of the post-IR IRSL dating protocols of K-feldspar. *Methods Protoc.* **2020**, *3*, 7. [CrossRef] [PubMed]

9. Hu, Y.; Li, B.; Jacobs, Z. Single-grain quartz OSL characteristics: Testing for correlations within and between sites in Asia, Europe and Africa. *Methods Protoc.* **2020**, *3*, 2. [CrossRef] [PubMed]

10. Sawakuchi, A.O.; Rodrigues, F.C.G.; Mineli, T.D.; Mendes, V.R.; Melo, D.B.; Chiessi, C.M.; Giannini, P.C.F. Optically stimulated luminescence sensitivity if quartz for provenance analysis. *Methods Protoc.* **2020**, *3*, 6. [CrossRef] [PubMed]

11. Tsukamoto, S.; Takeuchi, T.; Tani, A.; Miyairi, Y.; Yokoyama, Y. ESR and radiocarbon dating of gut strings from early plucked instruments. *Methods Protoc.* **2020**, *3*, 13. [CrossRef] [PubMed]

12. Zhang, J.; Qiu, W.; Hu, G.; Zhou, L. Determining the age of terrace formation using luminescence dating—A case of the Yellow River terraces in the Baode area, China. *Methods Protoc.* **2020**, *3*, 17. [CrossRef] [PubMed]

methods
and
protocols

Article

Sampling Methods for Luminescence Dating of Subsurface Deposits from Cores

Michelle Nelson [1],*, Tammy Rittenour [1,2] and Harriet Cornachione [2]

[1]   USU Luminescence Laboratory, North Logan, UT 84341, USA; tammy.rittenour@usu.edu
[2]   Department of Geosciences, Utah State University, Logan, UT 84322-4505, USA;
      harriet.cornachione@aggiemail.usu.edu
*   Correspondence: michelle.nelson@usu.edu

Received: 2 October 2019; Accepted: 19 November 2019; Published: 22 November 2019

Abstract: Study of subsurface deposits often requires coring or drilling to obtain samples for sedimentologic and geochemical analysis. Geochronology is a critical piece of information for stratigraphic correlation and rate calculations. Increasingly, luminescence dating is applied to sediment cores to obtain depositional ages. This paper provides examples and discussion of guidelines for sampling sediment core for luminescence dating. Preferred protocols are dependent on the extraction method, sedimentology, core integrity, and storage conditions. The methods discussed include subsampling of sediment in opaque core-liners, cores without liners, previously open (split) cores, bucket auger samples, and cuttings, under red lighting conditions. Two important factors for luminescence sampling of sediment core relate to the integrity of the natural luminescence signal and the representation of the dose rate environment. The equivalent dose sample should remain light-safe such that the burial dose is not reset (zeroed) by light exposure. The sediment sampled for dose rate analyses must accurately represent all units within at least 15 cm above and below the equivalent dose sample. Where lithologic changes occur, units should be sampled individually for dose rate determination. Sediment core extraction methods vary from portable, hand-operated devices to large truck- or vessel-mounted drill rigs. We provide recommendations for luminescence sampling approaches from subsurface coring technologies and downhole samplers that span shallow to deep sample depths.

Keywords: luminescence sampling; sediment cores; augered sediments; portable dark room

---

1. Introduction

Luminescence dating is a technique that provides an age estimate for the last time sediment or cultural material was exposed to sunlight or high heat which resets the luminescence signal [1]. Luminescence ages are calculated by dividing the equivalent dose of radiation the sample received during burial (D_E) by the dose rate environment of the surrounding sediments (D_R). Special sampling and handling methods are required for luminescence samples to prevent light exposure. Routinely, this involves a light-proof metal, black polyvinyl chloride (PVC), or polyethylene (PE) tube that is pounded horizontally into an exposure of sediments. The collection of sediment surrounding the D_E sample for D_R calculation is equally important for accurate age determination. Exposures of sediment from erosional escarpments and human-made excavations (i.e., roadcuts, trenches, soil pits) are commonly limited in availability, requiring the use of mechanical collection of cores through augering and drilling to characterize, describe, and date buried stratigraphy. Literature describing the best practices for luminescence sampling are mostly focused on settings where samples can be collected from exposures [2]. Guidelines for luminescence sampling related to core or auger samples are limited, although specific sediment coring technology and luminescence sampling of sand dunes has been introduced [3]. Additionally, subsampling of stored Ocean Drilling Program sediment cores for luminescence dating has also been well-described [4]. Here, we build on these contributions and describe best practices for collection of luminescence samples from a range of core and subsurface sediment collection methods. We also review common coring methods (augering, drilling) and how they should be outfitted to protect the natural luminescence signals.

Luminescence dating utilizes trapped charge (electrons) that accumulate in defects in quartz or feldspar minerals (sand or silt grain) due to exposure to ionizing radiation to calculate the last time that sediment was exposed to sunlight or heat [1,5]. Following burial or removal from heat, the grain acquires trapped charge proportional to the duration of burial and the radioactivity of the surrounding sediments, plus incident cosmic radiation [1]. The applicable age range for luminescence dating is dependent on the dose rate conditions and mineral properties and typically ranges from 100 years to ≥200,000 years for optically stimulated luminescence (OSL) dating of quartz and up to 500,000 years for infrared stimulated luminescence (IRSL) dating of potassium feldspar [6], however older ages can be obtained using other luminescence methods [7]. OSL and IRSL are advantageous in many settings given that quartz and feldspar are present in most surficial deposits. Moreover, these methods directly relate time with sediment deposition, unlike other methods that use radiometric decay of affiliated material (e.g., radiocarbon dating of charcoal).

Important factors for reducing uncertainty in luminescence dating are adequate sunlight exposure prior to sediment deposition, limited post-depositional sediment mixing, and stable dose rate conditions. Partial bleaching is the incomplete solar resetting of a luminescence signal and it is typical of high-turbidity water columns [8], glacially-sourced sediment [9], or subaqueous reworking processes [4]. When overlooked, partial bleaching can lead to depositional age overestimation. Bioturbation can cause mixing of different-aged sediments and/or deviations from a stable dose rate environment over time, limiting the accuracy and precision of luminescence ages [1]. Pedogenic alterations, such as oxidation of iron and buried organic material, translocation of particles, dissolution and precipitation of evaporites, and shrink-swell processes, also lead to sediment mixing and dose rate heterogeneity. Sediment with clear signs of bioturbation or pedogenic alteration should be avoided for luminescence sampling. Beware that in sediment cores and under subdued lighting, these features may be difficult to identify. Disturbance from coring methods and subsurface extraction can cause additional mixing, uncertainty in dose rate conditions, and loss of the of the original luminescence signal.

Subsurface samples have further complexity in their dose rate environment due to changes in groundwater levels and diagenetic processes. In the vadose zone, water content fluctuations impact the effective dose rate to which the sample was exposed, as water does not emit radiation and attenuates the dose absorbed by the grain [1]. Water saturation can also lead to mobilization of soluble radioelements and disequilibrium detected in the U-series decay chain, causing the dose rate environment to change over time [10]. Further complications in varying dose rate are found in lacustrine and marine settings where sediment compaction in the upper 5 m exponentially decreases sediment density and water content [11]. All of the factors mentioned here can impact the luminescence age estimate, and sampling details are provided for mitigating and controlling for unwanted sample material from sediment cores.

2. Drilling and Augering Methods

Sediment character and site conditions will dictate the optimum auger or drilling setup and recovery. Sedimentologic considerations are burial depth, grain size (cohesion), and compaction (water content and induration) of the sediment being extruded, as well as site accessibility and driller/operator availability [12]. If solid core liners are available, then a light-proof (opaque) core liner/tube/barrel should be selected such as aluminum or steel (for deeper core depths, >7 m) or dark PCV or PE for shallower cores (1–2 m). For best sediment core recovery in subaqueous settings, the core length to core diameter ratio should be ≥6:1 to maintain enough internal friction to keep the core intact [13]. Larger diameter cores are preferred for luminescence sampling after extraction because of the greater volume of sediment available for dating after removal of the outer sediment adjoining the core liner.

Generally, drilling method choice will depend on geologic setting, desired core length, and sample depth. Coring mechanisms well-suited for collecting samples specifically for luminescence dating include hand-augering (human or mechanized-power), vibracoring, sonic and percussion drilling, in addition to rotary drilling in limited use. Removal of the core or cuttings will be aided by a drill stem liner, core barrel, bucket, bailer, fluid, or air. For deeper cores that require flight extension and further penetration, casing, mud, or water may be used as means to stabilize the open borehole. Some of these methods may be combined or modified to fit specific sedimentologic, hydrogeologic, and depth objectives [13–16].

Key goals for successful coring and luminescence dating of core sediment are related to the preservation of original stratigraphy and the burial dose (natural luminescence signal). Sedimentary structures should be visible to help select the most suitable sediments for the D_E sample. Note that at least two D_R samples will be collected for each D_E sample (above and below) and undisturbed sediments are important for selecting intervals with intact sediments for all three sample intervals (two D_R and one D_E sample). In addition to D_R considerations, sediment disturbances can also mix different aged deposits. This can be minimized by limiting auger rotations or percussive drives. Minimizing loss of material during retrieval and post-extraction will also maintain the correct sample/core depth and allow accurate cosmic-dose contribution to the total D_R calculation. Recovery may be improved with the use of a vacuumed sample chamber (piston), or high internal friction (fine-grained sediment). Luminescence dating requires the sediment be kept in an opaque core liner, core-box, or bag until it can be subsampled in a darkroom laboratory. It is critical that the D_E sample be collected under safe lighting conditions, otherwise the sample is at risk of age underestimation due to loss of the natural signal [1]. We will discuss further methods for dealing with unlined and split (sunlight-exposed) cores in Section 3.

2.1. Augers

Shallow-depth subaerial deposits (<10 m) can be sampled by using hand-augering (rotating bit and core barrel) in non-indurated sedimentary environments such as sand dunes, fluvial terraces, sandy soil, and loess. Compared to more powerful drilling rigs, soil augers are generally cost-effective, transportable to remote field sites, and some require only one person to operate [3]. Additionally, hand-auger coring systems are desirable for use in sensitive sites, where minimally invasive sample extraction methods are required. Soil augers typically use human power to advance the bit by rotating a T-shaped handle to which rod extensions are attached. These serve to extend the auger bit to the desired sample depth. The main disadvantages of soil augers are the length of core section (~0.3 m), limited depth range, and the destruction of sedimentary structures. The van der Staay suction corer is specifically designed to sample saturated sand up to 30 m depth, and intact core section lengths recovered range from 2.5 to 5 m [17]. More powerful motorized augers can achieve extraction at greater depths (60 m), particularly when a mechanized hydraulic pump is attached to the auger head and a powered hoisting apparatus assists drill stem removal [3]. Truck or track-mounted hollow stem augers are commonly used in geotechnical or water well drilling through unconsolidated sediments. The diameter of the stem ranges from about 6 cm to 15 cm, and each drive is ~1.5 m long. An inner liner painted black or split spoon corer may be placed inside the hollow stem to extract undisturbed core sections for luminescence dating.

Auger Sampling Methods for Sand Dunes

Shallow sand-rich deposits are often the preferred environment for luminescence dating applications, and methods for hand-augering in shallow sandy deposits are described in detail. The sampling site should be determined in the field where bioturbation (i.e., burrowing, root zones, human disturbance) can be avoided. Loose, non-cohesive surficial sediment should be removed to form a platform that reduces the potential for surface grains to fall into the auger hole. The thickness of any removed sediment should be noted to maintain accurate sample depth records. A PVC pipe can be placed at the auger location to act as a casing and prevent contamination of samples by surface sediment falling into the hole, as seen in Figure 1. This will also provide a reference platform to determine sample depth for both the D_E and D_R components in the OSL/IRSL sample.

Figure 1. Hand-augering in Kanab Sand Dunes, southern Utah. Polyvinyl chloride (PVC) casing is inserted in top of auger hole to provide a reference platform for depth control and prevents influx of surface sediment downhole. The drill stem is advanced through the casing to the desired depth.

Bucket augers can be used to collect sediment samples as the auger hole is advanced to greater depths during exploration. Various designs ranging from a closed bucket, as seen in Figure 2A,B, to an open catcher, as seen in Figure 2C, are available and the best choice will be dependent on the sediment characteristics and environmental setting. Each bucket auger drive should be saved in stratigraphic order (by depth) as it may be used for part of the D_R sample, as seen in Figure 2C. For dose rate sediment, collect ~300 g of material from the auger-drive(s) within 15 cm above and 15 cm below the OSL sample interval. When the target depth for the OSL sample is reached, switch out the bucket auger head for the OSL sample head, as seen in Figure 2A. Sample collection heads should contain an inner opaque liner into which the OSL sample can be captured without light exposure. Use of foam inserts keep the sediment packed while it enters the OSL tube and act as a barrier when the sediment reaches the end (top) of the OSL tube. Once extracted, the sampler is removed from the drill stem and the inner metal tube is capped on both ends to keep the sample intact and to prevent light exposure.

Figure 2. Shallow-sediment auger set-up for luminescence sampling. (**A**) Optically stimulated luminescence (OSL) auger head with T-shaped handle. The enclosed OSL auger head is fitted with an aluminum liner to contain the OSL sample in a light-proof tube. (**B**) T-shaped handle with extension rods are used to drive bucket auger. (**C**) Each drive requires sediment removal from hole, placed on the surface in stratigraphic order (by depth) until desired OSL sampling depth is reached. This material may be needed as part of the D_R sample.

2.2. Vibracorers

Vibracoring utilizes steady and high-frequency vibrations to allow the sampler to move downward through subsurface deposits [13,18]. Vibracoring works best in fine-grained saturated sediments. Portability of vibracore devices can range from hand-held battery-operated corers to large vessel-mounted deep-water corers. Use of a steel core barrel is ideal for keeping the core light-proof for subsequent luminescence sampling. Undisturbed sediment cores greater than 10 m long may be obtained in finer-grained settings, while full recovery may be difficult in well-sorted water-saturated sands. Due to the vibration and rearranging of sediment particles, the amount of compaction should be calculated so that sample depth is accurate. For luminescence sampling of vibracored sediment, subsections of the core lengths may be cut and sent directly to the luminescence laboratory. Creation of a dark space is required if luminescence sampling occurs prior to shipment to the laboratory. Details for this in the field or remote dark room and subsampling are discussed below.

2.3. Piston and Gravity Corers

Gravity submersible and piston corers refer to similar coring methods and are gravity-driven, single-drive bottom penetrators. Obtainable core depth is based on the free-fall velocity and empirically calculated from water depth. Core length for gravity drivers is on the order of 1–5 m [12]. Piston corers can operate up to several hundred meters water depth depending on the drive mechanism [19]. The main advantage to piston-style corers is the partial vacuum created by the piston, which helps keep the sample intact [20]. Sediment does not enter the core barrel until the desired depth is reached [13]. These can operate via a rod or cables, noting that extension rods add weight and may limit the operational water depth [12]. Lake-bottom sediment corers are commonly used in limnological studies where preservation of stratigraphy and pollen are integral to sample analysis. The Livingstone-type drive rod piston corer is used for underwater operations and sediment sampling depths up to several meters are obtainable [21]. These techniques may be suitable for luminescence core sampling as well, preferably with core barrels and liners that are metal or black. If necessary, clear liners may be spray-painted black and core should be stored in a dark setting.

2.4. Percussion Drivers

Percussion drivers are a mechanized hammer that is restricted to vertical motion, typically used to advance the tube with core catcher into the sediment or rock by percussive force or direct push. These types of rigs are available as compact units for limited access situations up to industrial-sized track or truck-mounted rigs. For use in unconsolidated or semiconsolidated sediment above the water table, large diameter (15–25 cm) hollow-stem percussion drill rigs can drive through the subsurface to 30 m or more.

2.5. Rotary-Vibratory (Sonic) Drill

Sonic drilling combines rotary drilling and vibracoring mechanics in a dry drilling technique capable of extracting large diameter (30 cm) continuous core sections, up to 100 m depth. Core section length and recovery is dependent on the cohesiveness of drilled material. If clay content is high or the sediment is semi-indurated, then whole core sections on the order of 3 m per drive can be obtained [22]. For sonic drilling, the drill bit and auger stem are rotated as the auger is pushed downward [23], each flight is brought up to extrude the cored sediment. For luminescence dating, it is best for the extruded material to be stored in black plastic liners; however, the innermost core sediment may be sampled for the D_E even if cores and are stored in clear plastic liners if they have not been disturbed by liquefaction or desiccation, as discussed below.

2.6. Rotary Drill

Rotary drilling is essential for lithified or highly consolidated sediment and bedrock, and commonly produces rock cuttings which are not recommended for luminescence dating in most cases. Commonly, rotary drilling is used with water or mud to prevent borehole caving and to lift the cuttings out of the hole while the rotating auger head cuts through the rock. Borehole drilling that uses bentonite-based drilling additives should be avoided for luminescence dating. The major concern with type of additive is that it will be mixed with the dose rate material and will contribute an unknown amount of radiation to those measurements. Additionally, light-exposed grains or small clasts from further up the borehole could get mixed in the D_E sediment during retrieval and extraction.

Hydraulic rotary uses fluid circulation down the drill stem and return along the outside, while reverse rotary has fluid injected down the well and cuttings are brought up through the drill stem [24,25]. Each drive may be up to 10 m in length, allowing for drilling through several km of rock through either rod (conventional) or wireline extension methods [13]. Though not an endorsed application here, previous work has shown that cuttings may be large enough for luminescence characterization, particularly if there is a high content of silt and clay and drilling speed is reduced [26]. Telescoping casing can be used as a replacement for drilling mud to keep the borehole open, but at greater cost and decreasing core diameter with increasing depth [27]. An inner core barrel and special bit may be used if whole rock cores are desired for OSL dating or thermochronology [28]. OSL thermochronology is an innovative utility in neotectonics for quantifying rock exhumation and cooling rates through the Earth's crust [29]. The concept for sampling remains the same, that is, the innermost material which has not been exposed to light is used for luminescence measurements.

3. Core Sampling Methods for Luminescence Dating

The methods outlined here discuss strategies for sample collection from cores and auger samples of differing integrity. First we discuss how to sample for OSL/IRSL dating on whole (unsplit) intact sediment core, weighing such factors as sedimentary and stratigraphic interpretation prior to luminescence sampling and keeping the core intact and light-safe prior to and during sampling. Considerations and protocols for OSL sampling of cores previously exposed to light will also be discussed. Ideally, two cores immediately adjacent to each other should be collected. The first will be the pilot core, which will be analyzed under normal lighting conditions to fully describe the sedimentary units and select target depths for the luminescence samples. The second core would then be extracted in a darkroom setting (or at least at the D_E sample target depths) such that intact samples can be extracted and sent to the luminescence laboratory of choice. The sediment core used for luminescence sampling should be kept in a light-proof casing and split (opened) under safe lighting conditions in a dark lab, as recommended below.

3.1. Whole (Unsplit) Cores

Whole or unsplit cores stored in light-proof core liners are the most suitable for luminescence dating of sediment and rock cores. Figure 3A shows an example of a hollow-stem auger system that can accommodate core liners. Transparent liners may be helpful in cases where a companion core is not possible, and the target sample range is only a few cm thick (i.e., single-event strata). Cores retrieved in transparent liners may be sampled for luminescence dating following proper sampling protocol to remove exposed grains on the outer perimeter of the core [30]. The PVC or PE liner and core may be cut into smaller section lengths at desired sample targets and sent to the luminescence laboratory for D_E sample extraction under subdued lighting conditions (see below). The length of the core section sampled for D_E and D_R will depend on several factors, as seen in Figure 3. First, the D_E sample should contain quartz and/or potassium feldspar-rich sand (>20% in the 63–250 µm range). The length of the core section sampled should ensure at least 10 g of datable sediment will make it through mineralogic processing. Roughly 2–10 cm intervals will suffice in larger diameter cores, as shown in Figures 3 and 4, though greater length may be required in quartz and potassium feldspar-poor environments, such as evaporite rich or other fine-grained deposits. Water can be used to aid sample extraction in dry sands with little cohesion or indurated fine-grained sediment to ensure layers are removed uniformly and to help retain sample integrity. Unconformities should be avoided to help reduce the complexity of the dose rate environment and inclusion of different-aged depositional units in the D_E sample. The outer 1 cm of sediment in contact with the core liner should be removed prior to D_E sampling due to light exposure (in clear core liners) as well as the potential inclusion of sediment from different depths due to drag along the core liner. This outer material may be used for the D_R sample if there is minimal disturbance on the outer rim of the core. However, it should be discarded if there is evidence of friction and sediment smearing along the exterior of the core. Sampling for dose rate can become complex if multiple layers of varying grain size and/or lithology exist above or below the D_E sample. If different units are present within 15 cm above or below the D_E sample, each unit should be subsampled with distance to D_E sample noted, as seen in Figure 3C. A distance weighting will be used to average the radio-isotope contribution to the D_R because the sediments closest to the D_E sample will have a larger dose-rate contribution than sediments further away [1].

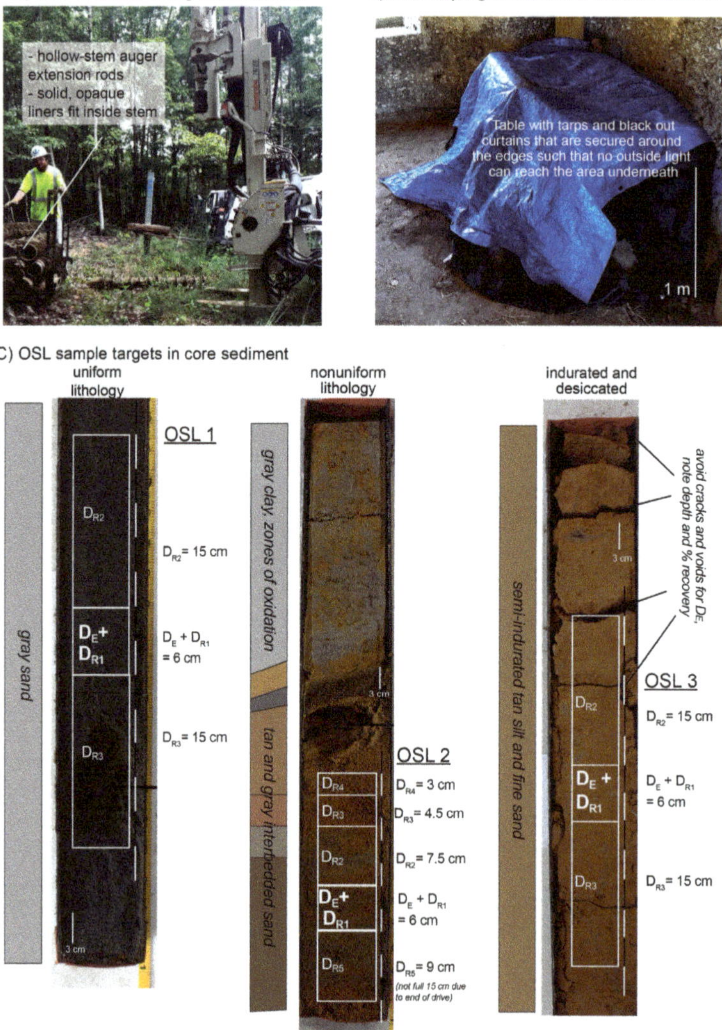

Figure 3. Whole core extraction and sampling with opaque liners. (**A**) Hollow-stem auger drill rig with flight extensions. Metal or black PVC tubing fits in opening for light-safe core sediment extraction. Photo credit: Abby Conklin-Muchnick. (**B**) Example of in-the-field dark space for subsampling core sediment. Tarps and blackout drapery are layered over folding table and edges are secured such that no outside light is visible under the material. Red LED headlamps can be used for lighting under the tarps. (**C**) Core photos of uniform, nonuniform sedimentology and stratigraphy, as well as indurated/desiccated core. Photo credit: Benjamin DeJong [16].

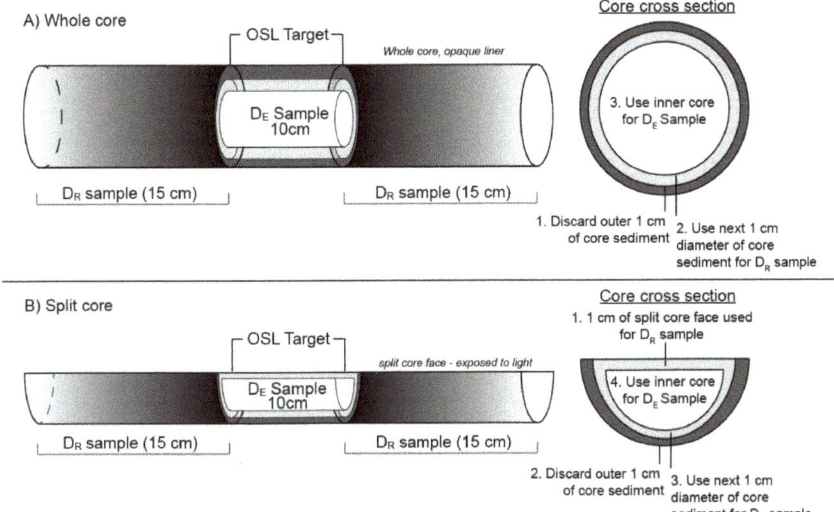

Figure 4. Idealized target sediment for D_E and D_R samples in luminescence dating applications to core sediment. (**A**) Whole core example preferably collected in opaque liner. The outer 1 cm is usually discarded as this may contain grains from unwanted sedimentary units above the target depth. The next 1 cm of core diameter used as dose rate sample for target depth. The innermost (light-proof) sediment is used for D_E sample processing. The overlying and underlying 15 cm of core is sampled for D_R. Subsamples may be required if sedimentology is vastly different from the target D_E sample. (**B**) Split core example. If the core face has been exposed to light, 1 cm of sediment at the split core face is used for the D_R sample. The outer 1 cm rim of core material is typically discarded as in example A. The inner core is used for D_E sample processing. The overlying and underlying 15 cm of core is used as the D_R samples. Subsamples may be required if sedimentology is vastly different from the target D_E sample. Desiccation cracks in the core and zones of liquefaction should be avoided.

3.2. Opened Cores and Cuttings Exposed to Light

Archived or stored sediment cores not originally intended for luminescence dating may also be sampled for OSL/IRSL if the following conditions are held. The core must remain in original stratigraphic order and depth should be known, along with percent of core recovery for proper cosmogenic dose rate calculation. If major desiccation has occurred while in storage, large cracks may form and can allow light to penetrate through the inner core and diminish the luminescence signal, as seen in Figure 3C. Additionally, the water content of a core in storage will not be representative of burial moisture conditions, a simple method for retrospectively estimating water content on desiccated core samples may be followed [31]. The outer 1–2 cm from each edge (exposed to light) will typically be removed in the dark lab, though it has been shown that <1 mm of sediment may be enough to shield the underlying material from light penetration in fine-grained sediments [4]. The exposed sediment can be added to the D_R sample material, which as noted above should come from units within 15 cm of the D_E sample, as seen in Figure 4B. When the outer core rim is contaminated with sediment from overlying units as indicated by smearing on outside of core, this material should be discarded. Recent exploratory work on soil and saprolite cores in the eastern US were successfully sampled for OSL/IRSL analyses [22], as seen in Figure 5A–E. For this work, three-meter sections of 10 cm diameter cores were extruded into clear plastic liners and stored outdoors. The methods for D_E and D_R subsampling included working with the core sections in the luminescence laboratory to:

1. Remove and discard 1–2 cm of clay smear from the outer diameter of the core.
2. Cut core in half to preserve one side for non-OSL analyses.

3. Collect sediment from the split face (inner material) of core for D_E processing, as seen in Figure 4B. In this case, the split core face material was safe to collect for the D_E sample as it was opened (split) in the dark room.

4. Collect remaining sediment for water content and D_R samples.

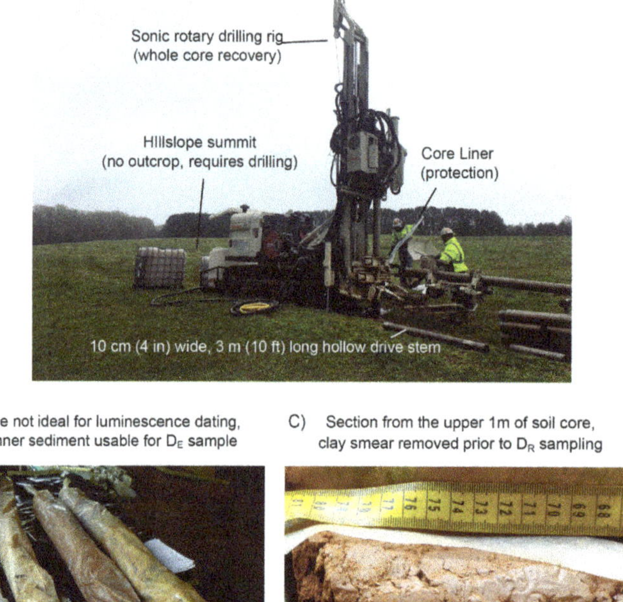

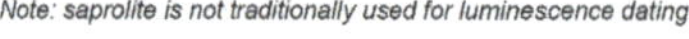

Figure 5. (**A**) Sonic-rotary drill rig at hillslope summit. Photo credit: Martha Eppes. (**B**) Yield, including intact soil and saprolite cores that were stored in plastic liners and outdoors. While not ideal, these conditions are not an impediment to luminescence geochronology [22]. (**C**) The upper 1 m of core shows signs of sediment and soil smearing on the sides, this material should be removed prior to dose rate sample collection. (**D**) Indurated saprolite, no smearing. (**E**) The dose rate sample will include the outer 1–2 cm from the sides, top, and bottom of the sample section. The inner most material is used for the equivalent dose sample.

At times subsurface sediment and rock are collected in the form of cuttings and chips through rotary drilling. Although not ideal, if these materials are large enough, they can be utilized for luminescence dating. For example, 0.25–1 cm cuttings made available through boreholes drilled for a basin analysis study were used for luminescence characterization using IRSL [26]. Given the very limited amount of material available, full processing and mineralogic refinement to isolate potassium feldspar or quartz was not feasible. Additionally, after removing the outer few mm of exposed grains there was not enough material to make the chemical measurements needed for dose rate calculation and thus apparent ages were reported [26].

3.3. In-the-Dark Field Sampling and Portable Luminescence Measurements on Core

Night-time subsampling or the construction of a field-based or off-site dark-room shelter may be required in some situations, for example if the core is needed for other analyses, or to reduce shipping costs. If night-time working conditions are not possible or suitable due to site restrictions based on safety, access or artificial lighting then cores should be taken to a windowless room in a building that is large enough to accommodate the core sections. For more remote field sites, a tent or some similar structure such as a very large cardboard box or folding table will suffice. This will support the layers of tarps and blackout curtains and sheeting needed to create the condition of total darkness required for selecting the luminescence samples, as seen in Figure 3B. Note that it can be ~5 °C (10–15 °F) warmer with little airflow under the blackout materials, so caution should be taken under extreme heat.

Once under night-time or darkroom conditions, only light sources with wavelengths of amber (~590 nanometers, nm) and red (~700 nm) light should be used to illuminate the work area. All other wavelengths of light, particularly those with shorter wavelengths (blue and white light or sunlight) will rapidly destroy the luminescence signal stored in the sediment. Additionally, near-infrared and infrared light sources will remove the luminescence signal stored in potassium feldspar (~890 nm). Common safe-light sources include red bike lights (only those with red light-emitting diodes (LEDs)). White LEDs with a red filter are not recommended for this purpose. It is important that these lights are only used as indirect light-sources and not directly shined on to the sediment core. Alternatively, if power is available at the sample site, a shop light with a red lightbulb (8 watt) can be used. Again, this light source should not be directed on the sediment core and should serve as an ambient light source for the space. It not necessary for the D_R sample to be collected under dark conditions, unless there is concern some D_R material might be needed for D_E processing due to a limited amount of datable mineral grains in the sample.

A well-designed, yet underutilized tool for rapid continuous-core stratigraphic analysis is the scanning luminescence reader [32]. The reader and sediment core are placed under total darkness, and the core moves under the photomultiplier reader by a motorized bed. The natural luminescence signal is measured along the entire length of the core [32], with an output similar to a gamma log. Advantages to this unique whole core analysis include: (1) it is nondestructive such that other analysis can take place after it is performed; (2) it can aid visualization of stratigraphic breaks at the mm scale; and (3) it displays zones with higher luminescence intensity to indicate better sedimentary targets for luminescence sampling (or those deposited under poor bleaching conditions as they too can have large luminescence response to stimulation) [32]. Note that this apparatus must be constructed individually. To our knowledge, it cannot be purchased as an assembled piece of equipment and has no professional installation option, requiring some knowledge of basic electrical engineering, blue or IR LEDs, a photomultiplier tube, a motorized core bed table (5 mm/sec), and sensor cards.

A portable luminescence reader has recently come on the market [33]. The portable unit utilizes much smaller sample sizes than those used in traditional luminescence dating, provides rapid luminescence measurements, and may be transported to core storage facilities. High-resolution subsampling (i.e., every 5–10 cm) of sediment cores can be run through the portable luminescence reader to rapidly identify stratigraphic changes and sedimentary targets to sample for full luminescence analysis. These measurements must be carried out in a darkroom facility if used on site.

4. Uncertainties and Mitigations

The systematic uncertainties in luminescence dating of core sediment are the same for noncore OSL/IRSL samples and are related primarily to instrument calibration and dose rate conversion factors [1]. Random uncertainties are sample specific and include pre-, syn- and post-depositional processes, OSL/IRSL and radioelement measurement uncertainties and sampling errors. For this reason, the random error on OSL/IRSL samples collected from cores may be greater than for traditional samples collected from well-exposed and carefully selected outcrop settings.

One contribution to D_E uncertainty is partial bleaching of a previously acquired burial dose that was not fully reset prior to deposition [34]. Poor luminescence zeroing can cause age overestimation and occurs in high-energy or extremely turbid depositional environments, such as subaqueous slumping, marine storm deposits, alluvial flood packages, proglacial, and subglacial deposits where sediment concentration is high and light penetration is low. Single-grain dating [35] and use of advanced statistical methods such as the Minimum Age Model [36] mitigate this effect by helping to identify the population of individual-grain D_E values that were fully reset by sunlight. Additional contribution to D_E uncertainty occurs with post-depositional mixing of different-aged sediment through liquefaction and resuspension of benthic sediments, bioturbation [37], and soil processes [38]. These features are often identifiable through detailed sedimentology (i.e., krotovina identification), chronostratigraphy (i.e., age reversals) and D_E statistics (i.e., overdispersion, skewness). Other sources of D_E uncertainty are inherent to the sediment and are less impacted by core sampling methods, including micro-dosimetry, and differences in luminescence sensitivity and signal components between grains from the same sample.

Dose rate uncertainty can be more complex for OSL/IRSL samples collected from sediment core than those collected from outcrop exposures. Water content in sediment pore spaces plays a key role in radiation attenuation, that is, as water content increases the dose rate is effectively reduced [1]. Samples in the vadose zone often have highly variable annual water content due to seasonal fluctuations in the water table, as well as longer-term climate-related factors. Additionally, soil-moisture retention factors like aspect, vegetation type, canopy cover, and geology play a key role in water content. *In situ* moisture content can be representative of average burial water content conditions; however, it is also likely that *in situ* samples over or underestimate average moisture conditions if there has been a recent precipitation event or long period of no precipitation. Additionally, sediment compaction will reduce pore spare and lower the *in situ* water content over time. Mean water state can be projected from soil moisture classification mapping to select value(s) of average water content for use in the D_R calculation [39]. Deposition and erosion impact paleo-moisture conditions by driving changes in base-level. Moisture samples from exhumed deposits, or coastal environments that have been through multiple cycles of marine transgression and regressions may not be representative of mean-state moisture content and might require a retrospective approach to estimate water content [16]. Dose rate modelling is required in cases like these or where a single value will not effectively represent the complex moisture history of a subsurface sample.

Further uncertainties in the dose rate measurements and calculation from sediment cores can arise from the presence of organic material, solubility of minerals, and pedogenesis. In the presence of large quantities of buried organic material, radioactivity is absorbed and attenuated [40]. For saturated sediments, decay chain disequilibrium can be an issue when radioelements are water soluble and mobilize, creating chemical excesses or deficits that do not reflect past burial chemistry [41]. Soluble minerals (i.e., calcite, halite) can precipitate out of solution in the vadose zone and change the overall dose rate environment over time, usually lowering the dose rate as they are not highly radioactive [42,43]. Additionally, pedogenic processes such as eluviation (leaching) and translocation add uncertainty in dose rate calculations for soil samples as deep weathering will chemically alter and physically move soluble and reactive elements and minerals down profile [44].

5. Summary

Sediment cores are essential for reconstructing Quaternary geologic history, and luminescence dating is at the forefront as a commonly used tool for building such geochronologies. Luminescence dated core sequences are used in (but not limited to) reconstructing flood chronologies, shoreline development, sea level change, dune field evolution, soil development, basin analysis, and paleoseismology. Coring mechanisms and devices are more readily available, and some are novice-user-friendly, in addition to being more compact for use in remote and ecologically-sensitive sites. It is possible to date sand or silt in previously collected sedimentary cores in storage using OSL/IRSL, given a few conditions are met. Knowledge of the basic uncertainties outlined here should help investigators target better D_E and D_R samples to meet the growing demand for better age resolution and more precise geochronology from sediment cores.

Author Contributions: Conceptualization, M.N. and T.R.; methodology, M.N., T.R., H.C.; writing—original and revised draft preparation, M.N., H.C., T.R.; writing—review and editing, H.C., T.R., M.N.; visualization, M.N., H.C.; supervision, T.R.

Funding: This research received no external funding.

Acknowledgments: Core samples shown in Figure 5 were provided by Martha Eppes, UNC-Charlotte. Core and drilling photos provided by Martha Eppes (UNC-Charlotte), Abby Cocklin-Muchnick (ECS Mid-Atlantic, LLC), and Benjamin DeJong (VHB). We are thankful for the suggestions made by three anonymous reviewers which helped shape the final version of this article.

Conflicts of Interest: The authors declare no conflict of interest.

References

1. Aitken, M.J. *An Introduction to Optical Dating: The Dating of Quaternary Sediments by the Use of Photon-Stimulated Luminescence*; Oxford University Press: New York, NY, USA, 1998; pp. 1–267.
2. Nelson, M.S.; Gray, H.J.; Johnson, J.A.; Rittenour, T.M.; Feathers, J.K.; Mahan, S.A. User Guide for Luminescence Sampling in Archaeological and Geological Contexts. *Adv. Archaeol. Pract.* **2015**, *3*, 166–177. [CrossRef]
3. Munyikwa, K.; Telfer, M.; Baker, I.; Knight, C. Core drilling of Quaternary sediments for luminescence dating using the Dormer Drillmite^{TM}. *Anc. TL* **2011**, *29*, 15–24.
4. Armitage, S.J.; Pinder, R.C. Testing the applicability of optically stimulated luminescence dating to Ocean Drilling Program cores. *Quat. Geochronol.* **2017**, *39*, 124–130. [CrossRef]
5. Huntley, D.J.; Godfrey-Smith, D.I.; Thewalt, M.L.W. Optical dating of sediments. *Nature* **1985**, *313*, 105–107. [CrossRef]
6. Rhodes, E.J. Optically Stimulated Luminescence Dating of Sediments Over the Past 200,000 Years. *Annu. Rev. Earth Plan. Sci.* **2011**, *39*, 461–488. [CrossRef]
7. Ankjærgaard, C.; Guralnik, B.; Buylaert, J.-P.; Reimann, T.; Yi, S.W.; Wallinga, J. Violet stimulated luminescence dating of quartz from Luochuan (Chinese loess plateau): Agreement with independent chronology up to ~600 ka. *Quat. Geochronol.* **2016**, *34*, 33–46. [CrossRef]
8. Jain, M.; Murray, A.S.; Bøtter-Jensen, L. Optically stimulated luminescence dating: How significant is incomplete light exposure in fluvial environments? *Quaternaire* **2004**, *15*, 143–157. [CrossRef]
9. Berger, G.W. Trans-arctic-ocean tests of fine-silt luminescence sediment dating provide a basis for an additional geochronometer for this region. *Quat. Sci. Rev.* **2006**, *25*, 2529–2551. [CrossRef]
10. Guibert, P.; Lahaye, C.; Bechtel, F. The importance of U-series disequilibrium of sediments in luminescence dating: A case study at the Roc de Marsal Cave (Dordogne, France). *Radiat. Meas.* **2009**, *44*, 223–231. [CrossRef]
11. Kadereit, A.; DeWitt, R.; Johnson, T.C. Luminescence properties and optically (post-IR blue-light) stimulated luminescence dating of limnic sediments from northern Lake Malawi- Chances and limitations. *Quat. Geochronol.* **2012**, *10*, 160–166. [CrossRef]
12. Frew, C. Coring Methods. In *Geomorphological Techniques (Online Edition)*; Cook, S.J., Clarke, L.E., Nield, J.M., Eds.; British Society for Geomorphology: London, UK, 2014; Chapter 4; Section 1.1; pp. 1–10.

13. Skilbeck, C.G.; Trevathan-Tackett, S.; Apichanangkool, P.; Macreadie, P.I. Chapter 5 Sediment Sampling in Estuaries: Site Selection and Sampling Techniques. In *Developments in Paleoenvironmental Research: Applications of Paleoenvironmental Techniques in Estuarine Studies (eBook)*; Weckström, K., Saunders, K.M., Gell, P.A., Skilbeck, C.G., Eds.; Springer: Dordrecht, The Netherlands, 2017; pp. 89–120.

14. Bristow, C.S.; Lancaster, N.; Duller, G.A.T. Combining ground penetrating radar surveys and optical dating to determine dune migration in Namibia. *J. Geol. Soc. Lond.* **2005**, *162*, 315–321. [CrossRef]

15. Bristow, C.S.; Duller, G.A.T.; Lancaster, N. Age and dynamics of linear dunes in the Namib Desert. *Geology* **2007**, *35*, 555–558. [CrossRef]

16. DeJong, B.D.; Bierman, P.R.; Newell, W.L.; Rittenour, T.M.; Mahan, S.A.; Balco, G.; Rood, D.H. Pleistocene relative sea levels in the Chesapeake Bay region and their implications for the next century. *GSA Today* **2015**, *25*, 4–10. [CrossRef]

17. Wallinga, J.; van der Staay, J. Sampling in waterlogged sands with a simple hand operated corer. *Anc. TL* **1999**, *17*, 59–61.

18. Lanesky, D.E.; Logan, B.W.; Brown, R.G.; Hine, A.C. A new approach to portable vibracoring underwater and on land. *J. Sediment. Res.* **1979**, *49*, 654–657. [CrossRef]

19. Chambers, J.W.; Cameron, N.G. A rod-less piston corer for lake sediments: An improved, rope-operated percussion corer. *J. Paleolimnol.* **2001**, *25*, 117–122. [CrossRef]

20. U.S. Army Corps of Engineers. *State-of-the-Art of Marine Soil Mechanics and Foundation Engineering*; U.S. Army Corps Engineers: Washington, DC, USA, 1972.

21. Myrbo, A.; Wright, H.E. An introduction to Livingstone and Bolivia coring equipment. *Limnol. Res. Cent. Core Facil. SOP Ser.* **2008**, *Draft v.3.1*, 12p. Available online: http://lrc.geo.umn.edu/laccore/assets/pdf/sops/livingstone-bolivia.pdf (accessed on 15 August 2019).

22. Nelson, M.S.; Eppes, M.C.; Rittenour, T.M. Luminescence signals from soil and saprolite in deeply weathered profiles—Intriguing new results from the piedmont of North Carolina (Invited Presentation). In Proceedings of the GSA Annual Meeting, Phoenix, AZ, USA, 22–25 September 2019. [CrossRef]

23. Geoprobe. Available online: https://geoprobe.com (accessed on 10 July 2019).

24. PeaceCorps. Well Construction: Hand Dug and Hand Drilled. Information Collection and Exchange 1982, Manual M-9. Available online: http://ces.iisc.ernet.in/energy/water/paper/drinkingwater/wellsconstruction/ (accessed on 15 August 2019).

25. State of Hawai'i. Section 5, Field Collection of Soil and Sediment Samples. In *TGM for the Implementation of the Hawai'i State Contingency Plan*; State of Hawai'i: Honolulu, HI, USA, 2016. Available online: http://www.hawaiidoh.org/TGM.aspx?p=0500a.aspx (accessed on 15 August 2019).

26. Hardy, F.; Lamothe, M. Quaternary basin analysis using infrared stimulated luminescence on borehole cores and cuttings. *Quat. Sci. Rev.* **1997**, *16*, 417–426. [CrossRef]

27. Preusser, F.; Radies, D.; Matter, A. A 160,000-Year record of dune development and atmospheric circulation in southern Arabia. *Science* **2002**, *296*, 2018–2020. [CrossRef]

28. Biswas, R.H.; Herman, F.; King, G.E.; Braun, J. Thermoluminescence of feldspar as a multi-thermochronometer to constrain the temporal variation of rock exhumation in the recent past. *Earth Planet. Sci. Lett.* **2018**, *495*, 56–68. [CrossRef]

29. King, G.E.; Guralnik, B.; Valla, P.G.; Herman, F. Trapped-charge thermochronometry and thermometry: A status review. *Chem. Geol.* **2016**, *446*, 3–17. [CrossRef]

30. Roberts, H.M.; Bryant, C.L.; Huws, D.G.; Lamb, H.F. Generating long chronologies for lacustrine sediments using luminescence dating: A 250,000year record from Lake Tana, Ethiopia. *Quat. Sci. Rev.* **2018**, *202*, 66–77. [CrossRef]

31. Lowick, S.E.; Preusser, F. A method for retrospectively calculating the water content for silt-dominated desiccated core samples. *Anc. TL* **2009**, *27*, 9–14.

32. Duller, G.A.T.; Li, S.H.; Musson, F.M.; Wintle, A.G. Use of infrared stimulated luminescence signal for scanning sediment cores. *Quat. Sci. Rev.* **1992**, *11*, 115–119. [CrossRef]

33. Sanderson, D.C.W.; Murphy, S. Using simple portable OSL measurements and laboratory characterization to help understand complex and heterogeneous sediment sequences for luminescence dating. *Quat. Geochronol.* **2010**, *5*, 299–305. [CrossRef]

34. Smedley, R.K.; Buylaert, J.-P.; Újvári, G. Comparing the accuracy and precision of luminescence ages for partially-bleached sediments using single grains of K-feldspar and quartz. *Quat. Geochronol.* **2019**, *53*, 11. [CrossRef]

35. Duller, G.A.T.; Bøtter-Jensen, L.; Murray, A.S.; Truscott, A.J. Single grain laser luminescence (SGLL) measurements using a novel automated reader. *Nucl. Instrum. Methods Phys. Res. B* **1999**, *155*, 506–514. [CrossRef]

36. Galbraith, R.F.; Roberts, R.G. Statistical aspects of equivalent dose and error calculation and display in OSL dating: An Overview and some recommendations. *Quat. Geochronol.* **2012**, *11*, 1–27. [CrossRef]

37. Bateman, M.D.; Frederick, C.D.; Jaiswal, M.K.; Singhvi, A.K. Investigations into the potential effects of pedoturbation on luminescence dating. *Quat. Sci. Rev.* **2003**, *22*, 1169–1176. [CrossRef]

38. Gliganic, L.A.; May, J.-H.; Cohen, T.J. All mixed up: Using single-grain equivalent dose distributions to identify phases of pedogenic mixing on a dryland alluvial fan. *Quat. Int.* **2015**, *362*, 23–33. [CrossRef]

39. Nelson, M.S.; Rittenour, T.M. Using grain-size characteristics to model soil water content: Application to dose-rate calculation for luminescence dating. *Radiat. Meas.* **2015**, *82*, 142–149. [CrossRef]

40. Polymeris, G.S.; Kitis, G.; Liolios, A.K.; Sakalis, A.; Zioutas, K.; Anassontzis, E.G.; Tsirliganis, N.C. Luminescence dating of the top of a deep water core from the NESTOR site near the Hellenic Trench, east Mediterranean Sea. *Quat. Geochronol.* **2009**, *4*, 68–81. [CrossRef]

41. Wintle, A.G.; Huntley, D.J. Thermoluminescence dating of a deep-sea sediment core. *Nature* **1979**, *279*, 710–712. [CrossRef]

42. Wintle, A. A thermoluminescence dating study of some Quaternary calcite: Potential and problems. *Can. J. Earth Sci.* **1978**, *15*, 1977–1986. [CrossRef]

43. Rodrigues, A.L.; Dias, M.I.; Valera, A.C.; Rocha, F.; Prudêncio, M.I.; Marques, R.; Cardoso, G.; Russo, D. Geochemistry, luminescence and innovative dose rate determination of a Chalcolithic calcite-rich negative feature. *J. Arch. Sci. Rep.* **2019**, *26*, 1–13. [CrossRef]

44. Fulop, E.C.; Johnson, B.G.; Keen-Zebert, A. A geochronology-supported soil chronosequence for establishing the timing of shoreline parabolic dune stabilization. *CATENA* **2019**, *178*, 232–243. [CrossRef]

Protocol

Transferring Grains from Single-Grain Luminescence Discs to SEM Specimen Stubs

Isa Doverbratt [1] and Helena Alexanderson [1,2,*]

[1]   Department of Geology, Lund University, Sölvegatan 12, SE-223 62 Lund, Sweden; isa.doverbratt@geol.lu.se
[2]   Department of Geosciences, UiT the Arctic University of Norway, PB 6050 Langnes, N-9037 Tromsø, Norway
*   Correspondence: helena.alexanderson@geol.lu.se; Tel.: +46-46-222-4483

Received: 1 October 2019; Accepted: 18 November 2019; Published: 21 November 2019

Abstract: The grain transfer protocol presents a step-by-step guide on how to successfully transfer positioned grains from a single-grain luminescence disc to a scanning electron microscope (SEM) specimen stub and how to transport them between laboratories. Single-grain luminescence analysis allows the determination of luminescence characteristics for individual sand-sized grains. By combining such luminescence data with other grain properties such as geochemical composition, shape, or structure also at single-grain level, it is possible to investigate factors controlling luminescence signals or study other material properties. The non-luminescence properties are typically measured in another instrument; thus, grains need to be transferred between machines and sample holders, and sometimes also between laboratories. It is then important that the position of each grain is known and stable so that the properties from the same grain are compared. By providing an easily observable orientation marker on the specimen stub, the hundred numbered grains from the single-grain disc can be transferred and later identified when analyzed in the SEM.

Keywords: luminescence; single-grain; SEM; sample holders

---

1. Introduction

With the development of single-grain luminescence analysis [1,2], it has become possible to measure the luminescence signals of individual quartz or feldspar grains. From a dating perspective, this has provided valuable information on dose distributions that can be used to evaluate, for example, sediment mixtures or incomplete bleaching [3,4], information that is partly lost in multigrain aliquots because of the averaging of grains that takes place [5]. Single-grain luminescence has also revealed the variability in luminescence characteristics between grains within a single sample [6]. While this may complicate age determination [7,8], it also offers possibilities to study and compare luminescence characteristics on the level of individual grains, with the aim of understanding causes and controls of luminescence of quartz or feldspar. However, this may require the additional analysis of grain properties that cannot be carried out in a single-grain luminescence reader, for example of geochemical composition, crystal structure, or grain shape. To do such analyses, grains need to be transferred from the luminescence reader sample holders (single-grain discs) to the sample holders of another instrument, while keeping track of which grain is which to be able to compare properties on the grain level. Potential applications of such comparative analyses include determining the K content in feldspar grains, identifying impurities in quartz or feldspar grains, studying the effect of hydrofluoric acid or other chemical treatments, evaluating bleaching for different transport pathways by analyzing grain shape and surface textures (of nonetched grains), and comparing luminescence signals or doses at the grain level. In some cases, it may also be interesting to simply review the grains in a scanning electron microscope (SEM) or optical microscope to determine grain integrity (missing grains, multiple grains per hole, grain appearance, etc.). In this paper, we present a protocol that allows for successful transfer of grains from single-grain discs to SEM specimen stubs.

2. Experimental Design

The main focus in this procedure is how to transfer grains from a single-grain disc (SG disc) to an SEM stub and transport the analyzed single grains, intact and in order, from one laboratory to another. The most crucial part in this procedure is sample orientation in order to identify grains, and noting that the grains on the SEM specimen stub will be the mirror image of the grains on the SG disc.

Four samples previously dated at the Lund Luminescence Laboratory, Lund University, Sweden, were used in this study. The grains were transferred from single-grain discs to SEM specimen stubs in the luminescence laboratory at the University of Sheffield, UK, and then transported by train, plane, and bus to Lund University, Sweden. Apart from the sample holders, the protocol requires few special materials and can be applied in most laboratories and with various transport distances.

2.1. Materials

The grains used for the transfer were of quartz in the 180–250 µm grain size fraction. The grains were extracted from Late Quaternary sediments from four sites in Sweden and Norway (Table 1) according to procedures described in [9].

Table 1. Sample information.

Lab. No Lund-	Site	Sediment	Disc	Reference
12057	Rauvospakka, N Sweden	glacilacustrine silty sand	4, 5	[10]
13031	Orsa, C Sweden	aeolian sand	1, 7, 8, 9	[11]
15012	Skorgenes, W Norway	glacifluvial sand	6	[12]
15064	Skogalund, SW Sweden	aeolian sand	2, 3	[9]

2.2. Equipment

- Instruments

  - Risø TL/OSL reader model DA-15 with a single-grain attachment [1] at the luminescence laboratory at the Department of Geography, University of Sheffield, United Kingdom
  - Tescan Mira3 High Resolution Schottky FE-SEM equipped with Oxford EDS at the Department of Geology, Lund University, Sweden
  - Stereo Microscope, Zeiss Stemi 2000-C, equipped with a camera, AxioCam ERc 5s, at the Department of Geology, Lund University, Sweden

- Sample holders

  - Single-grain aluminum discs, 9.7 mm in diameter and 1 mm thick, with one hundred 300 µm holes in a 10 × 10 grid
  - Specimen mount (stub), aluminum, 1/2″ slotted head, 1/8″ pin

- Other

  - Carbon adhesive tabs, 12 mm diameter, Electron Microscopy Sciences
  - Scissors
  - Small container, plastic or other material, about 0.5–1 cm higher than the stub height
  - Styrofoam, thickness about 1 cm (corresponding to height of stub rods)
  - Tweezers

3. Procedure

3.1. Preparing the Transportation Container

1. Find a container of appropriate size with a lid. An appropriate size is one that suits the number of samples and allows a safe transportation of the samples, that is keeping the samples upright.
2. Cut a piece of Styrofoam into the desired shape to fit the bottom of the container (Figure 1). Any other flexible but firm material may also be used.

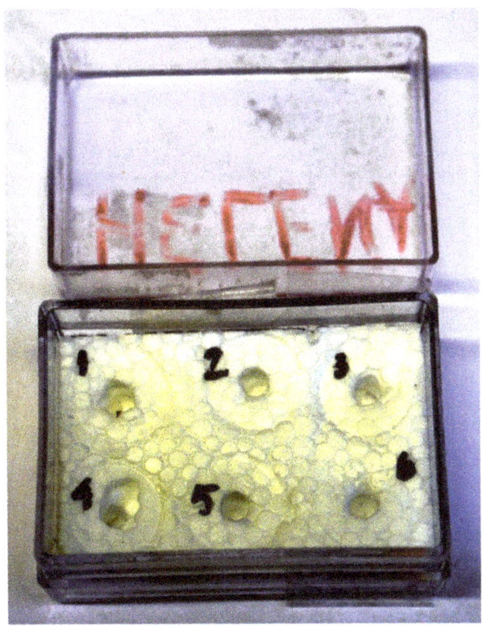

Figure 1. Container (3.5 × 5 cm) with lid, prepared with Styrofoam and holes for the scanning electron microscope (SEM) specimen stubs. This specific container can hold up to six stubs.

3. Prepare holes in the Styrofoam to fit the SEM specimen stubs (Figure 1). Hold the stubs upright (broad side up) and push down the rods into the Styrofoam. If necessary, use a sharp object (for example, a pen) to pierce the holes for the stub rods.
4. Mark each hole with a number, for example 1–6 (Figure 1).

3.2. Preparing the SEM Specimen Stubs

1. Use scissors to cut the carbon adhesive tabs so that there is one straight side to the previously circular piece.
2. Attach the precut tab centered on an SEM specimen stub, and leave the transparent plastic film on (Figure 2).
3. Place the stub in the container (Section 3.1) with the rod in one of the holes and the tape side up.

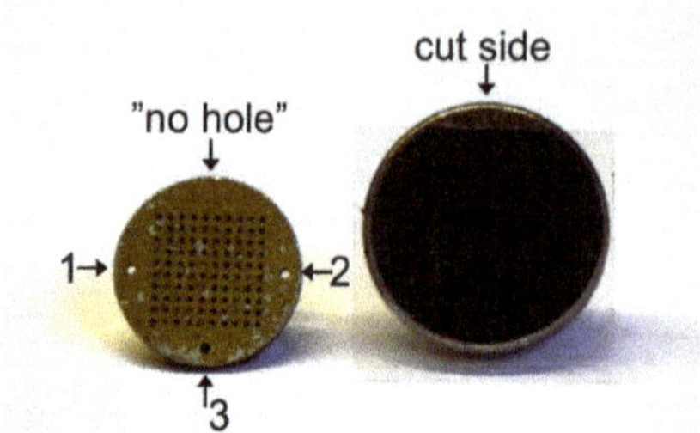

Figure 2. To the left is a single-grain disc with the "no-hole" side of the 10 × 10 grid in the uppermost part of the disc. Numbers 1–3 indicate the orientation of the holes, as registered in the luminescence reader software (SequenceEditor). To the right is an SEM specimen stub prepared with black, cut carbon adhesive tab and the transparent plastic film left on.

3.3. Transferring Grains from the Single-Grain Disc to the SEM Specimen Stub

1. After OSL measurements, take the wheel out of the reader and place it on a stable and horizontal surface.
2. Take one of the pre-prepared SEM specimen stubs out of the container and remove the transparent plastic film with tweezers.
3. Remember to note the orientation of the holes in the disc and the straight side of the carbon tape.
4. Hold the stub on its rod and carefully attach the sticky carbon tape to the SG disc, with the disc still in the wheel. Make sure that the stub is centered and placed with the cut tape side at the "no-hole side" of the SG disc (Figure 2). This enables you to keep track of the orientation, that is to be able to use the numbered grain positions from the reader software. Attention: The grain numbering on the carbon tape is the mirror image of the grains on the SG disc (Figure 3). The grains will also be upside-down compared to the SG disc.
5. When the SEM specimen stub is fitted to the SG disc, turn it into an upright position and tap the disc gently with the tweezers.
6. Remove the SG disc from the carbon tape using tweezers or your fingernails. You will now be able to see the grains on the carbon tape.
7. Place the stub carefully in the container prepared with Styrofoam. The rod should go into one of the rod holes, and the tape side should face upwards.
8. Make a note of which sample is in which hole (Figure 4).
9. Put the lid on the container, and, if necessary, tape it to keep it in place.

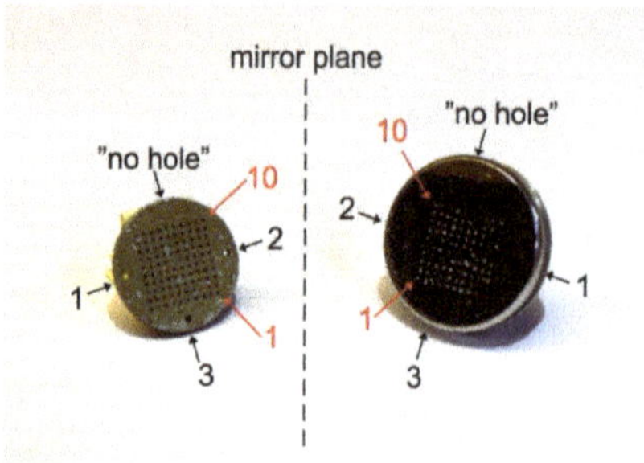

Figure 3. An SG disc (**left**) after the grains have been transferred to the carbon tape on the SEM specimen stub (**right**). The position of the grains on the stub is the mirror image of how the grains were positioned on the disc. Numbers 1–3 (in black) highlight the orientation of the holes, and numbers 1 and 10 (in red) highlight the grain positions on the disc as recorded in the OSL data file by the reader software.

Figure 4. Six SEM specimen stubs in the container prepared with Styrofoam (the two stubs to the left in this figure have a different appearance due to a Pd/Pt coating prior to SEM analysis).

3.4. Transport

During transport, make sure that the container stays upright and is packaged securely to ensure minimal movement of sample holders.

Additional Information

To orient the grid of grains once the sample is in the SEM, it is advisable to take an overview picture of the stub. The cut side of the carbon tape is clearly seen in the SEM (Figure 5), which allows the grains to be numbered (1–100). Pictures of each disc may also be taken with a stereo microscope equipped with a camera prior to SEM analysis.

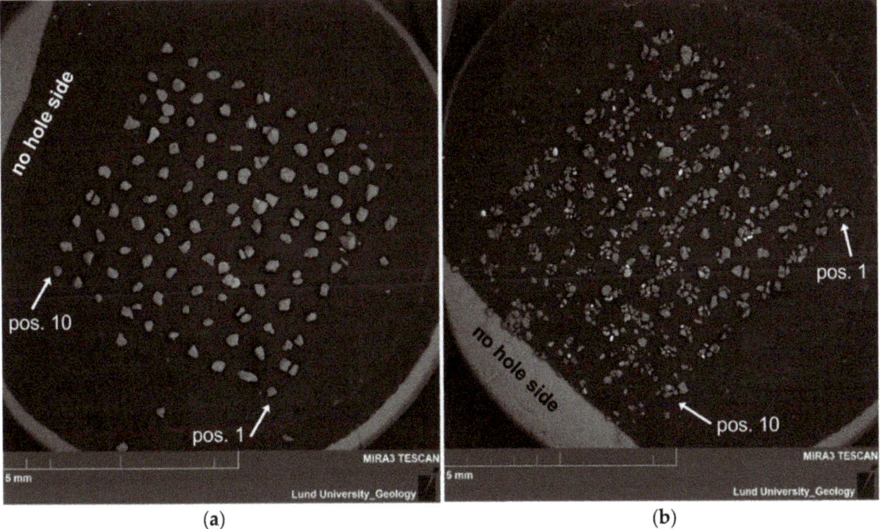

Figure 5. Examples of SEM images of transferred grains on stubs. The 10 × 10 grid can be seen clearly both where there is (**a**) mainly one grain per position (disc 2, sample Lund-15064) and (**b**) multiple grains per position (disc 6, sample Lund-15012).

4. Results

The success of the transfer was evaluated by (1) counting the number of empty holes in the single-grain discs by inspecting the OSL signals from each position; (2) counting the number of positions in the 10 × 10 grid on the SEM stubs that are filled with grain(s), as determined from an image taken in the SEM (Figure 5); and (3) comparing the numbers. On average, grains at 94% ± 6% of the positions were successfully transferred from the disc to the stub (Table 2). Some of the missing grains can, however, still be seen on the stub, but are not located in a grid position (Figure 5), either because they moved slightly within the grid or they ended up off the grid.

Even samples with multiple grains at almost all positions, such as the easily shattered quartz of sample 15012, retained the grid pattern (Figure 5b) and made it possible to identify position, though in some cases the piles of grains at neighboring positions merged. It is, therefore, fairly straightforward to identify grains that have moved, and we recommend that grains with a position off the grid are excluded from further analysis, since their connection to a specific single-grain luminescence measurement would be uncertain.

Table 2. Number of positions on the discs and stubs, respectively, that are filled with grains.

Disc	#Filled Positions on Disc	#Filled Positions on Stub	Transfer Success (%)
1	100	90	90.0
2	97	98 [1]	101.0 [1]
3	100	97	97.0
4	99	90	90.9
5	100	93	93.0
6	100	100 [2]	100.0 [2]
7	98	96	98.0
8	100	95	95.0
9	100	83 [3]	83 [3]

[1] There were no OSL signals from three SG positions, but observed in the SEM, only two positions are empty; [2] The transfer of sample 15012 from the disc to the stub was the most successful. However, the quartz was quite brittle and shattered easily. Therefore, there are several smaller grains in each position; [3] This is the least successful transfer during which grains have been moved around, and many grains do not logically fit in a specific position.

Concerns that may arise during SEM analysis of grains transferred according to the protocol presented here include the identification of individual grains, the orientation of grains, and contamination from the SG disc. Regarding the identification of individual grains, we found that having an overview image where (at least) grains in grid corners have been numbered (compare Figure 5) greatly facilitates finding specific grains based on their location and appearance, also at high magnification.

The second concern, that the same surface is not exposed for imaging or analysis before and after transfer (because of the grains being upside-down), may or may not be an issue depending on what type of analysis is carried out. Some analyses, including luminescence, retrieve information also from the interior of grains, and for these it should be less of a concern. This is something that could be studied further, as could the risk of contamination from the SG disc to the grains. If the discs are visually inspected in a microscope prior to use, and damaged or dirty discs are discarded, we consider the risk to be negligible, but we have not studied this specifically. Any such contamination would also be surficial and would likely not significantly influence measurements that include deeper parts of the grains, or whole grains.

Author Contributions: Conceptualization, H.A.; methodology, I.D. and H.A.; validation, I.D. and H.A.; investigation, I.D. and H.A.; data curation, H.A.; writing—original draft preparation, I.D. and H.A.; writing—review and editing, I.D. and H.A.; visualization, I.D. and H.A.; project administration, H.A.; funding acquisition, H.A. and I.D.

Funding: The Geological Survey of Sweden (grant No. 36-2021/2016) and the Royal Physiographic Society in Lund funded this research.

Acknowledgments: We would like to thank Mark Bateman at the Department of Geography, University of Sheffield, for his warm welcome, informative discussions and for lending us his single-grain machine, a.k.a. Scooby; Carl Alwmark for helping with the SEM; Amber Hood, Martin Jarenmark and the journal reviewers for comments on the manuscript.

Conflicts of Interest: The authors declare no conflict of interest. The funders had no role in the design of the study; in the collection, analyses, or interpretation of data; in the writing of the manuscript, or in the decision to publish the results.

References

1. Duller, G.A.T.; Bøtter-Jensen, L.; Kohsiek, P.; Murray, A.S. A high-sensitivity optically stimulated luminescence scanning system for measurement of single sand-sized grains. *Radiat. Prot. Dosim.* **1999**, *84*, 325–330. [CrossRef]
2. Murray, A.S.; Roberts, R.G. Determining the burial time of single grains of quartz using optically stimulated luminescence. *Earth Planet. Sci. Lett.* **1997**, *152*, 163–180. [CrossRef]
3. Duller, G.A.T. Single grain optical dating of glacigenic deposits. *Quat. Geochronol.* **2006**, *1*, 296–304. [CrossRef]
4. Bateman, M.D.; Boulter, C.H.; Carr, A.S.; Frederick, C.D.; Peter, D.; Wilder, M. Detecting post-depositional sediment disturbance in sandy deposits using optical luminescence. *Quat. Geochronol.* **2007**, *2*, 57–64. [CrossRef]
5. Duller, G.A.T. Single-grain optical dating of quaternary sediments: Why aliquot size matters in luminescence dating. *Boreas* **2008**, *37*, 589–612. [CrossRef]
6. Duller, G.A.T.; Bøtter-Jensen, L.; Murray, A.S. Optical dating of single sand-sized grains of quartz: Sources of variability. *Radiat. Meas.* **2000**, *32*, 453–457. [CrossRef]
7. Bailey, R.M.; Arnold, L.J. Statistical modelling of single grain quartz D_e distributions and an assessment of procedures for estimating burial dose. *Quat. Sci. Rev.* **2006**, *25*, 2475–2502. [CrossRef]
8. Rhodes, E. Quartz single grain OSL sensitivity distributions: Implications for multiple grain single aliquot dating. *Geochronometria* **2007**, *26*, 19–29. [CrossRef]
9. Alexanderson, H.; Bernhardson, M. Late glacial and Holocene sand drift in south-central Sweden. *GFF* **2019**, *141*, 84–105. [CrossRef]
10. Sigfúsdóttir, T. *A Sedimentological and Stratigraphical Study of Veiki Moraine in Northernmost Sweden*; Lund University: Lund, Sweden, 2013.
11. Alexanderson, H.; Bernhardson, M. OSL dating and luminescence characteristics of aeolian deposits and their source material in Dalarna, central Sweden. *Boreas* **2016**, *45*, 876–893. [CrossRef]
12. Anjar, J.; Alexanderson, H.; Larsen, E.; Lyså, A. OSL dating of Weichselian ice-free periods at Skorgenes, Western Norway. *Nor. J. Geol.* **2018**, *98*, 301–313. [CrossRef]

Article

Using DOSIVOX to Reconstruct Radiation Transport through Complex Archaeological Environments

Amber G.E. Hood [1,*] **and Edmund G. Highcock** [2]

[1] Department of Geology, Lund University, 223 62 Lund, Sweden
[2] Research & Development, Greenbyte, Östra Hamngaten 16, 411 06 Gothenburg, Sweden;
     edmund@greenbyte.com
*   Correspondence: amber.hood@geol.lu.se

Received: 8 October 2019; Accepted: 26 November 2019; Published: 11 December 2019

Abstract: The DOSIVOX programme is used to reconstruct radiation transport in complex depositional environments. Using two archaeological case studies from ancient Egypt, the burial environments for a selection of ceramic vessels are reconstructed using the DOSIVOX programme, allowing the simulation of the emission and transport of radiation throughout these burial environments. From this simulation we can extract the external dose rate of the archaeological samples, a measurement necessary for determine a luminescence age. We describe in detail how DOSIVOX can be used to best advantage at sites with complex depositional histories and highlight that DOSIVOX is a valuable tool in luminescence dating. This work illustrates that DOSIVOX is, at present, unparalleled in reconstructing a more accurate and detailed external gamma dose rate which can significantly improve upon simplistic scaled geometric models.

Keywords: optically stimulated luminescence dating; ceramics; DOSIVOX; radiation dose modelling; environmental dose rate; complex burial environments; Egyptian archaeology

---

1. Introduction

1.1. Overview

Luminescence dating, including optically stimulated luminescence (OSL) dating, is a valuable tool to archaeologists and geologists and remains arguably the best technique for the chronometric dating of archaeological or geological materials comprised of quartz and feldspar. In this manner, in archaeology, luminescence dating is widely applicable to pottery, minerogenic building materials and ceramic objects or sediments from archaeological contexts. While dating of recently excavated material can be routine (as in situ sampling can be done by a trained luminescence specialist aware of precisely what samples and measurements are required to achieve a luminescence age), working with more complex material, such as that recovered from museum contexts, can be more difficult. While working with recently recovered material suitable for OSL sampling is of course ideal, it is not always possible and in many cases OSL dating is not permitted, owing to strict archaeological laws which coincide with a lack of accessible laboratory facilities.

Archaeological materials housed in museums—most notably pottery vessels and ceramic objects—were often not collected with OSL dating in mind and in some cases were excavated and placed in collections well before OSL dating was even developed. As such, external dose rate ($\dot{D}_{ext}$) calculations can be incredibly difficult to reconstruct for museum material. Sometimes, for example if no external sediment from the original burial environment is forthcoming, it can be impossible to accurately reconstruct. However in some circumstances we can find museum material which is more favourable to OSL dating, for example, when original depositional sediment is attached to the vessels or is associated in some other way with the material. However, even then, the assumption of a uniform burial environment in archaeological luminescence dating is one that has often proved difficult for archaeologists to reconcile.

With the recent development of the DOSIVOX radiation transport modelling software [1], it has now become possible to reconstruct radiation transport throughout complex burial environments, where one must consider natural radiation from other archaeological materials, associated geological materials, and structural elements of the environment. Such modelling can vastly improve the accuracy of $\dot{D}_{ext}$ determinations for both museum specimens, as well as for complex burial environments recently excavated.

In this paper, we outline the general process for building a DOSIVOX model of an archaeological site or other complex depositional environments. This model, along with the DOSIVOX software, allows us to calculate $\dot{D}_{ext}$ from first principles and certain basic assumptions, using all the information we have at our disposal (assuming direct measurement is not possible). Thus, for complex depositional environments where direct measurement is not possible, the model represents the best possible estimate of $\dot{D}_{ext}$ using tools available at the time of writing.

In Section 1.2 we describe our approach to the measurement of $\dot{D}_{ext}$ starting from first principles. In Section 2 we begin by describing the application of the model to two complex sites from Early Dynastic Egypt where direct measurement of $\dot{D}_{ext}$ is not possible. We then, continuing into Section 3, compare the results to other possible methodologies for estimating $\dot{D}_{ext}$. In Section 4.1 we describe how to construct a DOSIVOX model based on known information about a site and in Section 4.2 we describe our approach to running DOSIVOX, including ensuring the quality of the simulations.

1.2. Theoretical Background

In our work, we start by observing that, owing to their very short attenuation depth, neither alpha nor beta radiation originating externally to the sample contribute to the equivalent dose. In other words, we assume $\dot{D}_{ext}$ is entirely composed of gamma radiation.

With that assumption, three key questions need to be answered when modelling $\dot{D}_{ext}$:

1. what was the quantity (flux) of gamma radiation arriving at the target site (the location of the sample) per unit of time;
2. what fraction of that radiation was absorbed;
3. what dose of laboratory-delivered beta radiation would be equivalent to the absorbed dose of gamma radiation?

1.2.1. Grain Absorption and Equivalent Beta Dose

We begin by tackling the second and third questions, which we will find, in our case, to be less interesting than the first. To estimate the fraction of incident radiation that is absorbed by a grain of quartz, it would be possible to model the grains themselves using DOSIVOX. However, given the extreme difference between the grain size and the gamma attenuation depth (roughly 1:2500) it is highly accurate to model the grains as thin targets, as is the usual practice [2]. This means that we assume that grains are irradiated evenly. Thus, we omit grain models and use DOSIVOX to calculate the average absorption in a uniform material of the right composition, as we will describe below.

In answer to the third question, we note first that the relationship between the luminescence caused by a dose of gamma radiation and the luminescence caused by the same dose of beta radiation remains impossible to determine from first principles. It would be possible to measure it experimentally by irradiating first with a gamma and then a beta source for the samples in question; however, it is in fact well established that the relationship between the two does not vary from one sample to another, and that it is accurate to assume that they are equal (Reference [3]: 47).

Thus in answer to the second and third questions, it is both standard practice and accurate to assume that the dose is evenly absorbed and the equivalent beta dose is equal to the gamma dose.

1.2.2. Incident Dose

Answering the first question will be the primary focus of this work. In general, the radiation incident on the grains can be directly measured using in situ gamma spectrometry, estimated using the infinite matrix assumption (IMA, see below), or, as in this work, calculated using DosiVox. We begin by briefly discussing gamma spectrometry, then examine in more detail the advantages and pitfalls of the IMA, and then lay out the new possibilities when using a first-principles method like DosiVox.

Gamma spectrometry

Gamma spectrometry is the direct measurement of the incident gamma flux at the site using a portable measuring device, that is, a gamma spectrometer (or a dosimeter, a gamma spectrometer may require an additional assumption about secular equilibrium). If it is possible to take a gamma spectrometer or dosimeter to the sample site and measure the incident dose directly, then that is undoubtedly the easiest method to pursue. However, there are reasons why this may not possible, for example:

- A portable gamma spectrometer is unavailable or cannot be transported to the site.
- The site has radically altered since antiquity; thus, the modern measurement of the gamma radiation intensity may not reflect what it was in the past.
- The context of the sample has already been changed during the current excavation, for example as a result of the removal of stratigraphic layers above.
- The artefact of interest was excavated in the past and its find spot either cannot be accessed or has been destroyed, for example, by backfilling.

Infinite Matrix Assumption Plus Geometric Model

Let us now turn to the case where direct in situ measurement of the incident gamma flux is not possible. A second method of determining the gamma flux begins by recovering samples of all materials within the gamma attenuation depth of the sample and determining the intensity of gamma radiation emitted by these materials. This may be through direct measurement using a lab-based gamma spectrometer, or by determining the radioisotopic concentrations within each sample using a form of mass spectrometery (e.g., ICP-MS). The gamma emissions of all commonly discovered radioisotopes are well-measured and readily available [4]. Thus, if we know the radioisotopic content of any material or material mix we know the amount of radiation emitted by that material (per unit mass). Regardless of which method is used, the amount of radiation emitted by each material can be determined.

This is a good start, but what we want to know is the amount of radiation absorbed by the sample (per unit mass). Happily there is one simple case in which one is equal to the other: the case of a homogeneous material of infinite extent. In which case, by symmetry all points in the material must absorb the same radiative energy, and by the conservation of energy, this absorbed amount must be equal to the emitted amount, which we know (given the elemental breakdown). In the case where the material is not infinite, but is nonetheless considerably larger in extent than the attenuation depth of the gamma radiation (typically $\sim$30 cm in archaeological and geological contexts), we may treat the material as being infinite, and give the absorbed dose as being the emitted dose: this is known as the *infinite matrix assumption* (Ref. [5]: 56).

The IMA is an excellent assumption if the sample is indeed embedded in a uniform material, as shown in Figure 1, but what if instead it is located at the boundary between two or more materials? Provided the materials can be reasonably approximated as occupying solid angles surrounding the sample, all radiation arriving at the sample only passes through one material before reaching the sample and the infinite matrix assumption still applies, weighted by the solid angles (see Figure 2). It is possible to prove this in some circumstances and we do so in Appendix A.

The convenient arrangement of materials in Figure 2 is unlikely to apply in a real case. However, in the absence of any other methodology we may assume that we can still use this method, approximating the solid angles that apply to each material as the volume occupied by that material in a sphere whose radius is the gamma attenuation depth and whose centre is at the sample, divided by the total volume of the sphere. The more complex the actual arrangement of materials, the more specious this approximation, as our results in Section 2.4 demonstrate. In the case of contexts with a simple arrangement of materials, this geometric approximation will frequently be acceptable (e.g., Reference [6]; see extended discussion in Reference [5]: 289). In the case of a complex archaeological context like a tomb, something more is needed.

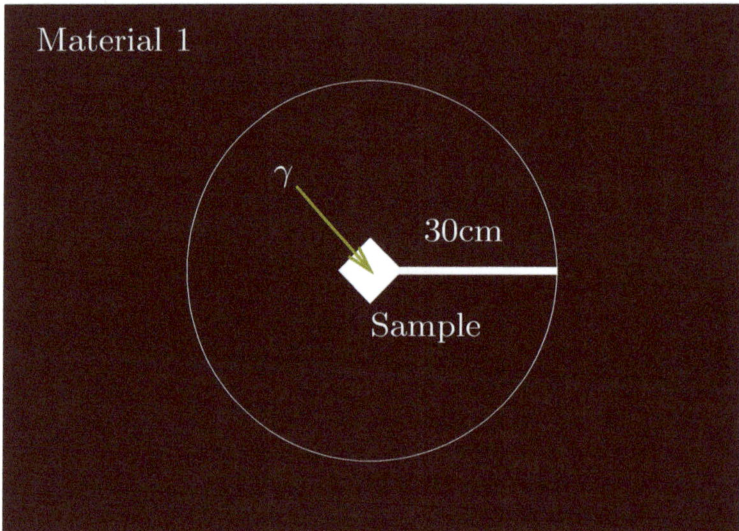

Figure 1. A sample in a simple context for which the infinite matrix assumption (IMA) is valid.

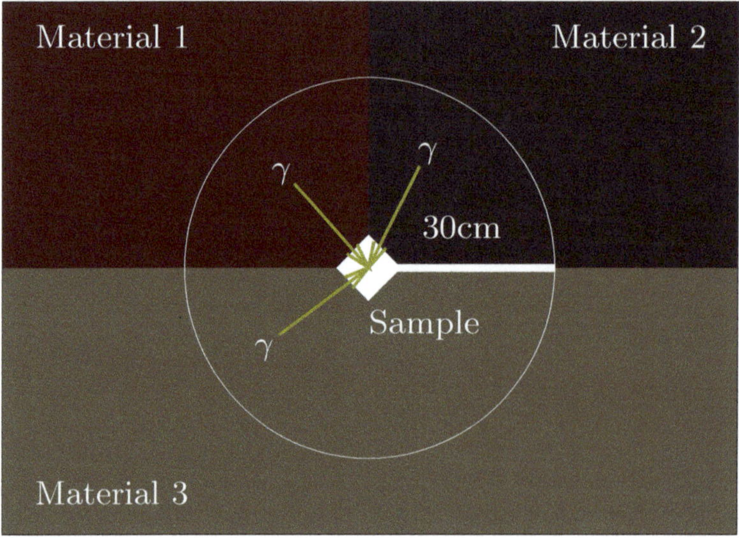

Figure 2. A sample in a context for which the infinite matrix assumption (IMA) in combination with geometric scaling is valid.

DOSIVOX

The large hadron collider (LHC) at CERN causes particles to collide with such high velocity that they break up into a shower of other particles, which are seen in the detector. Because the effects of these particles in the detector are indirect it is necessary to model the origin, transport and detection of these particles and then use statistical methods to determine the most likely sequence of events (e.g., Reference [7]).

One such model created and used at CERN is Geant-4, which simulates the creation, transport through matter and absorption of high-energy particles. Researchers at the Université Bordeaux Montaigne realised that this is exactly what was needed to model the dose rate that a luminescence sample experiences in its burial location, and thus the DOSIVOX software was developed [1].

As will be discussed further in Section 4.1, using DOSIVOX starts with making a model which includes the geospatial arrangement of all materials in the burial environment, as well as their composition. The composition of the materials includes their bulk composition, that is, the elements that make up the bulk of their mass (which affects the transport and absorption of the radiation) and the radioactive trace elements which emit the radiation (which are specified separately). We note that as well as specifying the composition and distribution of the materials, it is also possible to specify the grain-size of the sample; however, we omit this as we are only interested in the gamma radiation.

The dose absorbed by the sample location is the integral over time of the energy deposited by every gamma photon emitted from every radioisotope in the model. Since this integral is high-dimensional and expensive to calculate, DOSIVOX uses Monte-Carlo methods to approximate the integral, effectively allowing a given number of gamma particles to be generated at random from within the model and measuring the dose at the sample location from each particle. Thus, the greater the number of particles, the more accurate the calculation, as discussed in Section 4.2. For more details concerning the way DOSIVOX works the reader is referred to the original paper [1]. In this work, we wish to give a practical description of how DOSIVOX was applied, in the hopes that it may be useful to others wishing to use DOSIVOX in complex archaeological or geological contexts.

Other Gamma Sources

There are of course other sources of gamma radiation that might have affected the effective dose rate. Assuming that the sample was not exposed after excavation, the most notable of these is cosmic radiation. In our work here, we use the effective and well-established method of calculating the cosmic dose rate as a function of depth and global position of an archaeological/geological site that is provided by the Dose Rate and Age Calculator (DRAC) ([8]; following Reference [9]).

2. Results

2.1. Overview

To illustrate how DOSIVOX can be used to reconstruct radiation transport through complex burial environments, we look at two case studies from ancient Egypt. This research was initially carried out as part of Hood's doctoral thesis, which produced the first ever OSL dates obtained for ancient Egyptian ceramics. The results of this dating programme can be found in References [10,11]. Owing to a current law in Egypt which prevents the removal of archaeological material for scientific analysis, this research was carried out on material excavated at the turn of the 20th century which was stored in museums outside of the country. As a result the burial environment and thus DOSIVOX model, was established through analysis of the original excavation reports, and when possible, more recent archaeological examinations (as was possible in one case, see below).

2.2. Case Study 1: Bêt Khallaf

The first case study looks at twenty-four ceramics sampled from the Upper Egyptian site of Bêt Khallaf. The site was excavated by British archaeologist John Garstang in the early 1900s, and the intact ceramic material recovered from the site was removed from Egypt soon after excavation and dispersed across three museums: the Ashmolean Museum in Oxford, the Garstang Museum of Archaeology in Liverpool and the University of Pennsylvania Museum of Archaeology and Anthropology in Pennsylvania (Penn Museum). The assemblage is typologically characteristic of the late Early Dynastic Period of Egypt (c. 2650 B.C.E.).

The Bêt Khallaf assemblage is a perfect candidate for testing and comparing gamma dose rates obtained using DOSIVOX to those determined using the IMA. This is firstly because many of the vessel types found in the tombs at Bêt Khallaf are diagnostic (thus providing a perfect platform for using absolute dating to anchor the relative ceramic typology to a known calendrical system), and secondly because Garstang kept and published reports of high standard for that era. These reports were used to construct the DOSIVOX model. In addition to providing details of the tomb architecture (tomb size, depth etc.), Garstang also gives the location of the pottery deposit within each tomb, allowing an accurate find spot of the ceramics to be determined (Reference [12]: 8–16).

2.3. Case Study 2: Tomb of Djer, Abydos

The second case study was carried out using four vessels from the Tomb of Djer at Abydos. Excavated by Flinders Petrie in the early 1900s, this Abydene tomb was the burial place of King Djer, the 3rd king of the first Egyptian dynasty (c. 3000 B.C.E.). After excavation, the ceramic material was transported to the Ashmolean Museum, Oxford. Similarly to the Bêt Khallaf material, the Djer material was well-documented for its time and Petrie discusses the finding of the vessels and their precise location within the tomb. The DOSIVOX model was constructed based upon Petrie's excavation reports but was also supplemented by newly documented evidence from the tomb, which has recently been reexamined by the German Archaeological Institute. The combination of both sets of information allowed us to reconstruct the deposition of the vessels with a high degree of certainty and renders the Tomb of Djer a prime candidate for examining gamma dose rates and comparing the results obtained using DOSIVOX and the IMA.

2.4. Comparing Gamma Dose Rates

In both ancient Egyptian case studies, $\dot{D}_{ext}$ was determined using the U, Th and K concentrations from original depositional material found adhering to the vessels (under the rim and on a vessel handle). DOSIVOX was used to model the gamma dose rate received by the vessels from the surrounding burial environment (where the alpha and beta dose rates were determined using DRAC). The cosmic dose rate ($\dot{D}_{cos}$) was determined by using information reconstructed using the original excavation reports and Google Earth (where one can easily identify the tombs from which the vessels came from and thus their elevation, etc). Using the archival material discussed above, it was possible to reconstruct the burial environment for the examined ceramics. In both cases DOSIVOX modelling was considered necessary as the depositional environment for both sites was considered complex, without straightforward sample geometry. Complex sample geometry in luminescence dating occurs when a sample is not obtained from a (more or less) homogeneous depositional environment. In contrast, a simple sample geometry is what we see in Figure 1.

In the case of Bêt Khallaf, the ceramics were not fully buried, but rather 'strewn' about in a presumably haphazard manner. The precise manner of their deposition is unknown (i.e., they could potentially be sitting upright, lying on their side, or halfway out of the tomb fill; as described in Section 4.2.1 these variables were tested to see their effect upon the determined gamma dose rate and have been presented below), but from the tomb reports we were confident that they were not fully buried and thus would be affected by complex sample geometry, consisting of the pot itself, the tomb fill/debris on the ground as well as the tomb walls.

In the case of the Tomb of Djer, Petrie recorded the exact find spot of these vessels in the tomb, and the unpublished work of the German Archaeological Institute further informed us that these vessels were positioned upright and buried up to their mid-point in the surrounding sediment of a chamber in the north west corner of the mud-brick tomb and covered by a mud-brick staircase. These vessels also contained contents which were a mixture of organic and inorganic material. All of these known details helped to reconstruct the complex sample geometry for these vessels and to make the resulting DOSIVOX model.

Through reconstruction of the burial environment our resulting DOSIVOX models were able to present a more realistic gamma dose rate measurement for the OSL sample than if we simply assumed 4π geometry or determined a 'best guess' geometry based on assumed percentage contributions from each material measured.

Tables 1 and 2 present the calculated gamma dose rates for the Bêt Khallaf ceramics (Table 1) and those from the Tomb of Djer (Table 2). For each set of ceramics the gamma dose rate for each sample has been calculated in one of three ways: firstly by using a 4π geometric assumption using the IMA; secondly using a scaled IMA based on a 'best guess' of the complex sample geometry and; thirdly by using the gamma dose rate modelled using DOSIVOX. In the case of Bêt Khallaf, a sample 'best guess' of 2π is displayed. This was derived by following the principles outlined in Section 1.2.2, as the samples were known to be on the floor of the tomb (i.e., fill below and air above); in the case of the Tomb of Djer, the pots contained organic residue and were more than half buried in fill. Thus, a 'best guess' of 3π fill and 1π pot contents is displayed.

Table 1. A comparison of the DOSIVOX gamma dose rate and the gamma geometries (full 4π geometry and a 'best guess' scaled geometry of 2π, obtained using DRAC) for the tombs at Bêt Khallaf.

Sample ID	DRAC $\dot{D}_\gamma$ (4π Geometry) (Gy ka^{-1})	DRAC $\dot{D}_\gamma$ Error (4π Geometry) (Gy ka^{-1})	DRAC $\dot{D}_\gamma$ (2π Geometry) (Gy ka^{-1})	DRAC $\dot{D}_\gamma$ Error (2π Geometry) (Gy ka^{-1})	DosiVox $\dot{D}_\gamma$ (Gy ka^{-1})	DosiVox $\dot{D}_\gamma$ Error (Gy ka^{-1})
X4114	0.523	0.033	0.262	0.017	0.507	0.099
X4115	0.523	0.033	0.262	0.017	0.600	0.132
X4116	0.523	0.033	0.262	0.017	0.582	0.091
X4117	0.523	0.033	0.262	0.017	0.575	0.087
X4118	0.523	0.033	0.262	0.017	0.551	0.088
X5458	0.600	0.039	0.300	0.020	0.546	0.107
X5459	0.459	0.029	0.230	0.015	0.556	0.085
X5460	0.523	0.033	0.262	0.017	0.526	0.094
X5461	0.523	0.033	0.262	0.017	0.591	0.121
X5462	0.523	0.033	0.262	0.017	0.565	0.090
X5463	0.619	0.039	0.310	0.020	0.560	0.086
X5464	0.523	0.033	0.262	0.017	0.524	0.100
X5465	0.523	0.033	0.262	0.017	0.559	0.089
X5466	0.523	0.033	0.262	0.017	0.638	0.109
X5467	0.523	0.033	0.262	0.017	0.573	0.091
X5468	0.523	0.033	0.262	0.017	0.564	0.090
X5470	0.523	0.033	0.262	0.017	0.538	0.102
X5472	0.523	0.033	0.262	0.017	0.552	0.096
X5473	0.523	0.033	0.262	0.017	0.537	0.087

Table 2. A comparison of the DOSIVOX gamma dose rate and the gamma geometries (full 4π geometry and a 'best guess' scaled geometry of $3\pi + 1\pi$, obtained using DRAC), for the Tomb of Djer.

Sample ID	DRAC $\dot{D}_\gamma$ (4π Geometry) (Gy ka^{-1})	DRAC $\dot{D}_\gamma$ Error (4π Geometry) (Gy ka^{-1})	DRAC $\dot{D}_\gamma$ (Scaled Geometry) (Gy ka^{-1})	DRAC $\dot{D}_\gamma$ Error (Scaled Geometry) (Gy ka^{-1})	DosiVox $\dot{D}_\gamma$ (Gy ka^{-1})	DosiVox $\dot{D}_\gamma$ Error (Gy ka^{-1})
X5477	0.174	0.012	0.212	0.01	0.15	0.02
X5478	0.174	0.012	0.212	0.01	0.452	0.06
X5479	0.174	0.012	0.212	0.01	0.423	0.06
X6114	0.458	0.029	0.229	0.015	0.504	0.07
X6115	0.324	0.021	0.162	0.011	0.469	0.06
X6116	0.439	0.028	0.220	0.014	0.496	0.06
X6120	0.174	0.012	0.158	0.008	0.489	0.08

3. Discussion

We can see from the results presented in Tables 1 and 2 that the gamma dose rates obtained by using DOSIVOX differ from those obtained by assuming 'best guess' scaled geometries. In the case of Bêt

Khallaf the outcome saw DOSIVOX produce a gamma dose rate far more in keeping with an assumed 4π geometry than in our 'best guess' scaled geometry, illustrating how DOSIVOX can provide a result which is not subject to inbuilt biases and assumed geometries determined by the luminescence practitioner.

It can, of course, well be argued that the 'best guess' was in fact 4π, because the roof of the tomb was of similar composition to the floor and the dry air of the tomb would only slightly attenuate the radiation. We do not dispute this; we merely observe that all of these sweeping arguments and assumptions can be replaced by a more rigorous procedure using DOSIVOX. One benefit of this rigorous procedure is that there are many hidden assumptions in the 4π IMA estimate that DOSIVOX makes explicit and testable (position of the pot, etc). More importantly, however, it is only having run DOSIVOX (i.e., with hindsight) that we can know that the 4π IMA assumption was acceptable. Even though it might seem a reasonable assumption about the transport of gamma radiation *a priori*, with DOSIVOX the transport and absorption of gamma radiation is modelled in a way (using Geant-4) that has been extensively tested against highly accurate particle physics experiments; thus we can be much more confident in its results and we can eliminate at least one big assumption from our chain of reasoning.

In the case of the Tomb of Djer the DOSIVOX gamma dose rate was significantly different to the IMA and 'best guess' geometry dose rates. The complex depositional environment and the array of different elements affecting the dose rate was such that the gamma dose rate achieved using DOSIVOX modelling varied significantly enough from assumed geometries that final dose rate determinations varied considerably (which in turn affects the final luminescence age calculations). (It is beyond the scope of this paper to present the final age calculations for this material. If interested, the reader is referred to References [10,11], noting that final ages therein obtained using DOSIVOX were in good agreement with associated archaeological and historical chronologies.)

In simple geometric cases, for example, when a sample is known to have full, 4π geometry, DOSIVOX may be too significant an undertaking to apply to every luminescence sample and indeed will, in all likelihood, give a similar result. However, the more informative analysis and more accurate gamma dose rates achievable by DOSIVOX modelling in more complex burial environments, for example, that seen in the Tomb of Djer, illustrates how using the programme can improve luminescence age determination. In the case of complex sample geometries, it is therefore recommended that DOSIVOX modelling be incorporated into luminescence work to improve dose rate calculations and, in turn, age determinations.

One limitation of this research is that from the tomb reports we know that the ceramics at both sites were found as caches, that is multiple vessels were found in the same deposit. However, it was impossible at either site to determine the order or precise location of the ceramics relative to one another. In an ideal situation, if this information were recorded at the time of excavation, it would be possible to improve upon the DOSIVOX model even further by adding in each individual ceramic into the model so that the gamma dose rate determined for each vessel would also factor in the gamma dose being emitted by the surrounding vessels. This could be easily achievable for those wanting to use DOSIVOX modelling of complex deposits found in current excavations or fieldwork situations.

DOSIVOX is a valuable new tool when dealing with samples that have complex geometry and where in situ gamma measurements are not able to be made. Using DOSIVOX in conjunction with current projects could yield gamma dose rates far more accurate than those achievable through a simple incorporation of the IMA. This is because, in most cases, all the elements required to construct a DOSIVOX model would be readily available and, as a result of the more detailed model and more limited (and explicit) assumptions, a higher degree of confidence could be placed upon such a model, thus potentially yielding more accurate gamma dose rates than the IMA alone. We recommend that DOSIVOX modelling is routinely carried out at locations where in situ measurements are not available.

DOSIVOX modelling can be used in both helping to determine the gamma dose rate at recently excavated sites, those sites excavated in the past (i.e., where the resulting material has been stored in museums) and for both geological and archaeological situations. We are of the opinion that the incorporation of DOSIVOX into luminescence research can be of significant value in a number of situations, and believe that more accurate gamma dose rates can be achieved using DOSIVOX in cases of complex sample geometry. As such, DOSIVOX has the ability to greatly improve the reconstruction of dating profiles at geological and archaeological sites worldwide and assist in more accurate chronologies being achieved.

4. Materials and Methods

4.1. Constructing a DOSIVOX Model

A DOSIVOX model broadly consists of two parts:

1. a model of the composition and layout of the materials in the burial environment, and
2. a similar but more detailed model of the target site (the place where the sample is drawn from).

Before discussing how those parts are created it is helpful define the following terms:

- *Material*—the model is divided into regions that are filled with materials that are broadly uniform in composition: for example, a layer of sand, or a mud-brick wall.
- *Component*—materials are described by their density, their water content and the relative quantities of their chemical components (e.g., SiO_2 or TiO).
- *Voxel*—the whole burial environment is broken up into a grid of cubes which are called voxels (effectively "3D pixels"). Each voxel contains one material.
- *Subvoxel*—in the mode of DOSIVOX used in this paper, one voxel in the model may be broken down into a fine grid made up of subvoxels. This voxel is referred to as the 'sub-voxelized voxel'. The sub-voxelized voxel is used when one part of the model needs to be described in more detail. In this work, we have the case of a ceramic vessel in a tomb. The tomb itself is described at lower resolution using the voxels; one of these voxels (the sub-voxelized voxel) contains the vessel, which is described at higher resolution using subvoxels. This is elaborated below in Section 4.1.1.

4.1.1. The Model Layout

The volume that is being modelled (e.g., a tomb) must first be reconstructed, and then represented in terms of a 3D regular grid of voxels. Reconstructing the volume may be done using anything from field reports and excavation maps, to the latest 3D mapping tools like *GIS*. Here we discuss our own work using excavation reports from the early 20th century.

As discussed above, a major benefit of the Bêt Khallaf and Djer material is that Garstang's and Petrie's excavation records [12,13] were of excellent quality for the early 20th century and they allow us to reconstruct with a reasonable degree of confidence the exact burial environment of each vessel, which is further strengthened by finding parallels for similar temporal contexts and recent excavations (i.e., in the case of the Tomb of Djer).

Regardless of the source of information about the burial environment, what must eventually be constructed is a series of evenly spaced layers (i.e., horizontal sections) each of which is composed of an identical regular grid of voxels. In our model, the voxels were sized 30 cm × 30cm × 15 cm, with the total number of voxels varying with the size of the tomb being modelled. Thus, once the tomb shapes and sizes were reconstructed, each tomb layout was replicated using voxels in a three-dimensional grid.

Every voxel interior in the grid must be labelled with a number identifying a particular material. One of the voxels on one of the layers must be chosen as the sub-voxelized voxel—in our case, the location of the vessel of interest (as far as could be known, as we discuss below). Within the sub-voxelized voxel the whole process is repeated again on a finer scale. The voxel is divided into layers, which are divided into rectangular cells. Each cell, that is, each *subvoxel*, is labelled by a material representing a number (this process allows a higher spatial resolution in a given section of the tomb (i.e., the section that contains the vessel, without the cost of having such a high spatial resolution across the whole tomb ([14])).

A variety of possibilities are available when constructing the sub-voxelized voxel. As discussed in Reference [14], there are many 3D scanning tools which may be used to make a 3D image of the vessel, which can be converted into a series of subvoxel layers by, for example, associating colours with materials. In our work, we constructed the sub-voxelized voxel manually from drawings of the vessels. While we were careful to preserve the overall scale and thickness of the vessel, we determined it was unnecessary to precisely specify the shape of the vessel, owing to the long attenuation depth of the gamma rays. Accordingly, each jar-shaped vessel was simulated using a simplified model of the vessel composed of 1 cm × 1 cm × 1 cm sub-voxels forming a regular rectangular shape (and for the bowls and the pot stand other approximate shapes were used).

Figure 3 presents a set of schematic representations of the DOSIVOX simulation model used for vessel X5472 in tomb K2 at Bêt Khallaf. Figure 4 presents a similar set for vessel X5476 in the tomb of Djer. Supplementary Materials Video S1 presents a representative 3D visualisation of a ceramic vessel in a tomb.

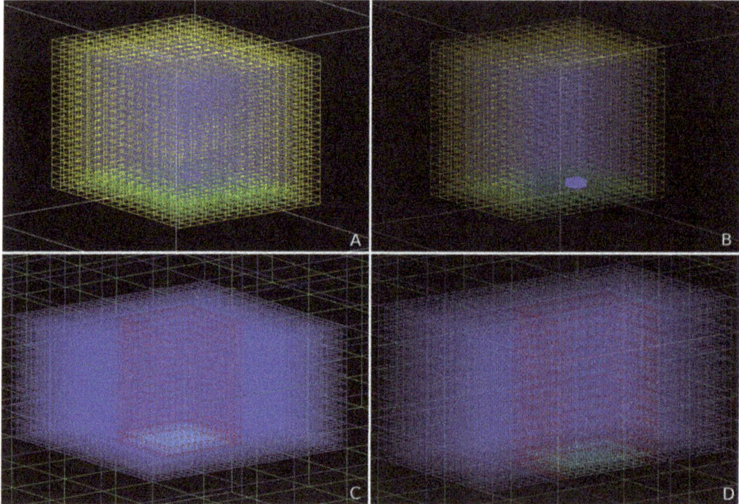

Figure 3. Wire frame representations of the DOSIVOX model for X5472. The whole tomb is subdivided into voxels which are filled with different components: yellow = gebel/bedrock, green = fill, blue = air, red = vessel walls, cyan = vessel base where the dose is recorded. (**A–D**) represent different views of the model: (**A**) whole tomb; (**B**) whole tomb cut away to reveal the detector (the sub-voxelised voxel which contains the vessel); (**C**) close up view of the detector, (**D**) detector cut away to reveal the vessel. NB these figures represent an unburied vessel.

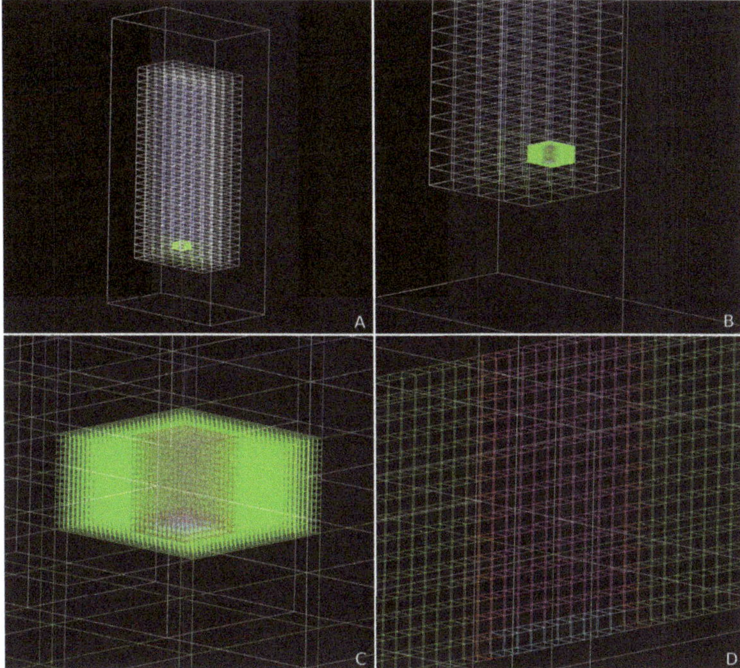

Figure 4. Wire frame representations of the DOSIVOX model for X5476 in the Tomb of Djer. The whole tomb is subdivided into voxels which are filled with different components: white = mud-brick, green = fill, blue = air, red = vessel walls, cyan = vessel base, orange = sediment attached to vessel wall (seen in (**D**) only), magenta = pot contents/residue. (**A**–**D**) represent different views of the model: (**A**) whole tomb; (**B**) whole tomb cut away to reveal the detector (the sub-voxelised voxel which contains the vessel); (**C**) close up view of the detector, (**D**) detector cut away to reveal a cross-section of the vessel and surrounding fill.

4.1.2. Describing Materials

Having labelled all voxels and subvoxels as containing a particular material, we now need to define these materials. For example, the materials of the tombs at Bêt Khallaf included air (within the tomb chamber), a sand-based fill (on the floor of the tomb), and gebel—Egyptian bedrock—which comprised the walls and roof of the tomb.

To define a material, we require its density, its bulk chemical composition, the concentrations of radioisotopes, and the water content. As with the tomb dimensions, the methodology for determining this information for each material will vary depending on the project. In an ideal situation, a large sample would be obtained of each material, which would then be sent away for analysis using, for example, a form of mass spectrometry. However, in most cases, all that is possible is to incorporate all available information, and to make assumptions where necessary. The assumptions themselves can then be tested, as we demonstrate below.

Below we present a complete list of all the materials used in this project, the methods used to analyse them and necessary assumptions.

- *Air*—assumed to be typical atmospheric concentration, with water content based on a specified temperature.
- *Fill at Bêt Khallaf*—bulk chemical composition and radioisotopes determined from analysis of sediment attached to the outside of the vessels; density assumed to be that of loose sand.

- *Gebel at Bêt Khallaf*—chemical composition and water content assumed to be the same as the fill; density assumed to be that of sandstone;
- *Fill at Djer*—radioisotopes determined from analysis of sediment attached to the outside of one vessel; bulk concentrations assumed to be similar to those at Bêt Khallaf;
- *Clay (vessel fabric)*—measurement of the bulk composition is discussed below in Appendix B.1; radioisotope concentrations were measured separately for every vessel using ICP-MS analysis of samples extracted using the minimum extraction technique [15] (the sub-voxels within the vessel were classified as two separate materials—'Clay' and 'ClayBase'—so that the average dose rate was recorded for the base (i.e., the OSL sample location) and the rest of the vessel separately. However, these two clay sub-voxel types were identical in their composition).
- *Pot contents at Bêt Khallaf*—vessels at Bêt Khallaf were assumed to be half-filled with fill.
- *Pot contents in the Tomb of Djer*—the vessels in the Tomb of Djer contained the organic/minerogenic residue of their original contents; this residue had been analysed as part of another project [11,16,17].
- *Mud-brick walls in the Tomb of Djer*—no direct samples or measurements were available of the mud-brick composition. Accordingly, the composition of the mud-brick was assumed to be similar to that of Nile silt clay.

The full chemical breakdowns of each material that resulted from these analyses and assumptions are detailed in Appendix B.

4.1.3. Specifying Radioisotope Concentrations

While the bulk chemical composition is specified only once per material, the radioisotope concentration is specified on a per-voxel (or per-subvoxel) basis. This allows the same material to have different radioisotope concentrations in different places. Accordingly, as well as a grid specifying which material is in which voxel (and another for subvoxels), we must also provide grids specifying the radioisotope concentration in each voxel/subvoxel. DOSIVOX calculates the dose from each radioactive element separately, and so a separate grid must be provided for each one.

4.2. Running and Resolving DOSIVOX

Once all the necessary data is assembled, it is time to input that data into DOSIVOX and then use it to calculate the dose rates. Running DOSIVOX consists of the following steps.

1. Obtaining DOSIVOX.
2. Generating a DOSIVOX pilot file.
3. Running DOSIVOX.
4. Extracting the dose rates for each voxel and subvoxel.
5. Repeating the previous two steps with larger and larger numbers of particles, up to the point where the dose rate is known to the required precision.

For the first step readers are referred to the DOSIVOX documentation [14]. For carrying out the second step, DOSIVOX provides a graphical tool. However it is also perfectly possible to manually construct a pilot file based on the included files. In this work, a template pilot file was manually constructed for each sample studied. From that point, an automated tool called CODERUNNER [18,19] was used to

1. set the radioisotope concentrations in the pilot file (for U, Th and K in turn);
2. run steps 3 and 4 above repeatedly, increasing the number of particles until the dose rates converged (that is, until the dose rates are known to the required precision);
3. parallelize the process by running multiple identical simulations concurrently, as detailed towards the end of this section.

Because of the Monte-Carlo methods used by DosiVox, there is no concept of time in the simulation. What is available in the DosiVox output data is the total dose emitted (E_γ) and the total dose absorbed (D_γ) by each voxel and subvoxel during the course of the simulation. In order to convert that total absorbed dose into a dose *rate* ($\dot{D}_\gamma$), we must scale it using the emitted dose rate $\dot{E}_\gamma$ (a known quantity given the concentration of the particular radioisotope, obtainable from, for example, Reference [4]). Thus, the absorbed dose rate from a given voxel or subvoxel is

$$\dot{D}_\gamma = \frac{D_\gamma}{E_\gamma} \dot{E}_\gamma. \tag{1}$$

This conversion was carried out automatically by the CodeRunner tool.

The output D_γ is provided voxel by voxel and subvoxel by subvoxel. However, it is also provided as an average per material. As mentioned earlier in Section 4.1.2, we define the material of the base of the vessel (where the samples were taken in every case) separately to the walls. Thus in every case we are calculating the average dose rate in the base of the vessel.

Given the complexity of the model and the relatively small volume occupied by the target site (the base of the vessel), a very large number of particles were required to achieve a sufficiently low uncertainty in the dose rates. Since the simulations were carried out on a laptop (albeit one with an up-to-date hyperthreaded 4-core CPU) total CPU time was limited. Nonetheless, even with the limited simulation time available, almost 550 simulations were carried out for this project (using approximately 5000 CPU hours). High-resolution simulations used 8 million particles; lower-resolution simulations used 2 or 4 million particles. For each case, six individual simulations were run in parallel (i.e., the same case, and hence the same input file, was run in parallel). As DosiVox uses a Monte Carlo method it was possible to take the mean of the six simulations and combine these to increase the simulation resolution to 48 million particles per high-resolution case, that is, six parallel simulations $\times$ 8 million particles (Pers. Comm. L. Martin, 2015).

4.2.1. Testing Assumptions in the Reconstruction

As touched on above, the information provided about the burial environments were incomplete, and several assumptions had to be made when constructing the DosiVox models. A DosiVox model allows us to obtain the best possible estimates given available sources of information. However, importantly, it allows us to assess the size of the error that may result from these assumptions. Thus, DosiVox allows us to be quantitative about the effect of missing information from the archaeological or geological record.

In this section we detail several of the necessary assumptions that were made for this project, and assess their impact on the accuracy of our results. These are assumptions concerned with:

1. The orientation of the vessel: the pot can be either standing (upright or upside down), or lying on its side. Since the OSL sample is drilled from the base in all instances in this project, the orientation of the pot in its burial environment will affect $\dot{D}_\gamma$ (i.e., was the sample location submerged in fill or exposed to air?)
2. Whether standing or lying on its side, the pot can be buried from 0% to 100% in fill, with the remainder surrounded by air (or other material).
3. The vessels can be filled from 0% to 100% with either fill or other contents (e.g., organic bulk residue).
4. The density of the vessel's clay matrix.
5. The density of the fill.
6. The vessel's proximity to the tomb walls, that is, its location within the tomb. Was it in the centre, corner, or along a side of the tomb?
7. The moisture content of the burial environment.

With regard to the position and burial situation of the vessels found within chamber tombs (i.e., where the tomb is not entirely filled up with fill, such as in the cases of Bêt Khallaf and the Tomb of

Djer), parallels at recently excavated sites can help inform of the most likely way in which the vessels were found buried. Figure 5A provides a schematic representation of how a ceramic vessel cache may look in an Early Dynastic chamber tomb (based on Hood's personal observations in the field). As can be seen, the vessels in such caches can be positioned in a number of ways and Figure 5B–E show the four most likely of these.

Figure 5. (**A–E**): A: schematic representation of a ceramic cache in a chamber tomb; (**B–E**): the four most common ways in which a ceramic vessel is deposited in a cache—the vessel is standing upright in the fill (**B**), the vessel is lying on its side partially submerged in fill (**C**), the vessel is lying on its side and resting on top of the fill (**D**), or the vessel is upside down in the fill (**E**). The red crosses seen in (**B–E**) are illustrative of OSL sample locations as it is considered best practice to take these samples from the base of a vessel wherever possible.

Owing to time constraints caused by the significant simulation time needed to test each case, this project limited itself to testing the first five assumptions listed above at high resolution, as these were deemed the most likely to affect $\dot{D}_{ext}$. Of the final two assumptions (assumptions 6–7), assumption 6 was considered, but at low resolution (the attenuation of radiation through air with low moisture content is such that it should not matter where the vessel is located as it will, in theory, still receive radiation from all four walls of the tomb). Assumption 7 was not tested as the literature consulted seemed unanimous that the correct approximate value for water content within the burial environment was 3% $\pm$ 2% ([20–23]). One situation that can not be tested for in these case studies is the proximity of other vessels to the case vessel. It was impossible, for lack of information, to determine the order or arrangement in which the vessels were located in the tombs. Thus, each vessel was modelled to be the only vessel in the modelled tomb (see Section 3).

A full description of the tests carried out for these seven assumptions and the effect of their changes on $\dot{D}_\gamma$ is given in Table 3 and illustrated in Figure 6.

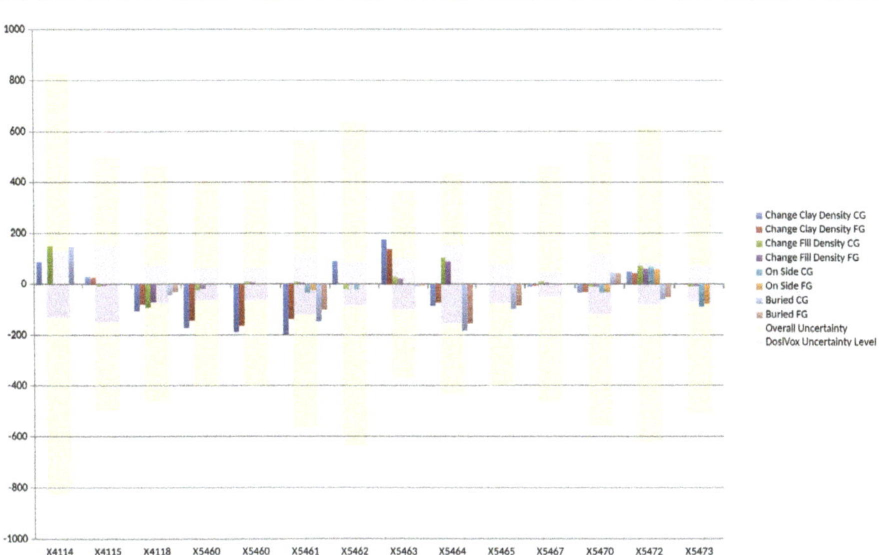

Figure 6. Graph depicting the effects of changes to the assumptions made in the DosiVox modelling. The change in the final age in years as a result of changing each assumption is plotted (using internal dose rate and equivalent dose rate data from Reference [10]). For each assumption we change, we display the effect on the age as determined from the coarse grain fraction (CG) and the fine grain fraction (FG) from the sample (NB a final age would result from combining these two values). The inner set of larger bars represent the magnitude of the error in years (at the 1σ level) resulting from the uncertainty in the DosiVox result for the reference case. It can be seen that changing all assumptions results only in errors which are comparable to or less than the typical error in the DosiVox simulations (which is itself small compared to the overall error). This means that there is insufficient evidence to reject the null hypothesis (the null hypothesis being that there is no effect of changes, or more precisely, that the effect of changes is smaller in magnitude than the uncertainty resulting from the DosiVox simulations). The outer set of larger bars represents the magnitude of overall uncertainty of the measured sample age in years. We can see that the DosiVox-derived uncertainty is itself only a minor contribution to the overall uncertainty, and that since the null hypothesis states that the effect of changes in assumptions are small, i.e., undetectable, with respect to the DosiVox-derived uncertainty, we can therefore conclude that *the effect of changing the assumptions is negligible.*

Overall it was found that changing these assumptions only had a weak effect (i.e., if there was an effect its magnitude was not statistically significant). This somewhat unexpected result is certainly worthy of further study. It is likely to be a consequence of the particular circumstances being considered here (specific dosimetries and so on). However, it may be hypothesised that it could be due to the characteristics of the emission spectra of the radionuclides. The emission spectra of U and Th are dominated by lower-energy, short-range gamma rays. Since the concentrations of U and Th are larger in the clay than in the surrounding fill, it might be expected that the dose rate from these radionuclides is dominated by the dose rate coming from the clay of the vessel (effectively the internal gamma dose rate) and will thus be unaffected by most of the assumptions tested here. By contrast the K spectrum is dominated by one high-energy, long-range gamma emission, for which the attenuation rate in any of the components of the model of the tomb will be low. Therefore, whatever the orientation of the vessel or its immediate surroundings, the absorbed K dose rate will effectively be an average of the tomb as a whole, i.e., it will be affected by whatever material is within 30 cm of it, or 30 cm of the material which is separated from the pot by air (as the gamma radiation will travel directly through air with low moisture content). Thus, even though the incomplete archaeological literature produced unavoidable uncertainty in the model, this did not give rise to significant uncertainty in the calculated $\dot{D}_\gamma$.

Table 3. The assumptions tested by DosiVox, illustrating the possible scenarios/assumptions that could be made (left column) and the outcome of the DosiVox modelling for each assumption (right column).

Modelling Unknown	Effect on Modelled Dose Rate of the Bêt Khallaf Assemblage
1. The pot can be (1) standing (upright or upside down), or (2) lying on its side. Since the OSL sample is drilled from the base in all instances, it also matters whether the sample location was on the upper or lower side of the deposited vessel (i.e., was the sample location submerged in fill or exposed to air)?	Two scenarios were considered: the pot standing on its base or lying on its side. These were considered the two limiting cases and as no measurable change was observed, it was concluded that the orientation of the pot and the location of the sample had no significant effect. To further examine the seeming minimal effect of vessel orientation on the measured $\dot{D}_{ext}$, it would be possible in the future to determine $\dot{D}_{ext}$ for a set of samples taken from different locations on the same vessel.
2. Whether standing or lying on its side, the pot can be buried up to 100% in fill, or be unburied and thus exposed to air up to 100%.	Two scenarios were considered: the vessel lying on the surface of the fill or buried ∼15 cm, which for bowls means they are completely buried and for storage jars, they are buried up to half their height. Again, no measurable change was observed.
3. The vessels can be filled from 0% to 100% with either fill or other contents (e.g., organic bulk residue).	Two scenarios were considered: the vessel was filled with fill or it had no fill inside it. Again, no measurable change was observed.
4. How does the density of the vessel's clay matrix affect the external $\dot{D}$?	Although we were not able to measure the clay density of all vessels, a single sherd (X5460) was measured as having a density of 1.4 g/cm^3. In the visual analysis of the clay fabric that accompanied each of the vessels the density of the vessels was determined as being 'high', 'medium', or 'low'. Independently of the density measurement carried out upon X5460, the vessel this sherd came from was deemed to have a 'low' density. Thus, based on this information, we were able to extrapolate that the density for vessels deemed to have a medium density was likely to be ∼1.6 g/cm^3, and for vessels with a 'high' density, ∼1.8 g/cm^3. Two scenarios were considered—the clay density was either set to its known value or to a reference value of 1.8 g/cm^3. No measurable change was observed between the two scenarios and thus it was concluded that although the clay densities are not known precisely for each vessel, variation within the likely range of values will not affect the result.

Table 3. *Cont.*

Modelling Unknown	Effect on Modelled Dose Rate of the Bêt Khallaf Assemblage
5. How does the density of the fill affect the external $\dot{D}$?	Two different fill densities were considered—1.2 g/cm^3 and 1.7 g/cm^3 and no measurable change was observed. Thus while the density of the fill is an unknown quantity, given that the fill density is visibly less than that of the low density clay whose density was measured (as above), the value of 1.2 g/cm^3 was used in final dose rate calculation. Thus, while the exact density of the Bêt Khallaf tomb fill can not be reconstructed accurately, simulation of two different scenarios illustrates that within the likely range of fill densities which could be present at Bêt Khallaf, variation within this range will not affect the modelled dose rate.
6. The vessel's proximity to the tomb walls, that is, its location within the tomb. Was it in the centre, corner, or along a side of the tomb?	A limited study of this was carried out at low resolution and no measurable changes were observed. Given the very low attenuation rates of gamma rays in air with low moisture content, this is to be expected.

Supplementary Materials: The following are available online at http://www.mdpi.com/2409-9279/2/4/91/s1, Video S1: 3D expanded view of a DOSIVOX model.

Author Contributions: This research was implemented and conceptualised by A.G.E.H., who also designed the project methodology and carried out all luminescence measurements and archaeological analysis. E.G.H. carried out the computational aspects and validation of the DOSIVOX modelling, with A.G.E.H. compiling the data required for modelling. E.G.H. constructed the derivation of the geometric approximation. A.G.E.H. and E.G.H. equally contributed to the writing and preparation of this manuscript.

Funding: This research was funded by the Clarendon Fund; the G.A. Wainwright Fund; Merton College, Oxford and Brasenose College, Oxford.

Acknowledgments: The Ashmolean Museum, Oxford, the Garstang Museum, Liverpool and the University of Pennslyvania Museum of Archaeology and Anthropology, Philadelphia provided the ceramics discussed in this paper and particular thanks are extended to Liam McNamara, Jenn Houser-Wegner and Steven Snape for their support of this research. Jean-Luc Schwenninger first brought DOSIVOX to the attention of Hood during her doctoral research and it was on his suggestion that this research was instigated. The authors extend their thanks to him also for his valuable insight and discussions on this project and for his comments and suggestions on the earliest version of this work appearing in Hood's doctoral thesis. The authors also extend their thanks to Christiana Köhler for her helpful discussions on ceramics in Early Dynastic Egyptian contexts which aided our consideration of complex depositional environments of ceramic materials. Many thanks go to the German Archaeological Institute, specifically to Martin Sählhof, for sharing his recent unpublished re-examinations from the Tomb of Djer at Abydos which were invaluable in assisting in the reconstruction of the tomb. The authors also extend their thanks to Loïc Martin, lead developer of the DOSIVOX programme for all his invaluable discussions, suggestions and trouble-shooting he provided during the course of this work. Finally thanks to Julie Durcan for her valuable insight and discussions on DRAC and dosimetry calculations.

Conflicts of Interest: The authors declare no conflict of interest.

Abbreviations

The following abbreviations are used in this manuscript:

OSL	Optically stimulated luminescence
IMA	Infinite matrix assumption
DRAC	Dose Rate and Age Calculator

Appendix A. Infinite Minute Matrix Assumption Plus Geometric Assumption

We wish to determine the radiation absorbed per unit time in an infinitesimal volume δV in the situation illustrated in Figure A1. In the case where Material 2 is identical to Material 1, the infinite matrix assumption (IMA) applies and the radiation absorbed is equal to $E_{\gamma 1}\delta V$ where $E_{\gamma 1}$ is the volumetric emissivity of Material 1. However, we may also write it as

$$E_{\gamma 1}\delta V = \int_0^{30} \delta r \int \delta\phi r^2 \Gamma\left(r,\phi\right), \tag{A1}$$

where r is the radius and $\Gamma(r,\phi)$ is the dose rate in δV that originates at radius r and solid angle ϕ.

Now let us consider the case where Material 2 is not identical to Material 1. Let us first define $\Gamma_1^a(r,\phi)$ to be the dose rate from material a where only one collision occurs (equivalent to path γ_1 in Figure A1). If we can neglect all intermediate (Compton) collisions, then we can write down the dose rate in δV as

$$\int_0^{30} \delta r \int_{\text{Material1}} \delta\phi r^2 \Gamma_1^1\left(r,\phi\right), + \int_0^{30} \delta r \int_{\text{Material2}} \delta\phi r^2 \Gamma_1^2\left(r,\phi\right) = f_1 E_{\gamma 1}\delta V + f_2 E_{\gamma 2}\delta V \tag{A2}$$

where f_1 and f_2 are the fractions of the total volume of the white sphere occupied by Materials 1 and 2 respectively. The right hand side of Equation (A2) is precisely the geometric approximation discussed in Section 1.2.2.

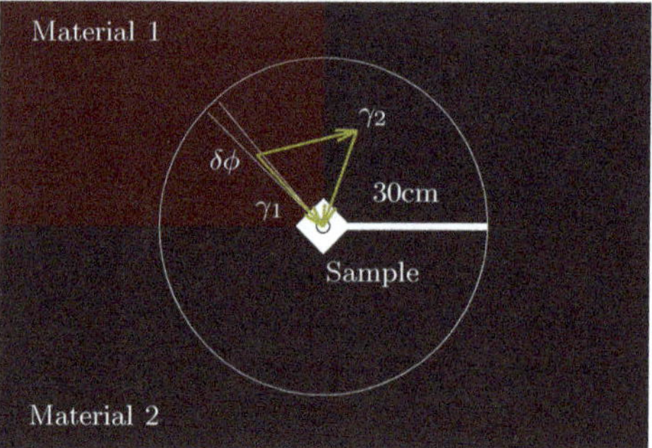

Figure A1. A sample surrounded by two materials; a solid angle $\delta\phi$ is projected onto the 2d plane; the small black circle represents an infinitesimal volume δV; gamma photons may arrive in δV directly (γ_1) or after one or more intermediate collisions (γ_2).

What about if intermediate collisions cannot be neglected? In fact the direct path γ_1 typically accounts for only half the dose (Ref. [5]: 289). In this case the geometric assumption still holds approximately for Figure A1 provided the attenuation of gamma radiation is similar for Materials 1 and 2. This is likely to be the case if, for example, Materials 1 and 2 have similar bulk compositions, differing chiefly in their radioisotope profiles. If, on the other hand, the materials have different attenuation properties, then the geometric assumption is not expected to apply.

Lastly, we note that in the case where the sample does not lie at the interface between two materials, one may use approximations like those described in Ref. [5]: 291.

Appendix B. Determining Chemical Compositions

Appendix B.1. The Bulk Composition of the Clays

The composition data for the Nile silt clay was provided by five bulk samples initially tested by a commercial laboratory, which in addition to the elemental analysis, provided a breakdown of the main components of the clay in percentages. With only five samples analysed it was not possible to know the specific chemical breakdown of each individual vessel from the Bêt Khallaf assemblage. Instead, based on these five samples, an average breakdown was determined (See Table A1). Table A2 gives the chemical composition breakdown of marl clay and Nile silt clay used in the DOSIVOX simulations (for more detailed discussion see Ref. [11]).

Table A1. Chemical breakdown of five ceramic vessels from Bêt Khallaf obtained using ICP-MS.

Sample	SiO_2	Al_2O_3	Fe_2O_3	MnO	MgO	CaO	Na_2O	K_2O	TiO_2	P_2O_5	Vacuum	Total
X5460	57.62	13.01	8.2	0.124	2.49	3.58	2.11	2.2	1.52	0.6	8.54	91.46
X4112	51.49	14.6	8.5	0.251	1.87	6.07	1.41	1.62	1.661	0.64	11.89	88.11
X4114	67.02	11.98	8.17	0.209	2.26	3.54	1.96	1.46	1.524	0.52	1.36	98.64
X4118	61.49	11.96	7.9	0.131	2.4	4.68	2.18	1.85	1.587	0.72	5.09	94.91
X4119	65.15	11.99	7.79	0.123	2.11	3.53	1.7	1.86	1.462	0.56	3.74	96.26
Mean Value	60.55	12.71	8.11	0.17	2.23	4.28	1.87	1.80	1.55	0.61	6.12	93.88

Table A2. Chemical breakdown of Nile clay, take from Table A1, and marl clay, taken from [24], used in the DOSIVOX simulations.

Clay	SiO_2	Al_2O_3	Fe_2O_3	MnO	MgO	CaO	Na_2O	K_2O	TiO_2	P_2O_5	Vacuum	Total
Nile	60.55	12.71	8.11	0.17	2.23	4.28	1.87	1.80	1.55	0.61	6.12	93.88
marl	41.3	21.95	6.76	0.05	1.8	18.8	1.39	1.20	0.90	0.45	1.3	98.7

Appendix B.2. Determining Radioisotope Concentrations

Appendix B.2.1. Isotope Concentrations in the Clay and Pot Contents

The radioisotope concentrations in the clay of the vessels were also needed to calculate the internal dose rate, and thus were already available for inclusion in the DOSIVOX model. Owing to the samples being taken from museum specimens, the minimum extraction technique was used for nearly all sample collection [15]. The first $\sim\frac{1}{2}$ of the sample was analysed using ICP-MS. Special care was taken with sample collection and measurement owing to the small size of the sample [11,15].

Tables A3 and A4 give the measured concentrations of radioisotopes in the clay and (in the case of the Tomb of Djer) contents of the vessels from Bêt Khallaf and Tomb of Djer respectively.

Table A3. Concentrations of U, Th and K (to three significant figures) for each Bêt Khallaf ceramic sample, measured using ICP-MS analysis, used to determine $\dot{D}_{int}$.

Sample	K (%)	^{232}Th (ppm)	^{238}U (ppm)
X4114	0.987	4.32	1.03
X4115	1.26	5.82	1.44
X4116	0.977	7.30	1.99
X4117	1.25	7.49	1.85
X4118	1.42	5.79	1.48
X5458	0.972	6.70	1.82
X5459	1.01	7.69	2.45
X5460^0	1.42	5.79	1.48
X5461	1.51	5.56	1.31
X5462	1.32	4.90	1.19
X5463	1.45	5.95	1.24
X5464	1.33	5.78	1.27
X5465	1.41	7.06	1.70
X5466	1.59	17.6	3.68
X5467	1.48	5.63	1.15
X5468	1.04	6.95	1.88
X5470	1.27	6.05	1.14
X5472	1.37	5.15	1.43
X5473	1.22	5.64	1.77

Table A4. Concentrations of U, Th, and K for each sample from the Tomb of Djer, measured using ICP-MS analysis, used to determine $\dot{D}_{int}$. Samples beginning with X5 are the clay fabric of the vessel; samples beginning with X6 are pot contents from select vessels.

Sample	K (%)	^{232}Th (ppm)	^{238}U (ppm)
X5474	0.87	8.15	3.75
X5475	1.08	9.03	4.21
X5476	1.21	14.19	3.39
X5477	1.00	10.57	1.93
X5478	0.85	8.75	3.46
X5479	1.00	7.99	3.20
X6114	0.51	4.2	1.3
X6115	0.37	3.2	0.8
X6116	0.38	3.3	1.8
X6120	0.13	2.26	0.35

Appendix B.2.2. Radioisotope Concentrations in the Fill

For the Bêt Khallaf material, it was noticed during the sampling process that several vessels had material adhering to their interior surface (vessels X5458, X5458, X5463, X5467, and X4120), or lodged tightly up under the rim of the vessel (vessels X5458 and X5459). It was possible therefore to collect this material for ICP-MS analysis and use the radionuclide concentrations as a proxy for the entire assemblage (see Table A5).

Table A5. Concentrations of radioisotopes in sediment samples in direct association with several Bêt Khallaf ceramic vessels.

Sample	Location of Sample	Concentration of K (%)	Concentration of ^{232}Th (ppm) [1]	Concentration of ^{238}U (ppm)
X5458	Interior of vessel	1.38	4.4	1.3
X5459	Interior of vessel	0.88	3.9	1.5
X5463	Interior of vessel	0.88	4.6	1.8
X5467	Interior of vessel	0.41	1.38	0.45
X4120	Interior of vessel	0.81	3.2	2
X5458	Under rim	0.98	3.4	1.2
X5459	Under rim	0.71	3.2	1.4
Average		0.86 ± 0.11	3.44 ± 0.40	1.38 ± 0.19

[1] ppm = parts per million mass.

The ICP-MS values for this project that were used to determine $\dot{D}_{ext}$ are presented in Table A5, with the average value being used. It was decided that an average value would be the best value to use for $\dot{D}_{ext}$ calculation, even though the values varied slightly between the individual vessels yielding external sediment. This was because the long attenuation depth of the external gamma radiation means that $\dot{D}_{ext}$ is affected by a larger body of material than the sediment in direct contact with the vessel. In this case, an average value will be more representative than an individual measurement as it will compensate for the variation in composition of the surrounding sediment.

References

1. Martin, L.; Incerti, S.; Mercier, N. DosiVox: Implementing Geant 4-based Software for Dosimetry Simulations Relevant to Luminescence and ESR Dating Techniques. *Anc. TL* **2015**, *33*, 1–10.
2. Brennan, B.J.; Lyons, R.G.; Phillips, S.W. Attenuation of Alpha Particle Track Dose for Spherical Grains. *Int. J. Radiat. Appl. Instrum. Part D Nucl. Tracks Radiat. Meas.* **1991**, *18*, 249–253. [CrossRef]
3. Aitken, M.J. *An Introduction to Optical Dating: The Dating of Quaternary Sediments by the Use of Photon-Stimulated Luminescence*; Oxford University Press: Oxford, UK, 1998.
4. Guérin, G.; Mercier, N.; Adamiec, G. Dose-rate Conversion Factors: Update. *Anc. TL* **2011**, *29*, 5–8.

5.   Aitken, M.J. *Thermoluminescence Dating*; Academic Press: London, UK, 1985.

6.   Rhodes, E.J. Optically Stimulated Luminescence Dating of Sediments Over the Past 200,000 years. *Annu. Rev. Earth Planet. Sci.* **2011**, *39*, 461–488. [CrossRef]

7.   Highcock, E.G. A Jet Sorting Algorithm for tt –> WWbb –> 6 Jets. Master's Thesis, University of Cambridge, Cambridge, UK, 2007.

8.   Durcan, J.A.; King, G.E.; Duller, G.A.T. DRAC: Dose Rate and Age Calculator for trapped charge dating. *Quat. Geochronol.* **2015**, *28*, 54–61. [CrossRef]

9.   Prescott, J.R.; Hutton, J.T. Cosmic Ray Contributions to Dose Rates for Luminescence and ESR Dating: Large Depths and Long-term Time Variations. *Radiat. Meas.* **1994**, *23*, 497–500. [CrossRef]

10. Hood, A.G.E.; Highcock, E.G.; Durcan, J.A.; Martin, L.; Dee, M.W.; Bronk Ramsey, C.; Köhler, E.C.; Schwenninger, J.L. The First Optically Stimulated Luminescence Dates for Ancient Egyptian Ceramics. **2019**, Submitted.

11. Hood, A.G.E. New Insights into Old Problems: The Application of a Multidisciplinary Approach to the Study of Early Egyptian Ceramic Chronology, with a Focus on Luminescence Dating. Ph.D. Thesis, University of Oxford, Oxford, UK, 2017.

12. Garstang, J.; Sethe, K. *Mahâsna and Bêt Khallâf*; Quaritch: London, UK, 1903.

13. Petrie, W.M.F. *The Royal Tombs of the First Dynasty*; Egypt Exploration Fund: London, UK, 1901.

14. Martin, L.; Mercier, N. *DosiVox Manual*; Université Bordeaux Montaigne: Bordeaux, France, 2015.

15. Hood, A.G.E.; Schwenninger, J.L. The Minimum Extraction Technique: A New Sampling Methodology for Optically Stimulated Luminescence Dating of Museum Ceramics. *Quat. Geochronol.* **2015**, *30*, 381–385. [CrossRef]

16. Hood, A.G.E.; Woodworth M.; Dee, M.W.; Schwenninger, J.L.; Bronk Ramsey, C.; Ditchfield, P.W. A Tale of Six Vessels: A Multidisciplinary Approach to the Analysis of Six Predynastic and Early Dynastic Vessels from Abydos, Ballas and Naqada—Preliminary Remarks. In *Egypt at its Origins 5: Proceedings of the Fifth International Conference "Origin of the State. Predynastic and Early Dynastic Egypt", Cairo, Egypt, 13–18 April 2014*; PEETERS: Leuven, Belgium, 2017.

17. Dee, M.W.; Wengrow D.; Shortland, A.J.; Stevenson, A.; Brock, F.; Bronk Ramsey, C. Radiocarbon Dating of Early Egyptian Pot Residues. In *Vienna 2—Ancient Egyptian Ceramics in the 21st Century*; Bader, B., Knoblauch, C.M., Köhler, E.C., Eds.; Peeters: Leuven, Belgium, 2016; Volume 245, pp. 119–125.

18. Highcock, E.G.; Mandell, N.; Barnes, M.; Dorland, W. Optimisation of confinement in a fusion reactor using a nonlinear turbulence model. *J. Plasma Phys.* **2018**, *84*. [CrossRef]

19. Highcock, E.G. CodeRunner Framework. 2009. Available online: http://coderunner-framework.github.io/coderunner/website/Welcome.html (accessed on 4 December 2019).

20. Sekkina, M.; El Fiki, M.A.; Nossair, S.A.; Khalil, N.R. Thermoluminescence Archaeological Dating of Pottery in the Egyptian Pyramids Zone. *Ceramics-Silikaty* **2003**, *47*, 94–99.

21. Bubenzer, O.; Hilgers, A. Luminescence Dating of Holocene Playa Sediments of the Egyptian Plateau, Western Desert, Egypt. *Quat. Sci. Rev.* **2003**, *22*, 1077–1084. [CrossRef]

22. Liritzis, I.; Vafiadou, A.; Zacharias, N.; Polymeris, G.S.; Bednarik, R.G. Advances in Surface Luminescence Dating: New Data from Selected Monuments. *Mediterr. Archaeol. Archaeom.* **2013**, *13*, 105–115.

23. Huyge, D.; Vandenberghe, D.A.G.; De Dapper, M.; Mees, F.; Claes, W.; Darnell, J.C. First Evidence of Pleistocene Rock Art in North Africa: Securing the Age of the Qurta Petroglyphs (Egypt) through OSL Dating. *Antiquity* **2011**, *85*, 1184–1193. [CrossRef]

24. Redmount, C.A.; Morgenstein, M.E. Major and Trace Element Analysis of Modern Egyptian Pottery. *J. Archaeol. Sci.* **1996**, *23*, 741–762. [CrossRef]

Article

Sedimentary Dosimetry for the Saradj-Chuko Grotto: A Cave in a Lava Tube in the North-Central Caucasus, Russia

Bonnie A. B. Blackwell [1,2,*], Mehak F. Kazi [2], Clara L. C. Huang [2], Ekaterina V. Doronicheva [3], Liubov V. Golovanova [3], Vladimir B. Doronichev [3], Impreet K. C. Singh [2] and Joel I. B. Blickstein [2]

[1] Department of Chemistry, Williams College, Williamstown, MA 01267-2692, USA

[2] RFK Science Research Institute, Glenwood Landing, NY 11547-0866, USA; mackazi101@gmail.com (M.F.K.); claralhuang@gmail.com (C.L.C.H.); impreet98@gmail.com (I.K.C.S.); joelblickstein@gmail.com (J.I.B.B.)

[3] ANO Laboratory of Prehistory, 199034 St. Petersburg, Russia; edoronicheva87@yandex.ru (E.V.D.); mezmay57@mail.ru (L.V.G.); labprehistory@yandex.ru (V.B.D.)

* Correspondence: bonnie.a.b.blackwell@williams.edu

Received: 26 November 2019; Accepted: 26 January 2020; Published: 26 February 2020

Abstract: Karst caves host most European Paleolithic sites. Near the Eurasian-Arabian Plate convergence in the Caucasus' Lower Chegem Formation, Saradj-Chuko Grotto (SCG), a lava tube, contains 16 geoarchaeologically distinct horizons yielding modern to laminar obsidian-rich Middle Paleolithic (MP) assemblages. Since electron spin resonance (ESR) can date MP teeth with 2–5% uncertainty, 40 sediment samples were analyzed by neutron activation analysis to measure volumetrically averaged sedimentary dose rates. SCG's rhyolitic ignimbrite walls produce very acidic clay-rich conglomeratic silts that retain 16–24 wt% water today. In Layers 6A-6B, the most prolific MP layers, strongly decalcified bones hinder species identification, but large ungulates inhabited deciduous interglacial forests. Unlike in karst caves, most SCG's layers had sedimentary U concentrations >4 ppm and Th, >12 ppm, but Layer 6B2 exceeded 20.8 ppm U, and Layer 7, >5 ppm Th. Such high concentrations emit dose rates averaging ~1.9–3.7 mGy/y, but locally up to 4.1–5.0 mGy/y. Within Layer 6, dose rate variations reflect bone occurrence, necessitating that several samples must be geochemically analyzed around each tooth to ensure age accuracy. Coupled with dentinal dose rates up to 3.7–4.5 mGy/y, SCG's maximum datable ages likely averages ~500–800 ka.

Keywords: ESR dating; sedimentary dosimetry; Saradj-Chuko Grotto (SCG), Russia; lava tube; Middle Paleolithic

---

1. Introduction

ESR (electron spin resonance) can establish absolute (chronometric) dates for fossils, like teeth and molluscs, which range from <10 ka to >4 Ma, depending on the radiation dose rates that the fossils experience in their depositional history. Like the other trapped charge dating methods, like thermoluminescence (TL), optically stimulated luminescence (OSL), and radioluminescence (RL), ESR measures the intensity of a radiation-induced signal recorded in the datable mineral compared with the rate at which that radiation irradiates that mineral when deposited in sediment's can date the hydroxyapatite in tooth enamel, carbonate minerals in molluscs and other invertebrates, and quartz in fault gouge and quartz-rich sediment. Used mainly in Quaternary geology and archaeology, ESR can date teeth and molluscs from some late Pliocene sites. Using teeth, ESR has been used to date many archaeological, human paleontological sites, and mammalian paleontological sites (see details in [1–6]).

In the hydroxyapatite in vertebrate tooth enamel, ESR dating uses the HAP signal at $g = 2.0018 \pm 0.0002$, which has a mean signal lifetime, $\tau_{HAP} \approx 10^{19}$ y ([3]; for all variables and their definitions, see Table A1). If a tooth experiences a low total radiation dose rate, $D_\Sigma(t)$, the minimum and maximum datable ESR age limits will be relative old, whereas a high $D_\Sigma(t)$ translates into much younger minimum and maximum age limits. By comparing the radiation doses accumulated in tooth enamel, $\mathcal{A}_\Sigma$, with $D_\Sigma(t)$ affecting the enamel, ESR dating can calculate an ESR age:

$$A_\Sigma = \int_{t_0}^{t_1} D_\Sigma(t)\mathrm{d}t \tag{1}$$

$$A_\Sigma = \int_{t_0}^{t_1} (D_{\mathrm{int}}(t) + D_{\mathrm{sed}}(t) + D_{\mathrm{cos}}(t))\mathrm{d}t \tag{2}$$

where

$\mathcal{A}_\Sigma$ = the total accumulated dose in the sample,
$D_\Sigma(t)$ = the total dose rate,
$D_{\mathrm{int}}(t)$ = the internal dose rate from within the tooth: U, its daughters, and other radioisotopes,
$D_{\mathrm{sed}}(t)$ = the external dose rate from sedimentary U, Th, K, and other radioisotopes,
$D_{\mathrm{cos}}(t)$ = the external dose rate due to cosmic radiation,
t_1 = the sample's age,
t_0 = today [7].

Ideally, 2–8 subsamples of enamel from each tooth are analyzed.

For subsample, the accumulated dose, $\mathcal{A}_\Sigma$, is determined with the additive dose method, in which 12–15 aliquots of pristine, powdered, well homogenized enamel receives added radiation doses with the highest added dose ~10 times higher than the accumulated dose measured in the 1–2 natural aliquot that receive no added irradiation. After measuring the ESR peak heights for both the naturally and artificially irradiated aliquots, the peak heights are plotted vs. their added doses to yield a growth curve whose x-intercept gives $\mathcal{A}_\Sigma$ (for example, see [6]).

In a tooth, the internal dose rate, $D_{\mathrm{int}}(t)$, depends on the U concentration in the dental tissues, and the U uptake rate at which U was absorbed into the enamel, dentine, dental cementum, and any immediately adjacent bone. In most environments, however, neither the external radiation dose rates, $D_{\mathrm{sed}}(t)$ nor $D_{\mathrm{cos}}(t)$, stay constant with time. For example, $D_{\mathrm{cos}}(t)$ drops as sediment or water accumulates above the tooth, but $D_{\mathrm{cos}}(t)$ rises again, if erosion exposes buried teeth. Meanwhile, $D_{\mathrm{sed}}(t)$ varies as deposition, erosion, cementation, or sedimentary diagenesis alter the sedimentary composition, its water concentration, and its thickness around a tooth. For most Middle Paleolithic teeth found in caves, however, low enamel and dentinal U concentrations often make $D_{\mathrm{int}}(t)$ low, while $D_{\mathrm{cos}}(t)$ tends to be small compared to $D_{\mathrm{sed}}(t)$. Together, these factors result in dating limits that range from ~10–20 ka to ~3–4 Ma and ESR ages whose accuracy and precision depend critically depend upon the accuracy and precision in the $D_{\mathrm{sed}}(t)$ measurements (see references in [1–6,8]).

In other depositional settings, however, the ESR age's dating limits, accuracy, and precision can be dramatically different. This study examines how the sedimentary radioactivity in a Paleolithic site found in a lava tube affects the ESR dating analyses by examining the dosimetry on Saradj-Chuko Grotto (SCG), a newly discovered Middle Paleolithic (MP) site occurring in a lava tube in the north-central Caucasus in Kabardino-Balkaria (Figure 1; Table 1).

Figure 1. The important stratified Middle Paleolithic sites in the Northern Caucasus, Russia. Nestled in the Saradj-Chuko Valley, the Saradj-Chuko Grotto sits close to the fluvial divide between rivers flowing northwest into the Black Sea and those flowing southeast to the Caspian Sea. Within the Kabardino-Balkaria Republic, in the central northern Caucasus Mt., Russia, this lava tube formed in the rhyolitic ignimbrites of the Gelasian-Pliocene Lower Chegem Fm. In the later Quaternary, the grotto hosted Mousterian, other Paleolithic, and Medieval peoples.

Table 1. Teeth from Saradj-Chucko Grotto in the Study.

Number					Location				Sample Type	
ESR	Field	Plan	Layer (Horizon)	Square	South, X (cm)	West, Y (cm)	Depth, Z (cm)		Species	Tooth
JT5	2018SCG35	4	6B (1)	P17	34	48	11		herbivore	cheek

2. ESR Dating in Archaeological and Human Paleontological Sites

ESR dating is often used to date the fossils in caves *sensu lato*, especially those containing Neanderthals, other early hominins, their tool assemblages, and/or vertebrate fossils. Although open-air sites do exist that have yielded Pleistocene hominin and other vertebrate fossils, the higher propensity for open-air sites to experience erosion before, during, and after deposition means that such sites tend to lack thick stratigraphical sequences, when compared to those in caves. Lava tubes, especially the smaller tubes, tend to collapse as they undergo diagenesis (for example, see [9]), likely making it harder to find them. For Neanderthals and their tool assemblages, the majority of the multi-layer stratified sites occur in caves or abris (rock shelters). Saradj-Chuko Grotto is one of the few lava tube caves in Europe that has yielded Mousterian assemblages, whereas hundreds of karst caves in Europe, western Asia, and the Mediterranean coast of northern Africa have yielded Neanderthals, or earlier hominins, and their characteristic assemblages. In Iceland, Hawaii, and other volcanic islands, several lava tube caves have yielded recent archaeological finds (for example, see [10]). In Africa, several caves in granite, ash beds, or lava tubes have yielded hominin fossils or tool assemblages, especially those near the East African Rift. These caves include Kitum [11], Leviathan [12], the Mau Mau Caves, and Makubike [13]. Again, however, many more important multi-layered sites yielding hominins, their artefacts, or Quaternary or Pliocene vertebrates have occurred in karst caves within Africa. Many important archaological and vertebrate sites known in Eastern Asia or India also occur in karst caves, although caves in lava tubes or igneous units are also known (for example, see [14]). Few studies have examined the sedimentary dosimetry in lava tubes, unlike that in karst caves.

Determining the External Dose Rates, $D_{ext}(t)$

Since Rink [1], Skinner [2], Blackwell [5], and Blackwell et al. [6] among other authors, have discussed the theory and application of ESR dating in detail, we will only discus the theory underlying the sedimentary dose rates herein. Radiation, α, β, and γ, from the sediment surrounding the dating sample generates $D_{sed}(t)$. In limestone caves, sediment may derive from aeolian and fluvial processes, mass wasting, seepage through cracks, stoping of the roof, non-human and human biogenic sources (Figure 2a). In a lava tube, like SCG, the sediment will lack or contain few, mostly carbonate deposits, while the more acidic sediment may destroy bone and carbonate fossils (Figure 2b). Thus, geological and archaeological sites have inhomogeneous ("lumpy") sediment sitting within sequences of several very thin horizons with mineralogically and biologically distinct geochemistries. This generates inhomogeneous radiation dose fields that must be measured precisely and accurately to generate reliable ESR dates.

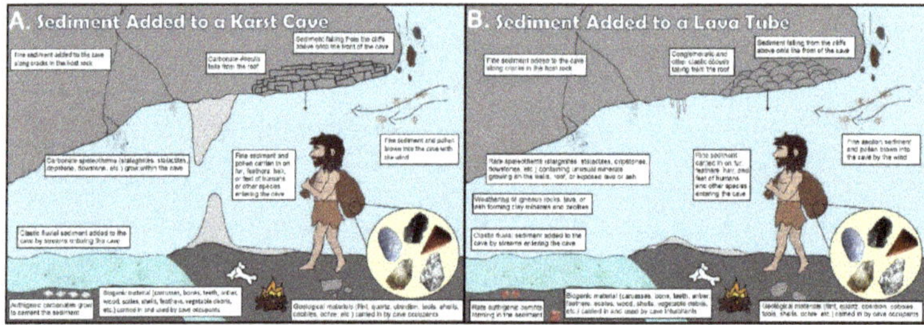

Figure 2. Sediment sources for caverns. Groundwater or surface water as fluvial or seeping in along cracks can transport sediment into caverns. Winds may add aeolian clastic or volcaniclastic sediment. Humans may contribute manuported minerals or fossils used as tools and ritual objects, but their fires also can add ash, burnt or charred wood and bone. Animals, including humans, accumulate their excretia, food debris, and their degradation products. As well as the dust and silt trapped on their feet, and in their pelts, feathers, or clothing, animals contribute their hunted or foraged prey carcasses. Plants add pollen. From the cave walls, roof, and the cliff face above the mouth come physical, chemical and biological weathering that contribute mass wasting products, including *éboulis*, as the cave stopes upward. Diagenetic processes and cementation can also add new authigenic minerals: (**A**) In karst systems, carbonate deposits, like stalagmites, stalactites, flowstone, and tufa, as well as authigenic carbonate cements usually form a significant sedimentary component. (**B**) In a grotto within a lava tube, carbonate sediment and cements will likely be insignificant compared to the clay generated by degradation of the igneous minerals or volcanic glasses.

To determine a reliable $D_{sed}(t)$ experienced by a tooth, the individual dose rate, $D_{sed,i,j}(t)$, for each mineral j, in each layer or horizon i within the sphere of influence that generates the total dose rate reaching the tooth must be measured. For the β radiation component in sediment, that sphere of influence has a radius of ~3 mm in sediment, but the γ radiation sphere's radius averages 30 cm in sediment. The radius for the sphere of influence represents the distance over which the radiation attenuates in sediment or its intensity drops below the level at which it produces a measurable effect in the sample. Radiation sources closer to the sample will produce a stronger effect on the sample than sources further away: This effect is modelled as a function of the distance squared between the source and the sample. Therefore, to assess the sedimentary β radiation dose rate, $D_{sed,\beta}(t)$, the fine sediment immediately attached to, or surrounding, and any other larger clasts, like *éboulis*, bone, or flint, within 3 mm of the dating sample is usually analyzed by neutron activation analysis (NAA) or other radiogeochemical analyses (i.e., fission track, X-ray fluorescence, TL dosimetry, β or γ spectrometry), while for the sedimentary γ radiation dose rate, $D_{sed,\gamma}(t)$, $D_{sed,\gamma,i}(t)$ measurements for all the layers or horizons within 30 cm of the dating sample must be calculated. Thus, since most archaeological and paleontological sites have inhomogeneous ("lumpy") sediment, finding the most representative value for $D_{sed}(t)$ requires collecting multiple samples immediately around the sample for sedimentary dosimetry up to 30 cm away from the tooth [5]. Brennan et al. [15] discussed the issues in lumpy sites. In most sites, using γ spectrometry and TL dosimetry usually requires that the dosimetry be completed before excavation removes any sediment. Because that sedimentary inhomogeneity produces significant variation in $D_{sed}(t)$ laterally and vertically for any tooth to be dated, γ spectrometry and TL dosimetry measurements often lack the precision and accuracy needed to get the highest reliability, if not supplemented by geochemical analyses.

Since water absorbs radiation, it reduces $D_{sed}(t)$, making it essential to correct $D_{sed,\gamma,i}(t)$ measurements, regardless of how they were measured, if the water content varies by >5 wt% [5]. In newly excavated sites, the modern water concentration, $[W_{sed}(0)]$, should be measured for each layer or horizon. Nonetheless, $[W_{sed}(0)]$ measurements do not guarantee that $[W_{sed}(t)]$ has been constant over time, because paleoclimatic and hydrological changes may have altered the time-averaged sedimentary water concentrations $\left[\overline{W}_{sed}(t)\right]$ in the past. Therefore, geomorphological, mineralogical and geochemical data should be used to model in order to find better $\overline{D}_{sed}(t)$ and $\overline{D}_{cos}(t)$.

To find the tooth's time and volumetrically averaged dose rate, $\overline{D}_{sed,\gamma}(t)$ derived from the γ radiation sources can then be calculated by integrating all $D_{sed,\gamma,i}(t)$ over the 30 cm sphere and over time [6,8,15,16]. is This includes correcting $D_{sed,\gamma,i}(t)$ for any absorption or leaching of radioactive elements from the various sedimentary components [16]. Analagously, $\overline{D}_{sed,\beta}(t)$ is integrated over the 3 mm sphere and over time.

3. Saradj-Chuko Grotto (SCG), South Russia

Lying at 934 m amsl in the central Caucasus Mt., south Russia, Saradj-Chuko Grotto sits ~35 m above the Saradj-Chuko River, which is a tributary of the Kishpek River that is, in turn, a tributary of the Baksan River within the Terek River Basin in Kabardino-Balkaria. Some 70–80 km NE of Mount Elbrus, Europe's tallest mountain, the grotto also sits 4 km south of Zayukovo and 20 km SE of Nalchik, the capital of the Kabardino-Balkaria Republic (Figure 1; [17–19]).

The cave formed in rhyolitic ignimbrites and tuff deposited in the Lower Chegem Formation, which erupted in the extension zone between the Laba-Malka monoclinal uplift and the Terek-Caspian Trough, in the continental convergence zone between the Saudi Arabian and Eurasian Plates (Figure 3). These Lower Chegem volcaniclastic deposits mainly comprise rhyolitic and liparitic tuffs, andesites, basalts, and rhyolitic ignimbrites ranging from a few tens to almost 500 m thick in the junction zone, although some extrusive domes, like Mt. Elbrus and Kazbek, also exist (Figure 3; [17–19]). Discovered in 2016, Saradj-Chuko Grotto contains 11 distinct geoarchaeological horizons ranging from modern, Medieval, Roman to Middle Paleolithic deposits (Figures 4 and 5). Averaging 5–6 cm thick, Layer 1 contains a grey sandy silt that dips to the NE. With few animal fossils and slag, mixed Medieval ceramic and several obsidian pieces suggest a bioturbated deposit. Layers 1A–C comprise yellow, sandy silt units mixed with charcoal and ash [17–19].

Dipping to the west, Layer 2 ranges from 11 to 24 cm of yellow, sandy silt (Figures 4 and 5). Although Layer 2 yielded a few mammal bones, it lacks artefacts. Also dipping to the west, Layer 3 averages 12–17 cm thick, with yellow sandy silt interspersed with three gruss horizons. Layer 3 yielded a few mammal fossils and obsidian artefacts, one of which was a Middle Paleolithic bifacial tool. Comprising 14–30 cm of grey-brown silt dipping to the west, Layer 4 also yielded obsidian artefacts and several mammal bones. Actually a tectonic fissure, Layer 5 produced flakes, but few fauna in a dark brown-black sandy silt [17–19].

Averaging 30–40 cm thick dipping to the west, Layer 6A comprises gray sandy silt with some tuff *éboulis* likely fallen from the roof (Figures 4 and 5). During the 2017 excavations, Layer 6A yielded 72 artefacts. Of these, most were made on obsidian, with only two on flint, and one on silicified limestone. With no formal tools or cores, eight flakes, 46 shatters, and 18 chips were identified, probably attributable to the Middle Paleolithic. Most of the obsidian tools and chips were made on Zayukovo (Baksan) obsidian [17–19].

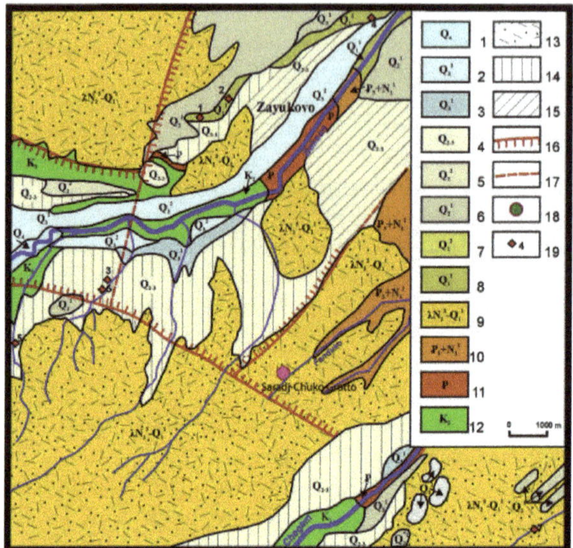

Figure 3. The Geology near Saradj-Chuko Grotto. Saradj-Chuko Grotto lies within the Lower Chegem Formation, which erupted in the extension zone between the Laba-Malka monoclinal uplift and the Tarak-Caspian Trough, during the collision zone between southern Euroasia and the Arabian Plate during the latest Pliocene and earliest Pleistocene (Gelasian). These volcaniclastic deposits extensively occur in the area. Zayukovo (Baksan) obsidian (Obsidian sites 1 and 2) was found in Layer 6B. Legend: 1–8: Quaternary units; 9: Early Gelasian-Late Pliocene Lower Chegem Fm.; 10: Middle-Upper Oligocene to Lower-Middle Miocene, Maikop Fm.; 11: Paleogene strata; 12: Upper Cretaceous strata; 13: Acidic volcanogenic deposits, mainly tuffs; 14: Undissected landslide, scree, deltaic, and fluvial units; 15: Fluvial deposits; 16: Emergent discharge, 17: Buried discharge; 18: Saradj-Chuko Grotto; 19: Obsidian and flint outcrops: 1–4: All Zayukovo (Baksan) obsidian; 5: Shtauchukua-1 flint; 6: HanaHaku-1 flint; 7: Kamenka-1 flint (adapted from Kizevalter and Karpinsky [20]).

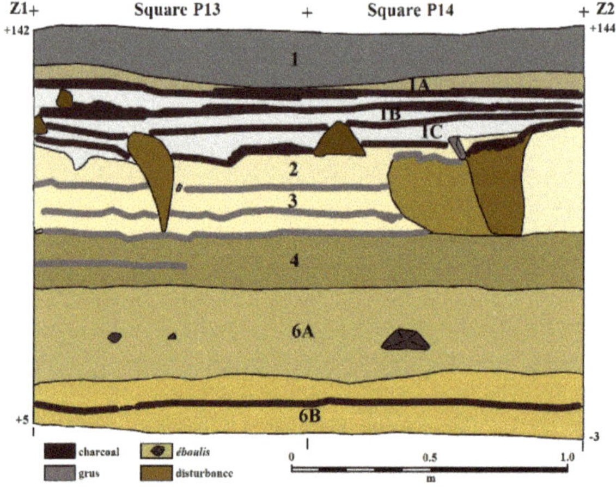

Figure 4. Profile Z1–Z2, Saradj-Chuko Grotto, Russia. In Squares Q13–Q14, nine geoarchaeological layers were identified. Cutting through Layers 2–3 in Square P17, the large disruption that begins just beneath the label for Layer 1C is likely a Medieval smelting pit or a slag discard pit.

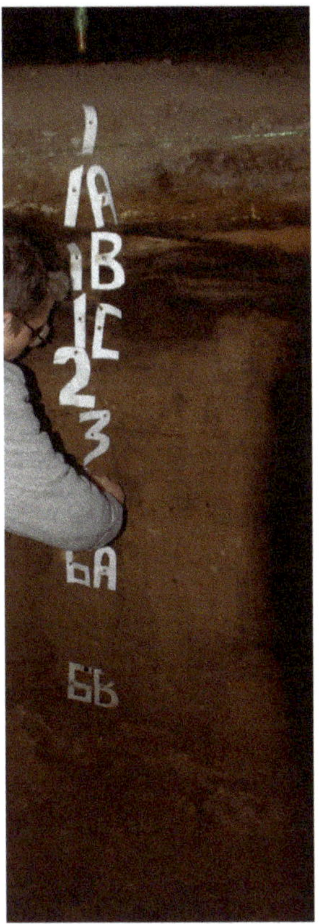

Figure 5. Sampling sediment, Saradj-Chuko Grotto, Russia. To find the sedimentary dose rates for ESR dating, 40 sediment samples were analyzed by NAA from 16 geoarchaeological horizons.

At 20 cm thick, Layer 6B contains dark brown silt with ignimbrite pebbles and tuff (Figures 4 and 5). Some organic matter contained secondary iron hydroxides, while some bones had decomposed to decalcified phosphate clusters. In 2017, Layer 6B yielded 1591 lithic artefacts, 98% of which were made on obsidian, again mainly Zayukovo obsidian [17–19].

Undoubtedly, hominins knapped local obsidian in SCG, but any flint artefacts had been imported ready to use. High quality grey, black, and pink flint was also available along the Baksan, Chegem, and Kamenka River valleys (Figure 3). The flint may have come from local sources, such as HanaHaku and Shtaucukua 5–7 km to the northwest, and from Kamenka to the southeast. Inhabitants rejuvenated flint artefacts at the site. In Layer 6B, the few retouched tools included simple side- scrapers, rare diagonal and transverse sidescrapers, convergent tools, and Mousterian points. Neanderthals likely made this Levallois-laminar Mousterian industry [17–19].

Where visible, Layer 7 ranges >20–30 cm thick (Figure 5), but it has not been fully excavated, iron hydroxide has cemented the archaeologically sterile ignimbrite. Excavations have yet not hit the grotto floor [17–19].

In Layers 6B and 7, pollen analyses found many tree and shrub grains, including *Alnus glutinosa*, *Ulmus campestris*, and *Castanea sativa*, with rare Compositae, Poaceaea, and Apiacae. These layers were deposited during an interglacial period, most likely Marine Isotope Stage (MIS) 5 when hazel-hornbeam-oak forests with oriental beech, chestnuts, and walnuts surrounded the cave. *Sula* and *Strobus* suggest that the minimum winter temperature locally reached −9 °C, while precipitation averaged 80–150 cm/y. These mixed deciduous and coniferous forests provided hominins with complex dietary choices [17–19].

Within Layer 6B, the very fragile bones ranged from whitish grey to yellowish-orangish red. In some, the original bone shape ("ghosts") could be discerned when first found, but they had been decalcified enough that they disintegrated into tiny flakes, spots, or dust if disturbed. Most intact bones had been broken or cracked before deposition. Thus, taxonomically or anatomically (typing) identifying many bones was impossible. Some bones, however, had been heated, charred, or calcined, hinting the animals had likely been hunted by Neanderthals [17–19].

4. Method

Since this study does not focus on the ESR dating age analyses, only a brief description of the tooth's preparation for ESR dating will suffice here. All samples were prepared using standard protocols for a Class 10,000 clean lab. To reduce cross-sample contamination, all glass- and plasticware were soaked in 6 M HCl(*aq*) and rinsed ≥15 times with doubly distilled, deionized water to remove all Cl⁻ ions [21].

After measuring their thicknesses with a CD-4C digital caliper, the tooth was split into four subsamples with a diamond-tipped Dremel drill, and each subsample cleaned of any attached sediment or dentine, which were removed and saved for NAA. After measuring enamel thicknesses in 30–50 places with a Mitutoyo IP-C112E micrometer, 20 μm were shaved off both enamel sides to remove the externally α-dosed enamel and thicknesses were remeasured. After powdering the enamel to 200–400 mesh (38–76 μm) in an agate mortar and pestle, it was split into 13–16 identical aliquots, each weighing 20.0 ± 0.1 mg [21].

Immediately after collection in the field (Figure 5), sediment samples were doubly sealed in ziplock bags. Samples with sharp gravel were triply bagged. Before powdering the sediment for NAA, ~30 g of sediment was sampled for $[W_{\text{sed}}(0)]$ immediately after opening the bags. To calculate $[W_{\text{sed}}(0)]$, the sediment mass was measured and gently heated at ~50 °C for several days until no mass changes were recorded to find the mass loss.

To measure $[U_{\text{sed}}]$, $[Th_{\text{sed}}]$, and $[K_{\text{sed}}]$, all associated sediment were powdered to ≤500 mesh and analyzed by neutron activation analysis (NAA). In JT5, 1–2 dentines/subsample and enamels/subsample were analyzed for their U concentrations, $[U_{\text{den}}]$ and $[U_{\text{en}}]$ respectively. After a 60.0 s irradiation and a 10.0 s delay, U was counted for 60.0 s in a delayed neutron counting system. Th and K were counted for 20.0 min in a γ counter. Th was counted after a 1.0 h irradiated and a 7.0 day delay. K had a 24–30 h delay after a 60.0 s irradiation. To ensure accuracy, all results were calibrated against NIST Standard 1633B [21].

Each $D_{\text{sed},i,j}(t)$ and its error for each mineral, *j*, in Layer *i*, was calculated from $[U_{\text{sed}}]$, $[Th_{\text{sed}}]$, $[K_{\text{sed}}]$, and $[W_{\text{sed}}(0)]$ [22], assuming no cosmic dose rate contribution, using both Rosy v. 1.4.2 [23] and Data [24]. Both these programs calculate $D_{\text{sed},i,j}(t)$ as seen by the sample sitting in a sphere of homogeneous sediment corrected for radiation backscattering and for β and γ, but not α, attenuation, due to the sedimentary and tissue water concentrations, tissue density, and thicknesses from sediment and tissues. $\overline{D}_{\text{sed}}(t)$ were calculated by using the program VolSed v. 2.2 that integrates all the applicable $D_{\text{sed},i,j}(t)$ values within 3 mm for the β dosimetry or 30 cm for the γ dosimetry respectively over their applicable sedimentary units, weighted by their volume and distance from the sample. Distances between sedimentary components and a tooth was determined visually in the field, from the total station data, and from the site photographs. For the saturation simulations and modelling, $\overline{D}_{\text{cos}}(t)$ were calculated assuming ramped box models [7] using Rosy v. 1.4.2, and integrated over time since the simulated deposition using the program AgeTimeAv v. 4.4. Comparisons between the other dosimetry methodologies, namely "instantaneous" γ and TL dosimetry, with the geochemical analyses described here, show that insignificant differences occurred between the three dosimetric methods (for example, see [5,6,25], and more references therein).

5. Results and Discussion

In Saradj-Chuko Grotto, water, U, Th, and K concentrations were measured for 16 different layers or horizons, from which $D_{\mathrm{sed},\beta}(t)$ and $D_{\mathrm{sed},\gamma}(t)$ were calculated (Table 2). To see the trends, means for the four concentrations and the two dose rates were plotted vs. depth (Figure 6).

Table 2. Sedimentary Radioactivity, Saradj-Chuko Cave, Russia.

| | | | Location | | | | Concentrations | | | | Sedimentary Dose Rates [1] | |
| | | | N–S, | E–W, | Depth, | | Water | U | Th | K | $D_{\mathrm{sed},\beta}^{BG}(t)^2$ | $D_{\mathrm{sed},\gamma}^{BG}(t)^3$ |
Sample	Layer	Square	X (cm)	Y (cm)	Z (cm)		(wt%)	(ppm)	(ppm)	(wt%)	(mGy/y)	(mGy/y)
a. Layer 1												
2018SCG11	1	Q17	72	98	+123		15.9	2.62	9.25	2.86	0.727	1.212
bulk sediment						±	0.2	0.02	0.18	0.07	0.064	0.016
b. Layer 1A												
2018SCG10	1A	Q14	40	97	+115		20.3	0.60	2.28	7.90	1.519	1.627
bulk sediment						±	0.2	0.02	0.09	0.20	0.142	0.037
c. Layer 1B												
2018SCG09	1B	Q14	35	97	+112		22.9	2.45	8.09	6.44	1.290	1.688
bulk sediment						±	0.1	0.02	0.15	0.18	0.119	0.033
d. Layer 1C												
2018SCG08	1C	Q14	43	97	+106		23.3	3.04	7.61	4.88	1.019	1.429
bulk sediment						±	0.2	0.02	0.17	0.13	0.093	0.024
e. Layer 2												
2018SCG06	2	Q15	4	96	+106		15.6	4.87	16.01	4.54	1.193	2.055
bulk sediment						±	0.4	0.02	0.27	0.12	0.105	0.029
f. Layer 3A												
2018SCG07	3A	Q17	23	97	+83		14.6	4.81	19.74	4.31	1.185	2.193
bulk sediment						±	0.1	0.02	0.35	0.11	0.104	0.027
2018SCG19	3A	Q17	58	13	+78		16.4	5.93	21.93	4.00	1.148	2.277
bulk sediment						±	0.1	0.02	0.37	0.10	0.099	0.026
bulk sediment	3A	Q17			+86		15.5	5.37	20.84	4.16	1.166	2.238
mean (*n* = 2)					− +78	±	0.9	0.56	1.09	0.16	0.072	0.019
g. Layer 3B												
2018SCG17a	3B	Q17	64	17	+79		17.5	4.28	17.82	3.84	1.014	1.889
bulk sediment						±	0.1	0.02	0.31	0.10	0.089	0.023
2018SCG18	3B	Q17	54	13	+78		17.2	4.42	17.13	3.74	1.000	1.861
bulk sediment						±	0.1	0.02	0.31	0.10	0.088	0.023
2018SCG20	3B	Q17	60	17	+72		17.8	4.31	16.99	3.96	1.029	1.873
bulk sediment						±	0.2	0.02	0.31	0.10	0.090	0.024
bulk sediment	3B	Q17			+79		17.5	4.34	17.31	3.85	1.014	1.875
mean (*n* = 3)					− +71	±	0.2	0.06	0.36	0.09	0.051	0.014
h. Layer 4												
2018SCG05	4	Q13	91	95	+69		16.5	4.60	17.42	3.92	1.054	1.943
bulk sediment						±	0.3	0.02	0.30	0.11	0.093	0.027
i. Layer 6A												
2018SCG22	6A	Q17	61	27	+17		19.7	4.66	12.91	2.62	0.735	1.436
bulk sediment						±	0.1	0.02	0.23	0.07	0.064	0.017
2018SCG23	6A	Q17	60	27	+10		18.5	3.89	13.59	2.77	0.754	1.447
bulk sediment						±	0.1	0.02	0.25	0.07	0.066	0.018
bulk sediment	6A	Q17			+17		19.1	4.28	13.25	2.70	0.745	1.441
mean (*n* = 2)					− +10	±	0.6	0.39	0.34	0.07	0.046	0.012
j. Layer 6B1a												
2018SCG29	6B1a	P16	65	53	+5		17.7	9.89	18.28	2.82	1.015	2.218
bulk sediment						±	0.1	0.02	0.31	0.07	0.084	0.019
2018SCG25	6B1a	P16	60	55	+3		18.8	8.42	17.11	3.36	1.044	2.108
bulk sediment						±	0.1	0.02	0.31	0.09	0.088	0.022
2018SCG27	6B1a	Q15	90	20	−3		21.8	5.76	14.42	2.76	0.783	1.575
bulk sediment						±	0.1	0.02	0.26	0.07	0.067	0.017
2018SCG26	6B1a	P16	60	55	−5		19.0	8.85	15.32	2.84	0.945	1.968
bulk sediment						±	0.1	0.02	0.28	0.08	0.079	0.020
2018SCG30	6B1a	P16	65	52	−5		17.8	6.98	14.76	2.71	0.867	1.781
bulk sediment						±	0.1	0.02	0.26	0.07	0.073	0.017
bulk sediment	6B1a	P16			+5		19.0	7.98	15.98	2.90	0.931	1.930
mean (*n* = 4)		-Q15			− −5	±	1.5	1.45	1.48	0.24	0.096	0.230

Table 2. *Cont.*

Sample	Layer	Square	N–S, X (cm)	E–W, Y (cm)	Depth, Z (cm)		Water (wt%)	U (ppm)	Th (ppm)	K (wt%)	$D^{BG}_{sed,\beta}(t)^2$ (mGy/y)	$D^{BG}_{sed,\gamma}(t)^3$ (mGy/y)
k. Layer 6B1b												
2018SCG32	6B1b	P18	36	90	+9		19.4	3.68	17.43	2.91	0.786	1.594
bulk sediment						±	0.1	0.02	0.31	0.08	0.068	0.021
2018SCG34	6B1b	P18	98	89	+4		18.1	3.25	13.63	2.94	0.769	1.431
bulk sediment						±	0.1	0.02	0.26	0.08	0.068	0.020
2018SCG40	6B1b	P17	23	99	−1		17.9	3.51	15.73	3.05	0.814	1.568
bulk sediment						±	0.1	0.02	0.29	0.08	0.071	0.020
2018SCG33	6B1b	P18	37	89	−2		19.9	3.53	15.53	3.01	0.783	1.512
bulk sediment						±	0.1	0.02	0.27	0.09	0.070	0.020
bulk sediment	6B1b	P17			+9		18.9	3.51	15.64	2.98	0.789	1.530
mean (*n* = 4)		-P18			−2	±	0.3	0.12	0.91	0.04	0.068	0.040
l. Layer 6B2												
2018SCG41	6B2	Q16	2	30	−1		20.8	18.43	15.13	3.15	1.303	2.810
bulk sediment						±	0.1	0.02	0.28	0.08	0.104	0.019
2018SCG42	6B2	P17	23	99	−11		21.9	14.33	13.13	2.49	1.012	2.213
bulk sediment						±	0.1	0.02	0.23	0.07	0.082	0.016
2018SCG61	6B2	Q17	40	36	−20		21.2	29.73	13.56	2.14	1.481	3.531
bulk sediment						±	0.1	0.02	0.25	0.06	0.112	0.016
bulk sediment	6B2	Q16			−1		21.3	20.83	13.94	2.59	1.276	2.851
mean (*n* = 3)		-P17			−20	±	0.5	6.51	0.86	0.42	0.339	0.539
bulk sediment	6B2	Q16			−1		21.4	16.38	14.13	2.82	1.160	2.532
mean (*n* = 2)		-P17			−20	±	0.6	2.05	1.00	0.33	0.015	0.032
m. Layer 6B3a												
2018SCG45	6B3a	P15	60	30	−9		20.9	5.30	16.94	2.90	0.819	1.682
bulk sediment						±	0.1	0.02	0.30	0.08	0.070	0.019
2018SCG44a	6B3a	P15	47	25	−10		20.4	6.15	19.00	2.53	0.802	1.795
bulk sediment						±	0.1	0.02	0.33	0.07	0.068	0.019
2018SCG44b	6B3a	P15	47	25	−10		20.4	14.19	15.90	2.82	1.108	2.421
bulk sediment						±	0.1	0.02	0.30	0.07	0.089	0.018
bulk sediment	6B3a	P15			−9		20.6	8.55	17.28	2.75	0.910	1.966
mean (*n* = 3)					−10	±	0.2	4.01	1.29	0.16	0.140	0.325
bulk sediment	6B3a	P15			−9		20.7	5.79	17.99	2.71	0.810	1.742
mean (*n* = 2)					−10	±	0.2	0.43	1.03	0.19	0.010	0.014
n. Layer 6B3b												
2018SCG54	6B3b	Q16	45	97	−10		22.2	16.49	13.27	2.56	1.093	2.408
bulk sediment						±	0.1	0.02	0.24	0.07	0.088	0.016
2018SCG49	6B3b	Q17	40	30	−11		28.2	-	-	-	-	-
bulk sediment						±	0.2	-	-	-	-	-
2018SCG58	6B3b	Q17	70	45	−16		23.6	15.62	12.69	2.06	0.949	2.175
bulk sediment						±	0.1	0.02	0.23	0.05	0.075	0.013
2018SCG57	6B3b	Q17	4	15	−17		21.0	10.66	15.09	2.81	0.972	2.058
bulk sediment						±	0.1	0.02	0.26	0.07	0.080	0.017
2018SCG62	6B3b	Q17	17	65	−19		22.4	15.40	15.05	2.43	1.040	2.353
bulk sediment						±	0.1	0.02	0.26	0.06	0.083	0.016
2018SCG55	6B3b	Q16	80	79	−19		25.2	21.12	15.73	2.51	1.195	2.771
bulk sediment						±	0.1	0.02	0.28	0.07	0.094	0.016
bulk sediment	6B3b	Q16			−10		23.8	15.86	14.37	2.47	1.050	2.353
mean (*n* = 5 or 6)		-Q17			−19	±	2.4	3.33	1.17	0.24	0.089	0.244
bulk sediment	6B3b	Q16			−10		23.0	14.54	14.03	2.47	1.014	2.249
mean (*n* = 4)		-Q17			−19	±	0.9	2.28	1.07	0.27	0.070	0.140
o. Layer 6B3c												
2018SCG46	6B3c	P15	47	22	−19		20.8	5.76	17.24	2.91	0.839	1.739
bulk sediment						±	0.1	0.02	0.31	0.08	0.071	0.019
2018SCG59	6B3c	Q17	3	10	−26		21.1	7.74	14.68	2.44	0.801	1.715
bulk sediment						±	0.1	0.02	0.26	0.06	0.067	0.015
2018SCG63	6B3c	Q17	43	39	−30		21.7	9.19	16.17	2.50	0.862	1.896
bulk sediment						±	0.1	0.02	0.27	0.07	0.072	0.017
bulk sediment	6B3c	P15			−19		21.2	7.56	16.03	2.62	0.834	1.783
mean (*n* = 3)		-Q17			−30	±	0.4	1.41	1.05	0.21	0.025	0.080
p. Layer 7												
2018SCG04	7	Q13	75	95	+5		-	10.18	27.20	3.20	1.164	2.714
bulk sediment						±	-	0.02	0.44	0.08	0.097	0.027
2018SCG03	7	Q13	75	95	−6		17.1	8.00	22.90	2.78	0.976	2.249
bulk sediment						±	0.3	0.02	0.37	0.07	0.081	0.023
bulk sediment	7	Q13			+5		17.1	9.09	25.05	2.99	1.070	2.481
mean (*n* = 2)					−6	±	0.3	1.09	2.15	0.21	0.094	0.233

Table 2. *Cont.*

Sample	Layer	Square	Location N–S, X (cm)	E–W, Y (cm)	Depth, Z (cm)	Water (wt%)	U (ppm)	Th (ppm)	K (wt%)	$D^{BG}_{sed,\beta}(t)^2$ (mGy/y)	$D^{BG}_{sed,\gamma}(t)^3$ (mGy/y)
q. Grotto walls											
2018SCG01	-	debris	-	-	-	-	11.05	31.64	3.97	1.317	3.011
weathered roof		fall				±	-	0.02	0.51	0.11	0.147	0.209
2018SCG02	-	debris	-	-	-	-	5.81	25.34	5.60	1.621	2.963
unweathered roof		pile				±	-	0.02	0.42	0.13	0.142	0.033
roof rock	-		-	-	-	-	8.43	28.49	4.79	1.469	2.987
mean (*n* = 2)						±	-	2.62	3.15	0.82	0.152	0.024

[1] Abbreviations: $D^{BG}_{sed,\beta}(t)$ = the bulk sedimentary dose rate from β sources; $D^{BG}_{sed,\gamma}(t)$ = the bulk sedimentary dose rate from γ sources; Concentrations and 1 σ errors were calculated for the closest tooth; [2] Calculated using the nearest tooth's thicknesses and clastic sediment density, ρ_{sed} = 2.66 ± 0.01 g/cm³; enamel density, ρ_{en} = 2.95 ± 0.02 g/cm³; enamel density, ρ_{den} = 2.85 ± 0.02 g/cm³; [3] Calculated using cosmic dose rate, $D_{cos}(t)$ = 0.00 ± 0.00 µGy/y; [4] Values below detection limits: Assumed to be 0.00 ± 0.00 ppm for calculations.

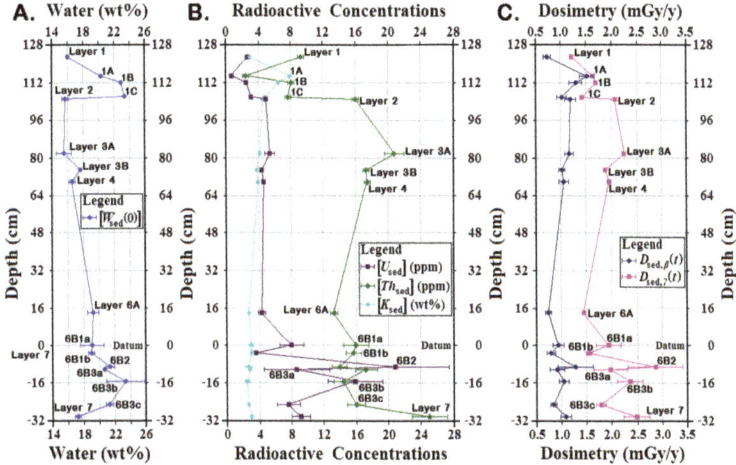

Figure 6. Factors affecting sedimentary dosimetry vs. depth, Saradj-Chuko Grotto (SCG), Russia. Mean factors plotted here show that significant changes occur with depth in the sediment at SCG: (**A**) Mean modern sedimentary water concentration, $\left[\overline{W}_{sed}(0)\right]$, vs. depth: Layers 1B, 1C, and 6B3b had highest mean sedimentary water concentrations, all of which exceeded 22 wt%, while Layers 3A and 2 had the lowest averaging <16 wt%. Layers 1, 4, and 7 also had water concentrations at 16–17 wt%. Within Layer 6, water concentrations ranged from 18.9 ± 0.3 to 23.8 ± 2.4 wt%, with higher uncertainties than seen in other layers. (**B**) Radioactive elemental concentrations vs. depth: For all but Layers 1A, 1B, 1C, and 2, mean $[K_{sed}]$ stayed below 4 wt%, with a low at 2.5 ± 0.2 wt%. Layer 1A had the highest $[K_{sed}]$ occurred at 7.9 ± 0.2 wt%, but the lowest $[Th_{sed}]$ and $[U_{sed}]$ at 2.28 ± 0.09 and 0.60 ± 0.02 ppm respectively. Below Layers 1–1C, all the mean $[Th_{sed}]$ exceeded 12 ppm, while the mean $[U_{sed}]$ exceeded 3.5 ppm. Within Layer 6, significant variations in both $[U_{sed}]$ and $[Th_{sed}]$ and their uncertainties occurred. (**C**) Sedimentary dosimetry vs. depth: While both the highest $D_{sed,\beta}(t)$ and $D_{sed,\gamma}(t)$ within Layer 6 occurred in 6B2, Layers 6B1b and 6A both had very low $D_{sed,\beta}(t)$ and $D_{sed,\gamma}(t)$. Since both $D_{sed,\beta}(t)$ and $D_{sed,\gamma}(t)$ showed substantial variation and high uncertainties from one horizon to the next, the means, $\overline{D}^{BG}_{sed,\beta,i,j}(t)$ and $\overline{D}^{BG}_{sed,\gamma,i,j}(t)$, could not be used to calculate the specific volumetrically averaged dose rates, $\overline{D}_{sed,\beta,i}(t)$ or $\overline{D}_{sed,\gamma,i}(t)$, near each tooth. Instead, several samples with 30 cm of each tooth must be measured, which will increase the cost to date each tooth.

5.1. Sedimentary Geochemistry

Throughout the layers at SCG, the sediment retained high modern water concentrations, $[W_{sed}(0)]$ (Table 2; Figure 6a). Layers 3A and 2 had the lowest mean $[W_{sed}(0)]$ at 15.5 ± 0.9 and 15.6 ± 0.4 wt% respectively, while Layer 1 averaged 15.9 ± 0.2 wt%, and Layer 4, at 16.5 ± 0.3 wt%. Probably due to its higher sand concentrations and less clay, Layer 7 had a water concentration at 17.1 ± 0.3 wt%. Within Layer 6, water concentrations ranged from 18.9 ± 0.3 to 23.8 ± 2.4 wt%, with higher uncertainties than seen in other layers. Meanwhile, Layers 1B, 1C, and 6B3b all had >22 wt% $[W_{sed}(0)]$. Within Layer 6, $[W_{sed}(0)]$ ranged from 18.9 ± 0.3 to 23.8 ± 2.4 wt%, with higher uncertainties than seen in other layers. In most karst cave sediment, $[W_{sed}(0)]$ tends to average 5–15 wt% [5,6,8,25–33]. At $[W_{sed}(0)]$ >12–15 wt%, the sediment feels damp, which tends to discourage both long-term human inhabitation and cave bear hibernation. Undoubtedly, the high clay concentrations in most of SCG's sedimentary layers ensures that they stayed wet, which also promoted higher degradation of bone and other the organic remains, as well as the igneous minerals found in the ignimbrite constituting the cave rocks. Whether under different climate regimes, the sediment may not have stayed as wet as it is now. Nonetheless, wet sediment might have discouraged hominin and cave bear inhabitation in the grotto during all but the driest seasons and the coldest glacial periods. With the high bone dissolution, the dental seasonal analyses needed to answer this question have yet to be completed.

The SCG sediment also contained high K concentrations, $[K_{sed}]$ (Table 2; Figure 6b). In both Layers 1B and 1C, $[K_{sed}]$ exceeded 6.5 wt%, but in Layer 6B3c, $[K_{sed}]$ averaged a low of 2.47 ± 0.24 ppm. Nonetheless, throughout SCG, $[K_{sed}]$ at 2.5–4.0 wt% more than doubles the typical $[K_{sed}]$ at 0.5–2.0 wt% seen in most karst cave sediment [5,6,8,25–33]. Thus, the high $[K_{sed}]$ also leads to higher $D_{sed,\beta}(t)$ (see below).

Except for Layer 6B1b, all the layers deeper than Layer 1C had sedimentary U concentrations, $[U_{sed}]$, >4 ppm (Table 2; Figure 6b). Layer 1A had the lowest U concentrations at 0.60 ± 0.02 ppm, but Layer 6B2's $[U_{sed}]$ exceeded 20.8 ppm U, with 2018SCG61 at 29.73 ± 0.02 ppm (Table 2l). In SCG, its $[U_{sed}]$ values average 4–6 ppm for those layers above Layer 6, but below, $[U_{sed}]$ generally exceeded 7.5 ppm, except for Layer 6B1b. In karst caves, the sediment tends to have $[U_{sed}]$ ≤4 ppm [2,5,6,8,25–33] namely ≤50% of that seen in SCG.

Below Layer 1C, all the mean Th concentrations, $[Th_{sed}]$, exceeded 12 ppm (Table 2; Figure 6b). The lowest $[Th_{sed}]$ occurred in Layer 1A at 2.28 ± 0.09 ppm, while Layer 7 had highest $[Th_{sed}]$ at 25.05 ± 2.15 ppm. In typical karst caves, $[Th_{sed}]$ usually ranges from 3 to 8 ppm, but the carbonate rocks often have $[Th_{sed}]$ near or below NAA $[Th_{sed}]$ detection limits [5,6,8,25–33]. Hence, SCG's $[Th_{sed}]$ ranged from 150% to >300% more than typically seen in karst cave sediment.

Interestingly, within Layer 6, large variations in both $[U_{sed}]$ and $[Th_{sed}]$ occurred (Table 2i–o; Figure 6b). Due to the high $[W_{sed}(0)]$, and its igneous source rocks that weather into sediment with high acidity, the SCG sediment contained up to 40–50% clay in many horizons. That acidity contributed to high dissolution rates for bone and other organic tissues, as shown by the prevalence of the bone ghosts seen in many layers. Once exposed to U in the sediment, however, bone, dentine, and dental cementum begin to scavenge U rapidly [34–39]. Certainly, samples with higher bone concentrations had more $[U_{sed}]$, coupled with somewhat lower $[Th_{sed}]$. These horizons also yielded the higher artefact numbers. Likely, in these horizons, the hominins contributed more bone and dental tissues to the sediment than in other horizons. Without a good proxy for the initial bone concentrations, however, estimating how much $[U_{sed,os}]$ derived from U scavenging by bone and dental issues from the groundwater compared to U scavenged from $[U_{sed,igrx}]$, including $[U_{sed,\ell b}]$, after igneous rock dissolution is difficult. Nor can we estimate how high $[U_{sed}]$ might have been for each horizon before the bones began to dissolve and release $[U_{sed,os}]$ again to the sediment, from where it may have been subsequently leached.

5.2. Sedimentary Dose Rates

Overall, the SCG's $D_{\text{sed},\beta}(t)$ ranged from 0.727 ± 0.064 to 1.519 ± 0.064 mGy/y, while $D_{\text{sed},\gamma}(t)$ varied from a low of 1.212 ± 0.016 mGy/y to a high of 2.851 ± 0.539 mGy/y (Figure 6c; Table 2). The highest $D_{\text{sed},\beta}(t)$ and $D_{\text{sed},\gamma}(t)$ occurred in Layers 2–4 and some horizons in Layers 6 and 7.

In Layer 1, although $[W_{\text{sed}}(0)]$ rose with depth, $D_{\text{sed},\beta}(t)$ reached a local maximum in Layer 1A due to high $[K_{\text{sed}}]$ and $D_{\text{sed},\gamma}(t)$ in Layer 1B due to $[Th_{\text{sed}}]$ (Figure 6; Table 2). In Layers 2–3, lower high $[W_{\text{sed}}(0)]$ coupled with high $[Th_{\text{sed}}]$ produced high $D_{\text{sed},\beta}(t)$ and $D_{\text{sed},\gamma}(t)$. In Layer 7, low $[W_{\text{sed}}(0)]$ mixed with both high $[U_{\text{sed}}]$ and $[Th_{\text{sed}}]$ again yielded high $D_{\text{sed},\beta}(t)$ and $D_{\text{sed},\gamma}(t)$.

Within Layer 6, Layers 6a, 6B1b, 6B3a, and 6B3c had lower $D_{\text{sed},\beta}(t)$ and $D_{\text{sed},\gamma}(t)$, while Layers 6B1a, 6B2, and 6B3b gave higher $D_{\text{sed},\beta}(t)$ and $D_{\text{sed},\gamma}(t)$ (Figure 6c; Table 2i–o). In Layers 6a and 6B1b, moderately low $[W_{\text{sed}}(0)]$ combined with low $[U_{\text{sed}}]$ to make low $D_{\text{sed},\beta}(t)$ and $D_{\text{sed},\gamma}(t)$. Despite higher $[W_{\text{sed}}(0)]$, the very high $[U_{\text{sed}}]$ in Layers 6B2 and 6B3b produced the highest and second highest $D_{\text{sed},\beta}(t)$ and $D_{\text{sed},\gamma}(t)$ within Layer 6.

Within Layer 6, $D_{\text{sed},\beta}(t)$, $D_{\text{sed},\gamma}(t)$, and their uncertainties varied greatly from horizon to horizon, due in part to $[U_{\text{sed},os}]$ associated with the sedimentary bone concentrations (Figure 6c; Table 2i–o). Since $[U_{\text{sed}}]$ strongly affected both $D_{\text{sed},\beta}(t)$ and $D_{\text{sed},\gamma}(t)$, the effect of $[U_{\text{sed},os}]$ was examined by plotting the means for both $D_{\text{sed},\beta}(t)$ and $D_{\text{sed},\gamma}(t)$ using all samples analyzed from the layer, as well as the means in the samples having higher amounts of bone, and those without the highest $[U_{\text{sed},os}]$ samples (Figure 7; Table 2i–o). In each of the three horizons, removing the sample with the highest $[U_{\text{sed}}]$ left a smaller mean with a significantly smaller uncertainty. Nonetheless, the mean for Layer 6B2 without the highest $[U_{\text{sed}}]$ still had much higher $D_{\text{sed},\beta}(t)$ and $D_{\text{sed},\gamma}(t)$ than for all other horizons and comparable those seen in Layer 7, which lacks hominin artefacts.

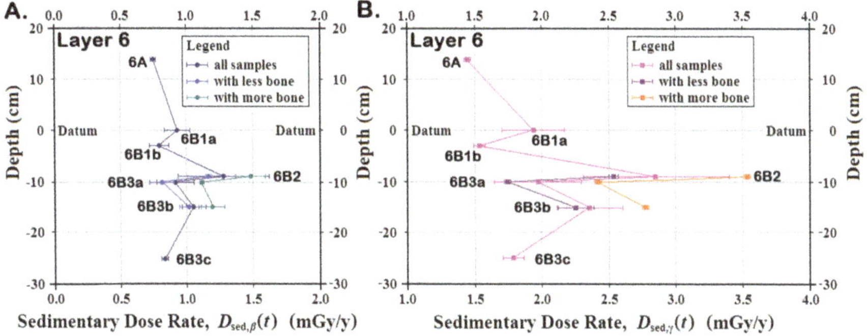

Figure 7. The effect of bone in the sedimentary dose rates, $D_{\text{sed}}(t)$, at Saradj-Chuko Grotto, Russia. Within Layer 6 at Saradj-Chuko, having more bone in the sediment produced: (**A**) higher β sedimentary dose rates, $D_{\text{sed},\beta}(t)$ (**B**) higher γ sedimentary dose rates, $D_{\text{sed},\gamma}(t)$ Bone-rich samples had up to 0.32 mGy/y for $D_{\text{sed},\beta}(t)$ and 1.0 mGy/y for $D_{\text{sed},\beta}(t)$. If bone-rich samples were removed from the means for $D_{\text{sed},\beta}(t)$ and $D_{\text{sed},\beta}(t)$, the resulting $D_{\text{sed},\beta}(t)$ and $D_{\text{sed},\gamma}(t)$ had lower mean rates with significantly smaller uncertainties. Thus, knowing the precise locations for bone-rich sediment will increase the precision for the dose rates.

In karst caves, $D_{\text{sed},\beta}(t)$ typically ranges from 100 to 500 µGy/y, while $D_{\text{sed},\gamma}(t)$ varies from 300 to 1000 µGy/y, partly due to the limestone and *éboulis* that tend to range at 30–60 µGy/y for $D_{\text{sed},\text{éb},\beta}(t)$ and 100–250 µGy/y for $D_{\text{sed},\text{éb},\gamma}(t)$ [5,6,8,16,25–33]. By comparison, at SCG, both $D_{\text{sed},\beta}(t)$ and $D_{\text{sed},\gamma}(t)$ range 3–15 times higher.

5.3. The Effects on the ESR Ages

With such high $D_{sed}(t)$ in several horizons and their high variations within Layer 6, the volumetrically averaged sedimentary dose rates, $\overline{D}_{sed}(t)$, will need to be calculated using the individual $D_{sed,i,j}(t)$ for each layer and sedimentary component within the 3 mm and 30 cm spheres of influence around each tooth. Using the means, $\overline{D}^{BG}_{sed,\beta,i,j}(t)$ and $\overline{D}^{BG}_{sed,\gamma,i,j}(t)$, for each horizon will not provide, $\overline{D}_{sed}(t)$ that will be accurate and precise enough to get the most reliable ESR ages. Thus, several individual sediment samples near each tooth must be tested by NAA. This could dramatically increase the costs for dating each tooth.

To test the dentinal dose rates, eight subsamples from JT5 were analyzed for its U concentrations, $[U_{den}]$ (Table 3). Not surprisingly, both $[U_{inden}]$ and $[U_{outden}]$ ranged from 136.12 to 162.38 ± 0.02 ppm, which produce $D_{den}(t)$ ranging from 3.751 ± 0.270 to 4.475 ± 0.322 mGy/y assuming an early U uptake model in the dentine. Again, $[U_{den}]$ tend to range 100–150 ppm higher than comparable $[U_{den}]$ values seen in dentine from karst caves [5,6,8,25–33,38,39]. JT5's $D_{den}(t)$ emits as much as ~2.5–4.0 mGy/y, significantly more than in the comparable dentine seen in Middle Paleolithic teeth collected from karst caves.

Table 3. Dental Radioactivity at Saradj-Chuko Grotto, Russia.

Sample		U Concentrations (ppm)		
		Enamel	Inner Dentine	Outer Dentine
		$[U_{en}]$	$[U_{inden}]$	$[U_{outden}]$
JT5, cheek tooth, Layer 6B1:				
JT5en1		-[1]	136.78	136.12
JT5en2		-[1]	149.66	151.08
JT5en3		-[1]	162.38	158.35
JT5en4		-[1]	156.85	161.71
Mean		-[1]	151.42	151.82
	±	-[1]	11.06	11.37
Typical concentration	~	0.01	0.01	0.01
uncertainties [2]	-	0.02	0.02	0.02
Typical isotopic	~	0.01	0.01	0.01
detection limits [2]	-	0.02	0.02	0.02
Typical water		0.02	0.05	0.05
concentrations (wt%) [2]	±	0.02	0.02	0.02

[1] Data not available. [2] Typical uncertainties, detection limits, and water concentrations depend on the tissue's mass, tissue type, and diagenetic state.

In humans, the lethal dose for 50% of people tested, LD_{50}, is ≤4 Gy of radiation, although a dose as low as 0.25 Gy produces measurable effects in the body [40]. With a combined $D_{sed}(t)$ averaging from ~1.9 to 3.7 mGy/y, but locally as high as 4.1–5.0 mGy/y, coupled with $D_{den}(t)$ as high as 3.7–4.5 mGy/y, hominins living in SCG received measurable effects after as few as ~26 years at the highest dose rates, assuming that no areas in the cave have a higher dose rate. Although they would not likely accumulate lethal doses in a lifetime, especially if the wet sediment discouraged long-term inhabitation, the effects on mutation and cancer rates likely affected people who inhabited SCG for short times periodically over many years or those visiting frequently.

Since both $D_{sed}(t)$ and $D_{den}(t)$ are so high, any tooth from SCG will have a much higher accumulated dose, $\mathcal{A}_\Sigma$, than a tooth of comparable age from a karst cave. Because the precision with which $\mathcal{A}_\Sigma$ can be calculated drops as $\mathcal{A}_\Sigma$ rises, the precision for the ESR ages for the SCG teeth also drops with the rising $\mathcal{A}_\Sigma$. For most teeth, full saturation of the HAP signal occurs at ~13–22 kGy, but must be tested in each. At the highest $D_{sed}(t)$ and $D_{den}(t)$ seen in SCG, teeth might reach their full saturation dose, $\mathcal{A}_{sat}$, within 250 ka of being deposited. More realistically, given that neither all the horizons within the spheres of influence around each tooth are likely to emit the highest $D_{sed}(t)$ nor are all $D_{den}(t)$ will have the highest concentrations, the maximum dating age could be <500–800 ka. If the teeth date <200–250 ka (i.e., Marine Isotope Stage, MIS, 7) or even <360–420 ka (i.e., MIS 11), reliable ages should be calculable. Given that the assemblages in the oldest layers resemble closely the Mousterian, one would expect their teeth to post-date 200 ka. If, however, the cave contains teeth deposited older than MIS 15, their calculated ages might represent minimum ages, because their teeth might have reached their $\mathcal{A}_{sat}$. Should more archaeological layers occur below Layer 7, the potential for encountering teeth that might have reached saturation increases.

At a depth of +11 cm, JT5 sat within Layer 6B1a, just above the boundary with Layer 6B1b. For the β dosimetry, both Layers 6b1a and 6B1b fell within JT5's sphere of influence, giving its preliminary $\overline{D}_{sed,\beta}(t) = 955 \pm 91$ µGy/y. For the γ dosimetry, 2 cm of Layer 4 and 18 cm of Layer 6A overlay JT5, while 18 cm of Layer 6B2 lay below Layer 6B1b. After estimating and correcting for the amount of bone and rooffall around JT5, its preliminary $\overline{D}_{sed,\gamma}(t) = 2000 \pm 109$ µGy/y. Final calculations must await excavation of nearby quadrants to assess the currently hidden sediment nearby and its inhomogeneous components. If JT5 was deposited during MIS 5 (i.e., ~74–128 ka), as the palynological analyses suggest [18], JT5's accumulated dose, $\mathcal{A}_{\Sigma}$, would be expected to lie between 580 and 1080 Grays, assuming a linear U uptake model (LU; Figure 8a), which would allow an definite age determination for JT5, rather than a minimum age estimate. Assuming LU, JT5 would likely reach $\mathcal{A}_{sat}$ at ages between 1.16 and 1.92 My after its initial deposition (Figure 8b). With the crystal damage rates produced by the high ionizing radiation fields bathing the teeth and the dentinal U concentrations (Table 3), however, the teeth might have uptaken U more frequently as the tooth aged. Thus, an RU model, with the uptake rate, $p > 0$, would better model the tooth's U uptake history. If so, JT5's MIS 5 ages would be older, and it would likely reach its $\mathcal{A}_{sat}$ if it had been deposited earlier in the Quaternary than would be generated by assuming LU.

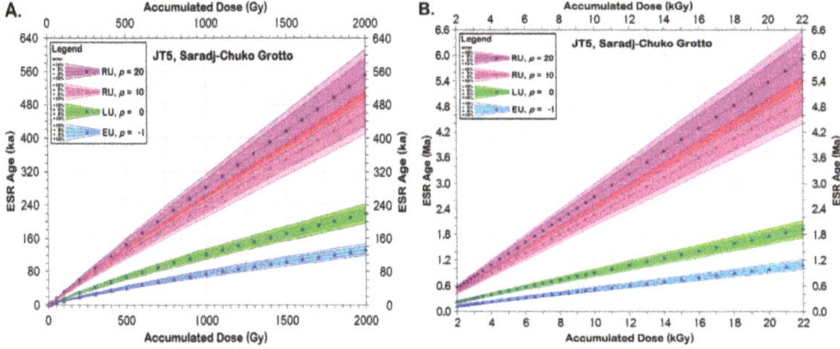

Figure 8. Calculated ages vs. accumulated doses, $\mathcal{A}_{\Sigma}$, JT5, Saradj-Chuko Grotto, Russia. Using the preliminary volumetrically averaged sedimentary dose rates, $\overline{D}_{sed,\beta}(t)$ and $\overline{D}_{sed,\gamma}(t)$, ESR ages were calculated for JT5 at: (**A**) $\mathcal{A}_{\Sigma}$ = 0.0–2.0 kGy (**B**) $\mathcal{A}_{\Sigma}$ = 2.0–22.0 kGy As $\mathcal{A}_{\Sigma}$ rises, so do the calculated ages, with a nearly linear function. Assuming that an early U uptake model (EU, $p = -1$) describes the U uptake rate into the tooth, JT5 would have $\mathcal{A}_{\Sigma}$ = 0.99–1.96 kGy if it dates to MIS 5. Under EU, JT5 would reach its saturation dose, $\mathcal{A}_{sat}$, after ~665–1100 ky following its initial deposition. Using a linear U uptake model (LU, $p = 0$), JT5 would have $\mathcal{A}_{\Sigma}$ = 0.58–1.06 kGy if it dates to MIS 5. Under LU, JT5 would reach its $\mathcal{A}_{sat}$ at ~1.16–1.92 My after its initial deposition. Assuming a recent U uptake (RU) model with an U uptake rate, $p = 10$, an age for JT5 dating to MIS 5 would have $\mathcal{A}_{\Sigma}$ = 280–520 Gy. Assuming $p = 10$ (RU), JT5 would likely reach its $\mathcal{A}_{sat}$ after ~2.92–4.94 My. Assuming $p = 20$, a tooth from MIS 5 would have $\mathcal{A}_{\Sigma}$ = 250–240 Gy, and reach its $\mathcal{A}_{sat}$ after ~3.59–5.90 My in the sediment. Thus, this analysis shows that JT5 will give definitive ESR ages regardless of its actual U uptake model. For teeth found in other layers with higher $\overline{D}_{sed,\beta}(t)$ and $\overline{D}_{sed,\gamma}(t)$, or for those found in thicker bone-rich horizons, however, their $\mathcal{A}_{\Sigma}$ may approach the saturation dose, $\mathcal{A}_{sat}$, too closely to distinguish its $\mathcal{A}_{\Sigma}$ from $\mathcal{A}_{sat}$. That would make it impossible to calculate anything other than a minimum age limit.

6. Conclusions

Unlike typical karst limestone caves, Saradj-Chuko Grotto nestles in rhyolitic ignimbrite. Due to its calcium carbonate, limestone buffers the sediment and its groundwater, but the rhyolitic ignimbrite at SCG lacks an effective geochemical buffer. Thus, in situ degradation of the ignimbrite has made the sediment very acidic. Thus, the acidic sediment contains high clay concentrations from the degraded silicates and obsidian in the lava. Acting as an aquatard, the clay also retains water well, producing $[W_{\text{sed}}(0)]$ higher than 16 wt% on average, but as high as ~24 wt%. Since clays adsorb Th, the SCG sediment contains 10–20 ppm more $[Th_{\text{sed}}]$ than those seen in most karst cave sediment. At SCG, $[K_{\text{sed}}]$ concentrations also far exceeded those in karst caves. Dissolution of bones that scavenged $[U_{\text{sed}}]$ from the local groundwater has produced $[U_{\text{sed}}]$ up to ~30 ppm, SCG's $[U_{\text{sed}}]$ averaged 5–10 times higher than $[U_{\text{sed}}]$ seen in most karst caves. In SCG, $\overline{D}_{\text{sed},\beta}(t)$ averaged from 0.727 ± 0.064 to 1.519 ± 0.142 mGy/y and $\overline{D}_{\text{sed},\gamma}(t)$ from 1.212 ± 0.016 to 2.987 ± 0.024 mGy/y. Namely, SCG's $\overline{D}_{\text{sed},\beta}(t)$ and $\overline{D}_{\text{sed},\gamma}(t)$ both exceed those seen in typical karst caves by 200–300%. In the fossils, the dentine and bone scavenged U from the uraniferous groundwater bathing the sediment, leading to high $D_{\text{den}}(t)$ measures. With high $[U_{\text{den}}]$ near 130–160 ppm, $D_{\text{den}}(t)$ also contribute radiation at 2–4 times faster than typical seen in karst caves. Therefore, for some SCG teeth found in the horizons rich in bone where the highest $\overline{D}_{\text{sed}}(t)$ occur, a viable maximum datable age may be as small as 0.25–0.8 Ma. Hominins living in SCG might have begun to experience medical effects from the high radiation rates within a few decades. People excavating in SCG should also be monitored with personal dosimeters.

Author Contributions: All authors contributed to all aspects of the publication. All authors have read and agreed to the published version of the manuscript.

Acknowledgments: The Russian Science Foundation Grant 17-78-20082 to E.V.D. "Human-nature interaction in ancient times in the central Caucasus: Dynamics of environmental change and technological innovations, and adaptations of subsistence strategies" supported the excavations and research at Saradj-Chuko Grotto (SCG). McMaster University Nuclear Reactor (MUNR), Williams College, RFK Science Research Institute (RFKSRI), and the ESR Foundation funded the ESR dating and NAA. Alice Pidruczny and her team at MUNR performed the NAA. The 2018 SCG excavation team assisted with the sediment collection in the field. Riyadh Ally and other 2019 RFKSRI crew helped with some ESR and NAA sample preparation. We thank the reviewers and the volume editor for their helpful suggestions.

Conflicts of Interest: The authors have no conflicts of interest.

Appendix A

Table A1. ESR symbols and abbreviations.

Symbol	Definition
ESR	electron spin resonance (also called electron paramagnetic resonance, EPR)
HAP	hydroxyapatite, the major constituent mineral in bone, enamale, dentine, and dentinal cementum
$\mathcal{A}_{\Sigma}$	the total accumulated radiation dose in the dated sample or subsample
$D_{\Sigma}(t)$	the total dose rate from all sources for the dated sample or subsample
$D_{\text{int}}(t)$	the dose rate from U, its daughters, and other radioisotopes inside the dated sample or subsample
$D_{\text{ext}}(t)$	the dose rate from all sources outside the dated sample or subsample
$D_{\text{sed}}(t)$	the dose rate from sedimentary U, Th, K, and other radioisotopes around the dated sample
$D_{\text{sed},\beta}(t)$	the dose rate from β particles from sedimentary U, Th, K, and other radioisotopes around the dated sample or subsample
$D_{\text{sed},\gamma}(t)$	the dose rate from γ radiation from sedimentary U, Th, K, and other radioisotopes around the dated sample or subsample
$D_{\text{cos}}(t)$	the dose rate from cosmic sources affecting the dated sample or subsample
t_1	the dated sample's or subsample's age
t_0	today
τ	the mean ESR signal lifetime for an ESR signal
τ_{HAP}	the mean ESR signal lifetime for the hydroxyapatite (HAP) signal in bone, enamel and other dentinal tissues
$[U_{\text{en}}]$	the uranium concentration in the enamel for a dated tooth
$[U_{\text{den}}]$	the uranium concentration in the dentine for a dated tooth
$[U_{\text{inden}}]$	the uranium concentration in the inner dentine for a dated tooth
$[U_{\text{outden}}]$	the uranium concentration in the outer dentine for a dated tooth
$[U_{\text{sed}}]$	the uranium concentration in the sediment around a dated sample

Table A1. *Cont.*

Symbol	Definition
$[U_{\text{sed},\acute{e}b}]$	the uranium concentration in the *éboulis* (clasts generated by mass waste as the cave roof stopes upward) within the sediment around a dated sample
$[U_{\text{sed,os}}]$	the uranium concentration in the other osseous components within the sediment around a dated sample, including in the bone, enamel, dentine, and/or dental cementum
$[U_{\text{sed,igrx}}]$	the uranium concentration in the igneous rock clasts (e.g., *éboulis*, fluvial, or aeolian clasts) in the sediment around a dated sample, including any igneous rocks, tephra, volcaniclastic deposits, etc.
$[Th_{\text{sed}}]$	the thorium concentration in the sediment around a dated sample
$[K_{\text{sed}}]$	the potassium concentration in the sediment around a dated sample
$[W_{\text{sed}}(0)]$	the modern water concentration measured now in the sediment around a dated sample
$[\overline{W}_{\text{sed}}(0)]$	the mean modern water concentration measured now in the sediment
$[W_{\text{sed}}(t)]$	the water concentration in the sediment as a function of time around a dating sample
$[\overline{W}_{\text{sed}}(t)]$	the time-averaged water concentration in the sediment around a dating sample
ρ_{en}	the density of the enamel in a dated tooth
ρ_{den}	the density of the dentine in a dated tooth
ρ_{sed}	the clastic sedimentary density around a dating sample
p	the U uptake rate (parameter) used in calculating $D_{\text{int}}(t)$
EU	the early U uptake model used in calculating $D_{\text{int}}(t)$ with $p = -1$
LU	the linear (continuous) U uptake model used in calculating $D_{\text{int}}(t)$ with $p = 0$
RU	any recent U uptake model used in calculating $D_{\text{int}}(t)$, often generally used with $p = 10$
$D_{\text{sed}}(t_0)$	the modern sedimentary dose rate measured now for a dated sample
$D_{\text{sed},i}(t)$	the sedimentary dose rate derived from Layer i
$D_{\text{sed},j}(t)$	the sedimentary dose rate derived from Component j
$D_{\text{sed},i,j}(t)$	the sedimentary dose rate derived from Component j in Layer i
$D_{\text{sed},\beta,i,j}(t)$	the individual sedimentary dose rate derived from β sources within Component j in Layer i
$D_{\text{sed},\gamma,i,j}(t)$	the individual sedimentary dose rate derived from γ sources within Component j in Layer i
$D_{\text{sed},\acute{e}b,\beta}(t)$	the sedimentary dose rate derived from β sources in the *éboulis* within the sediment
$D_{\text{sed},\acute{e}b,\gamma}(t)$	the sedimentary dose rate derived from γ sources in the *éboulis* within the sediment
$D_{\text{sed,os}}(t)$	the sedimentary dose rate derived from osseous components, including the bone, dentine, dental cementum
$D_{\text{sed,clay}}(t)$	the sedimentary dose rate derived from the clay minerals
$D_{\text{sed},\beta}^{\text{BG}}(t)$	the sedimentary dose rate from β sources derived from bulk sedimentary geochemical analyses
$D_{\text{sed},\gamma}^{\text{BG}}(t)$	the sedimentary dose rate from γ sources derived from bulk sedimentary geochemical analyses
$D_{\text{sed},\beta,i}^{\text{BG}}(t)$	the sedimentary dose rate from β sources derived from bulk sedimentary geochemical analyses in Layer i
$D_{\text{sed},\gamma,i}^{\text{BG}}(t)$	the sedimentary dose rate from γ sources derived from bulk sedimentary geochemical analyses in Layer i
$\overline{D}_{\text{sed}}(t)$	the time- and volumetrically averaged sedimentary dose rate for a dated sample
$\overline{D}_{\text{sed},\beta}(t)$	the time- and volumetrically averaged sedimentary dose rate for a dated sample from β sources
$\overline{D}_{\text{sed},\beta,i}(t)$	the time- and volumetrically averaged sedimentary dose rate for a dated sample from β sources in Layer i
$D_{\text{sed},\beta,i,j}(t)$	the individual sedimentary dose rate derived from β sources within Component j in Layer i
$D_{\text{sed},\gamma,i,j}(t)$	the individual sedimentary dose rate derived from γ sources within Component j in Layer i
$\overline{D}_{\text{sed},\beta,i,j}^{\text{BG}}(t)$	the mean sedimentary dose rate derived from bulk analyses due to β sources within Component j in Layer i
$\overline{D}_{\text{sed},\gamma,i,j}^{\text{BG}}(t)$	the mean sedimentary dose rate derived from bulk analyses due to γ sources within Component j in Layer i
$\overline{D}_{\text{sed},\gamma}(t)$	the time- and volumetrically averaged sedimentary dose rate for a dated sample from γ sources
$\overline{D}_{\text{sed},\gamma,i}(t)$	the time- and volumetrically averaged sedimentary dose rate for a dated sample from γ sources in Layer i
$\overline{D}_{\cos}(t)$	the time- and volumetrically averaged sedimentary dose rate for a dated sample
$\mathcal{A}_{\text{sat}}$	the accumulated radiation dose in the dated sample or subsample at signal saturation
LD_{50}	the lethal dose for 50% of humans tested
X	the north-south position within a square for a tooth or sediment sample relative to the (0,0,0) cave datum
Y	the east-west position within a square for a tooth or sediment sample relative to the (0,0,0) cave datum
Z	the depth for a tooth or sediment sample relative to the (0,0,0) cave datum
MIS	Marine (Oyygen) Isotope Stage

References

1. Rink, W.J. Beyond ^{14}C dating: A user's guide to long-range dating methods in archaeology. *Earth Sci. Archaeol.* **2001**, *27*, 975–1005.

2. Skinner, A.R. General principles of electron spin resonance (ESR) dating. In *The Encyclopedia of Scientific Dating Methods*; Rink, W.J., Thompson, J.W., Eds.; Springer: New York, NY, USA, 2015; pp. 246–255.

3. Skinner, A.R.; Blackwell, B.A.B.; Chasteen, D.E.; Shao, J.M.; Min, S.S. Improvements in dating tooth enamel by ESR. *Appl. Radiat. Isot.* **2000**, *52*, 1337–1344. [CrossRef]

4. Skinner, A.R.; Blackwell, B.A.B.; Chasteen, D.E.; Shao, J.M. Q-band ESR studies of fossil tooth enamel. *Quat. Sci. Rev. (Quat. Geochronol.)* **2001**, *20*, 1027–1030. [CrossRef]

5. Blackwell, B.A.B. Electron spin resonance (ESR) dating in karst environments. *Acta Cars.* **2006**, *35*, 123–147. [CrossRef]

6. Blackwell, B.A.B.; Skinner, A.R.; Blickstein, J.I.B.; Montoya, A.C.; Florentin, J.A.; Baboumian, S.M.; Ahmed, I.J.; Deely, A.E. ESR in the 21st Century: From buried valleys and deserts to the deep ocean and tectonic uplift. *Earth Sci. Rev.* **2016**, *158*, 125–159. [CrossRef]

7. Deely, A.E.; Blackwell, B.A.B.; Mylroie, J.E.; Carew, J.L.; Blickstein, J.I.B.; Skinner, A.R. Testing cosmic dose rate models for ESR: Dating corals and molluscs on San Salvador, Bahamas. *Radiat. Meas.* **2011**, *46*, 853–859. [CrossRef]

8. Blackwell, B.A.B.; Šalamanov-Korobar, L.; Huang, C.L.C.; Zhuo, J.L.; Kitanovski, B.; Blickstein, J.I.B.; Florentin, J.A.; Vasilevski, S. Hunting elusive sedimentary U and Th in an Upper Paleolithic-Middle Paleolithic (MP-UP) transition site: Increasing ESR tooth dating accuracy at Golema Pešt, Macedonia. *Radiat. Prot. Dosim.* **2019**, *186*, 92–114. [CrossRef]

9. Greeley, R. The role of lava tubes in Hawai'ian volcanoes. *US Geol. Surv. Prof. Pap.* **1987**, *1350*, 1589–1602.

10. Kennedy, J.; Brady, J.E. Into the nether world of Island Earth: A reevaluation of refuge caves in ancient Hawai'ian society. *Geoarchaeol.* **1997**, *12*, 641–655. [CrossRef]

11. Lundburg, J.; McFarlane, D.A. Speleogenesis of the Mount Elgon elephant caves, Kenya. In *Perspectives on Karst Geomorphology, Hydrology, and Geochemistry: A Tribute to Derek C. Ford and William B. White*; Harmon, R.S., Wicks, C., Eds.; GSA: Boulder, CO, USA, 2006; Volume 404, pp. 51–63.

12. Forti, P.; Galli, E.; Rossi, A. Minerogenesis of volcanic caves of Kenya. *Int. J. Speleol.* **1998**, *32*, 3–18. [CrossRef]

13. Willoughby, P.R.; Compton, T.; Bello, S.M.; Bushozi, P.M.; Skinner, A.R.; Stringer, C.B. Middle Stone Age human teeth from Magubike Rock Shelter, Iringa Region, Tanzania. *PLoS One* **2018**, *13*, e0200530. [CrossRef] [PubMed]

14. Crawford, R.L. The world's longest lava tube caves: Third revision. *J. Speleol. Soc. Korea* **1996**, *4*, 79–96.

15. Brennan, B.J.; Schwarcz, H.P.; Rink, W.J. Simulation of the γ radiation field in lumpy environments. *Radiat. Meas.* **1997**, *27*, 299–305. [CrossRef]

16. Blackwell, B.A.B.; Blickstein, J.I.B. Considering sedimentary U uptake in external dose rate determinations for ESR and luminescent dating. *Quat. Int.* **2000**, *68*, 329–343. [CrossRef]

17. Doronicheva, E.V.; Golovanova, L.V.; Doronichev, V.B.; Nedomolkin, A.G.; Shackley, M.S. The first Middle Paleolithic site exhibiting obsidian industry on the northern slopes of the central Caucasus. *Antiquity* **2017**, *91*, 1–6. [CrossRef]

18. Doronicheva, E.V.; Golovanova, L.V.; Doronichev, V.B.; Nedomolkin, A.G.; Korzinova, A.S.; Tselmovitch, V.A.; Kulkova, M.A.; Odinokova, E.V.; Shirobokov, I.G.; Ivanov, V.V.; et al. The first laminar Mousterian obsidian industry in the north-central Caucasus, Russia: Preliminary results of multi-disciplinary research at Saradj-Chuko Grotto. *Archaeol. Res. Asia* **2019**, *18*, 82–99. [CrossRef]

19. Doronicheva, E.V.; Golovanova, L.V.; Doronichev, V.B.; Shackley, M.S.; Nedomolkin, A.G. New data about exploitation of the Zayukovo (Baksan) obsidian source in northern Caucasus during the Paleolithic. *J. Archaeol. Sci.* **2019**, *23*, 157–165. [CrossRef]

20. Kizevalter, D.S.; Karpinsky, A.P. *Sheet K-38-II. The Geological Map of the USSR 1959, Scale 1:200,000*; USSR Geological Research Institute: Moscow, Russia, 1959.

21. Blackwell, B.A. Laboratory Procedures for ESR Dating of Tooth Enamel. *McMaster Univ. Dept. Geol. Tech. Memo* **1989**, *89.2*, 234.

22. Adamiec, G.; Aitken, M.J. Dose rate conversion factors: Update. *Anc. TL* **1998**, *16*, 37–50.

23. Brennan, B.J.; Rink, W.J.; McGuirl, E.L.; Schwarcz, H.P. β doses in tooth enamel by "one-group" theory and the Rosy ESR dating software. *Radiat. Meas.* **1997**, *27*, 307–314. [CrossRef]

24. Grün, R. The DATA program for the calculation of ESR age estimates on tooth enamel. *Quat. Geochronol.* **2009**, *4*, 231–232. [CrossRef]

25. Dibble, H.L.; Aldaeias, V.; Alvarez-Fernàndez, E.; Blackwell, B.A.B.; Hallett-Desguez, E.; Jacobs, Z.; Goldberg, P.; Lin, S.C.; Morala, A.; Meyer, M.C.; et al. New excavations at the site of Contrebandiers Cave, Morocco. *Paleoanthrop.* **2012**, *2012*, 145–201.

26. Blackwell, B.A.B.; Skinner, A.R.; Brassard, P.; Blickstein, J.I.B. U uptake in tooth enamel: Lessons from isochron analyses and laboratory simulation experiments. In *Proceedings of the International Symposiumon New Prospects in ESR Dosimetry and Dating*; Whitehead, N.E., Ikeya, M., Eds.; Society of ESR Applied Metrology: Osaka, Japan, 2002; Volume 18, pp. 97–118.

27. Blackwell, B.A.B.; Liang, S.S.; Golovanova, L.V.; Doronichev, V.B.; Skinner, A.R.; Blickstein, J.I.B. ESR at Treugol'naya Cave, northern Caucasus Mt., Russia: Dating Russia's oldest archaeological site and paleoclimatic change in Oxygen Isotope Stage 11. *Appl. Radiat. Isot.* **2005**, *62*, 237–245. [CrossRef]
28. Blackwell, B.A.B.; Yu, E.S.K.; Skinner, A.R.; Turk, I.; Blickstein, J.I.B.; Turk, J.; Yin, V.S.W.; Lau, B. ESR-datiranje najdišč a Divje babe I, Slovenija. In *Divje babe I: Paleotiscko Najdišče Mlajšega Pleistocena v Sloveniji (Divje babe I: Upper Pleistocene Palaeolithic Site in Slovenia), Vol. 1: Geologija in paleontologija (Geology and Paleontology)*; Turk, I., Ed.; Opera Instituti Archaeologici Sloveniae: Ljubljana, Slovenia, 2007; Volume 13, pp. 123–149.
29. Blackwell, B.A.B.; Yu, E.S.K.; Skinner, A.R.; Turk, I.; Blickstein, J.I.B.; Turk, J.; Yin, V.S.W.; Lau, B. ESR dating at Divje Babe I, Slovenija. In *Divje babe I: Paleotiscko Najdišče Mlajšega Pleistocena v Sloveniji (Divje babe I: Upper Pleistocene Palaeolithic Site in Slovenia), Vol. 1: Geologija in Paleontologija (Geology and Paleontology)*; Turk, I., Ed.; Opera Instituti Archaeologici Sloveniae: Ljubljana, Slovenia, 2007; Volume 13, pp. 151–157.
30. Blackwell, B.A.B.; Skinner, A.R.; Blickstein, J.I.B.; Golovanova, L.V.; Doronichev, V.B.; Séronie-Vivien, M.R. ESR dating at hominid and archaeological sites during the Pleistocene. In *The Sourcebook for Paleolithic Transitions*; Camps, M., Chauhan, P.R., Eds.; Springer: New York, NY, USA, 2009; pp. 93–119.
31. Blackwell, B.A.B.; Huang, Y.E.W.; Chu, S.M.; Mihailović, D.; Roksandic, M.; Dimitrijević, V.; Blickstein, J.I.B.; Skinner, A.R. ESR dating tooth enamel from the Mousterian Layers at Pešturina Cave, Serbia. In *Paleolithic and Mesolithic Research in the Central Balkans*; Mihailović, D., Ed.; Serbian Archaeological Society: Belgrade, Serbia, 2014; pp. 21–35.
32. Skinner, A.R.; Blackwell, B.A.B.; Mian, A.; Baboumian, S.; Blickstein, J.I.B.; Wrinn, P.J.; Krivoshapkin, A.I.; Derevianko, A.P.; Lundburg, J.A. ESR analyses on tooth enamel from the Paleolithic layers at the Obi-Rakhmat hominid site, Uzbekistan: Tackling a dating controversy. *Radiat. Meas.* **2007**, *42*, 1237–1242. [CrossRef]
33. Skinner, A.R.; Blackwell, B.A.B.; Martin, S.A.; Ortega, A.J.; Blickstein, J.I.B.; Golovanova, L.V.; Doronichev, V.B. ESR dating at Mezmaiskaya Cave, Russia. *Appl. Radiat. Isot.* **2005**, *62*, 219–224. [CrossRef]
34. Cherdyntsev, V.V. *Uranium-234 (Uran-234), Israel Program for Scientific Translations*; Schmorak, J., Ed.; Keter Press: Jerusalem, Israel, 1971; p. 234.
35. Gascoyne, M. Geochemistry of the actinides and their daughters. In *Uranium Series Disequilibrium: Application to Environmental Problems*, 2nd ed.; Ivanovich, M., Harmon, R.S., Eds.; Clarendon: Oxford, UK, 1992; pp. 34–94.
36. Ivanovich, M.; Harmon, R.S. (Eds.) *Uranium Series Disequilibrium: Application to Environmental Problems*, 2nd ed.; Clarendon: Oxford, UK, 2007; p. 910.
37. Osmond, J.K.; Ivanovich, M. Uranium-series mobilization and surface hydrology. In *Uranium Series Disequilibrium: Application to Environmental Problems*, 2nd ed.; Ivanovich, M., Harmon, R.S., Eds.; Clarendon: Oxford, UK, 1992; pp. 259–289.
38. Brassard, P.; Skinner, A.R.; Blackwell, B.A.B.; Blickstein, J.I.B. Specific sorption of uranium in modern and fossil dentine. In *International Conference on Luminescent & ESR Dating (LED)*; University ofNevada: Reno, NV, USA, 2002; p. 85.
39. Grün, R.; Taylor, L. Uranium and thorium in the constituents of fossil teeth. *Anc.TL* **1996**, *14*, 21–26.
40. ICRP. The Recommendations of the International Commissionon Radiological Protection. IRCP103. *Ann.ICRP* **2007**, *37*, 2–4.

Article

Testing Luminescence Dating Methods for Small Samples from Very Young Fluvial Deposits

Joel Q. G. Spencer [1],*, Sébastien Huot [2], Allen W. Archer [1] and Marcellus M. Caldas [3]

[1] Department of Geology, Kansas State University, Manhattan, KS 66506, USA; aarcher@ksu.edu
[2] Illinois State Geological Survey, Prairie Research Institute, Champaign, IL 61820, USA; shuot@illinois.edu
[3] Department of Geography and Geospatial Sciences, Kansas State University, Manhattan, KS 66506, USA; caldasma@ksu.edu
* Correspondence: joelspen@ksu.edu; Tel.: +1-785-532-6724

Received: 2 October 2019; Accepted: 4 December 2019; Published: 6 December 2019

Abstract: The impetus behind this study is to understand the sedimentological dynamics of very young fluvial systems in the Amazon River catchment and relate these to land use change and modern analogue studies of tidal rhythmites in the geologic record. Initial quartz optically stimulated luminescence (OSL) dating feasibility studies have concentrated on spit and bar deposits in the Rio Tapajós. Many of these features have an appearance of freshly deposited pristine sand, and these observations and information from anecdotal evidence and LandSat imagery suggest an apparent decadal stability. The characteristics of OSL from small (~5 cm) sub-samples from ~65 cm by ~2 cm diameter vertical cores are quite remarkable. Signals from medium-sized aliquots (5 mm diameter) exhibit very high specific luminescence sensitivity, have excellent dose recovery and recycling, essentially independent of preheat, and show minimal heat transfer even at the highest preheats. These characteristics enable measurement of very small signals with reasonable precision and, using modified single-aliquot regenerative-dose (SAR) approaches, equivalent doses as low as ~4 mGy can be obtained. Significant recuperation is observed for samples from two of the study sites and, in these instances, either the acceptance threshold was increased or growth curves were forced through the origin; recuperation is considered most likely to be a measurement artefact given the very small size of natural signals. Dose rates calculated from combined inductively coupled plasma mass spectrometry/inductively coupled plasma optical emission spectrometry (ICP-MS/ICP-OES) and high-resolution gamma spectrometry range from ~0.3 to 0.5 mGya^{-1}, and OSL ages for features so far investigated range from 13 to 34 years to several 100 years. Sampled sands are rich in quartz and yields of 212–250 µm or 250–310 µm grains indicate high-resolution sampling at 1–2 cm intervals is possible. Despite the use of medium-sized aliquots to ensure the recovery of very dim natural OSL signals, these results demonstrate the potential of OSL for studying very young active fluvial processes in these settings.

Keywords: Amazon; Rio Tapajós; quartz OSL dating; small samples; very high specific sensitivity; very young fluvial deposits

---

1. Introduction

1.1. Background

An important facet of the development of a geochronological technique is the investigation of potential age range. Much recent work in the luminescence field has focused on maximum achievable ages using high-temperature post-infrared infrared (pIRIR) signals from feldspars [1,2]. In contrast for quartz optically stimulated luminescence (OSL), the more efficient signal resetting coupled with environments where grain reworking is evident make it well suited to assessment of minimum achievable age. Notable examples are studies of young fluvial deposits [3–6] and dunes [7–11].

Regarding the application of OSL dating to fluvial sediments in the Amazon region, a number of studies have used the technique to try to constrain the origin and development of the drainage system, documenting Mid–Late Pleistocene ages [12–14], and OSL analyses have also been carried out to investigate the Late Pleistocene to Holocene development of fluvial bars [15].

The impetus behind this work was to investigate the feasibility of optically stimulated luminescence (OSL) dating of very young fluvial and shoreline landforms in the Amazon River catchment. The ultimate goal of the study is to use OSL to help understand the sedimentological dynamics of fluvial systems in the Amazon. This has relevance to the important issue of the anthropogenic effect of decades of land use and land cover change on the Amazon biome [16–18], that has impacted the stock of carbon and biodiversity [19,20] and resulted in erosion in many areas of the basin including along the rivers [21]. Furthermore, the Amazon is subject to significant marine tides, which propagate inland 1000 km from the mouth region, and OSL data have the potential to contribute to depositional models for modern analogues of ancient tidal rhythmites [22]. Initial OSL dating feasibility studies have concentrated on fluvial/shoreline features in the Rio Tapajós.

1.2. Geologic Setting

The Rio Tapajós is a major river system draining the Amazon basin, running ~1930 km from the Mato Grosso plateau (14°25′ S) north to the confluence with the Rio Amazonas at Santarém (2°25′ S) (Figure 1a). In the last ~160 km the Tapajós widens to 6–14 km, deepens considerably, and forms a ria (flooded river valley) (Figure 1b). To the south, deposits of pristine quartz-rich sand line the banks of the ria. These sands are primarily sourced by Cretaceous sandstone bedrock [23,24] that forms prominent bluffs as high as 90–120 m. Because of prevailing, equatorial tradewinds, out-of-phase peak discharge between Amazon mainstem (May–July) and Tapajós (March–May) [25], and potential tidal influence [24], the spits and bars exhibit an unusual pattern of upstream progradation. LandSat imagery indicates depositional systems have undergone only minor morphological changes in four decades (Figure 2). On the Amazonas mainstem at Santarém, a ~6 m seasonal oscillation of river level is documented [22]. Examples of both subaerially exposed and subaqueous spits and bars were identified in the Tapajós during fieldwork; the range in seasonal oscillation of the Tapajós is not as well documented but has been reported to be a similar order to the mainstem at ~5 m [24].

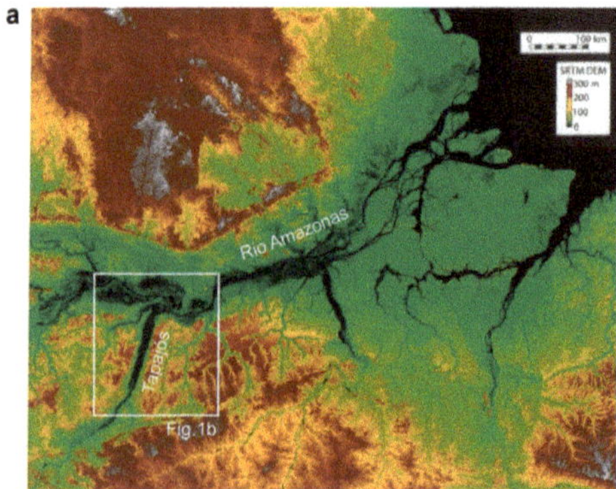
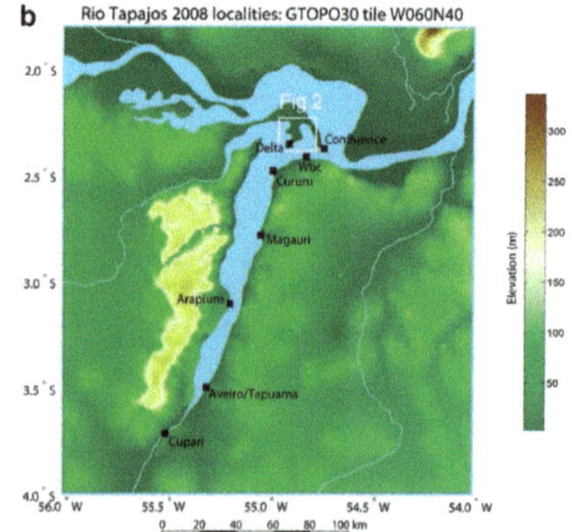

Figure 1. (**a**) Shuttle Radar Topography Mission Digital Elevation Model (SRTM DEM) image [26]. Box indicates study area in lower ~160 km stretch of the Rio Tapajós. (**b**) Detail of study area (USGS Global 30 Arc-Second Elevation data, GTOPO30, [27]) indicating sampling sites (black squares). The work described here investigated samples from Cupari, Tapuama and Arapiuns.

Figure 2. Historic LandSat imagery of bird's foot delta, northern Tapajós, indicates only minor morphological changes over the past four decades.

The drainage basin of the Tapajós is covered with dense rainforests on highly weathered, ancient shields, resulting in a clearwater river. Conversely, the mainstem of the Amazon has very high suspended sediment loads. Floodplain deposition along the Amazon has kept pace with Holocene rise in base level. Along the Tapajós, however, lack of sediment has resulted in a ria that is partially dammed along the Tapajós-Amazon confluence. The waters within these two disparate types of rivers maintain individual identity downstream of this confluence. A zone of mixing, very similar to the "meeting of the waters" at Manaus, occurs along the riverfront at Santarém.

2. Study Area and Sampling

Our study focused on shoreline features (spits, bars and dunes) that were accessed by boat and speedboat. The study area and sampling localities are shown in Figure 1b. Sands were sampled with a vertical push corer, with black spray-painted plastic sleeve inserts, allowing cores of ~65 cm in length by ~2 cm in diameter to be collected. The painted sleeves were examined carefully for complete paint coverage and tested to ensure bright white light was not visible through the painted exterior. Empirical luminescence tests of the light-tightness of the sleeves were not considered necessary but, as an added precaution, all core samples were promptly capped and immediately wrapped in thick black plastic when removed from the corer.

The work described in this study investigated core samples from three southerly sampling localities at Cupari, Tapuama and Arapiuns. At Cupari, duplicate samples (CUP-030808-01 and CUP-030808-02) separated by ~2–3 m were collected from a densely vegetated sand bar ~150 m from the riverbank. At the Tapuama locality, two samples were collected from a spit. The first (TAP-030808-03) from the unvegetated southerly distal end and the second (TAP-030808-04) from sands a few meters within the vegetated proximal end. At Arapiuns, a single sample (ARA-040808-05) was collected from the crest of a shoreline Aeolian dune directly behind a sandy beach. Sampling was conducted in the month of August, when fluvial discharge is roughly half-way between maximum (March–May) and minimum (September–December) flow periods [24]. The sampled cores were collected above the observable river water level, and probably above reach of capillary fringe influence (~0.2–0.3 m in sands [28]). Given the seasonal oscillation of the Tapajós, even if ~5 m (Section 1.2, [24]) is an overestimate, we anticipate the sampling sites oscillate between subaerial and subaqueous fluvial landforms. The altitude of all sampling sites was recorded as ~10 m above sea level with a hand-held GPS.

3. Luminescence Studies

3.1. Sample Preparation

Preparation procedures to produce 212–250 µm or 250–310 µm quartz grains for OSL analyses were carried out under low-intensity red safe lighting. Sediment at a core depth of ~63–65 cm was removed from the end of each core to exclude the possibility of analyzing grains that had been exposed to daylight during sample retrieval. The next ~5 cm from a core depth of ~58–63 cm was prepared for luminescence analyses. Standard preparation steps [29] included dry sieving, 10% HCl and 30% H_2O_2 pre-treatments to remove carbonates and organic matter, respectively, separation of heavy minerals (>2.70 gcm^{-3}) with lithium metatungstate (LMT) heavy liquid, and treating with 48% HF for 40 min to dissolve feldspar minerals and etch the surface of quartz grains to minimize luminescence due to ionization from external alpha particles. Initial test measurements indicated the dimmest natural OSL signals could only be recovered with use of medium-sized (5 mm diameter) aliquots. Potentially such small signals could be observed if a large number (e.g., ~100) of small (e.g., 1 mm) multi-grain aliquots were analyzed but given practical limitations of machine time we prepared medium-sized aliquots for all samples analyzed. Monolayers (5 mm circles; ~280 grains of 250–310 µm, and ~420 grains of 212–250 µm) of quartz grains were dispensed onto ~9.7 mm diameter stainless steel discs using silicone oil and a spray template.

The sediment at a core depth of ~63–65 cm was used for inductively coupled plasma mass spectrometry / inductively coupled plasma optical emission spectrometry (ICP-MS/ICP-OES) measurements and at a core depth of ~53–58 cm for high-resolution gamma spectrometry. After drying, these samples were pulverized in a Shatterbox ring and puck mill before sending for analysis.

3.2. Measurements

OSL measurements were carried out using a Risø TL/OSL-DA-20 reader [30], with optical stimulation of quartz provided by an array of blue light (470 nm, FWHM 20 nm) diodes, optical stimulation of feldspar with infrared (870 nm, FWHM 40 nm) diodes, a calibrated ^{90}Sr/^{90}Y beta source (~0.16 Gys^{-1}) to administer laboratory radiation doses, and a heating stage for thermal stabilization. All luminescence signals were detected in the ultraviolet (peak transmission ~340 nm) using 7.5 mm of Hoya U-340 filter with an EMI 9235QB photomultiplier tube.

Determination of the equivalent dose (D_e) was carried out using a single-aliquot regenerative-dose (SAR) protocol [31–33] with modifications. Continuous power or continuous wave OSL (CW-OSL) was conducted in all measurements. We routinely utilize post-infrared optically stimulated luminescence (post-IR OSL) measurement approaches [29,34–40], which in certain instances has been shown to improve dose recovery results even if infrared signals are negligible or absent [29]. Post-IR OSL was used to measure the luminescence from the quartz grains in this study. This procedure removes charge sensitive to infrared stimulation, commonly associated with remnant feldspathic minerals, before measuring OSL from the quartz grains. The post-IR OSL measurement comprised 40 s infrared-stimulated luminescence (IRSL) at ~117.9 mWcm^{-2} (22 Vishay TSFF5200 IR led's at 90% power) at a sample temperature of 50 °C, followed by 40 s OSL at 38.7 mWcm^{-2} (28 Nichia NSPB500S blue led's at 90% power) at a sample temperature of 125 °C. After measurement of the natural OSL in the first SAR cycle, regenerative doses in subsequent cycles were approximately 0.8, 1.6, 2.4, 0, and 0.8 Gy. The test dose administered for sensitivity correction was typically ~1.6 Gy, exceeding typical D_e values by a factor of ~10 to 400 (consistent with data from [7]). Test doses were heated to 160 °C prior to measurement. A hot bleach measurement of 40 s OSL at 280 °C was incorporated at the end of each SAR cycle [32].

All natural IRSL signals appeared negligible but natural OSL signals (Figure 3) were also very dim. There was little observable scaling in size of IRSL at the regenerative dose level suggesting IR contamination may not be a problem. However, for such small OSL signals, negligible IRSL may still be a source of overestimation and requires careful assessment if a post-IR protocol is not used.

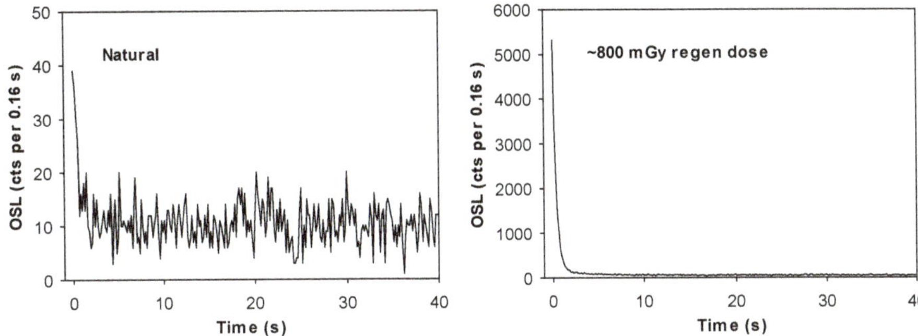

Figure 3. Typical optically stimulated luminescence (OSL) signals for an aliquot from sample Tapuama (TAP)-030808-03.

For all measurements we compared two approaches to define the net OSL signal: (1) late background subtraction where the signal was defined as the initial 0.8 s integral with subtraction of the final 8 s integral [31], and (2) early background subtraction with the same initial 0.8 s integral with subtraction of the following 0.8 to 2.72 s integral. The latter method has been assessed to optimize the contribution from the fast component [41]. The D_e value was estimated by interpolation of the natural OSL with a best-fit linear or saturating exponential curve fitted to regenerative OSL data. Uncertainty in D_e was estimated by combining error from counting statistics for the natural OSL, curve fitting, and instrumental systematic uncertainty [42].

Dose-rate measurements were conducted using the core portions described in Section 3.1. High-resolution gamma spectrometry was performed using a small-sample 2 g well geometry for assessment of U and Th. With the use of such a small sample for gamma spectrometry, a homogenous medium is assumed for accurate assessment of the radioactivity within a 30 cm radius sphere; the fluvial sand samples studied here are of uniform composition with well-sorted grain sizes, and thus a homogeneous medium is a good approximation. Li-metaborate fusion ICP-OES and ICP-MS were performed for K and Rb, respectively. These data were converted to annual dose rate using conversion factors [43]. Calculated beta dose was corrected using attenuation factors for grain size and HF etching (described in detail in [44] and references therein). In the absence of detailed imagery or documented evidence of the nature of subaerial-to-subaqueous cyclicity at the Tapuama and Cupari sites, attenuation of dose rate via moisture conditions over the burial time of the samples was calculated by using present day field moisture content with a maximum absolute error of 5% to allow for past changes. The dose rate from the ionizing cosmic ray component was calculated following [45]. For the purposes of this feasibility study, a constant overburden depth was assumed; we deliberately chose the deepest part of the cores for our sample selection in an attempt to minimize shallow gamma and hard cosmic corrections [46]. Finally, an estimate of an internal dose rate of 0.01 ± 0.002 mGya^{-1} [47] was incorporated into total dose-rate assessment.

3.3. Luminescence Characteristics

3.3.1. Specific Sensitivity of OSL Signals

An important consideration with dating features on an annual to decadal timescale is whether the luminescence is sufficiently sensitive to enable measurement of very small radiation doses. Quartz from the Tapajós shows weak natural OSL signals but very high specific sensitivity in response to a small regenerative dose (Figure 3). In Section 1.2 above, we suggest that the primary source of sand in the Tapajós is from Cretaceous sandstone bedrock that forms prominent bluffs. Given the high sensitivity of the quartz, the interplay of sources from Paleozoic [23] or even Paleoproterozoic [48] igneous and metamorphic rocks from the Amazon Craton must also be considered.

3.3.2. Effect of Preheating

The importance of investigating the influence of preheating for very young samples has been emphasized in previous studies [7]. We carried out D_e plateau tests and thermal transfer tests on all samples investigated; the latter tests were carried out using an optical bleach at ambient temperature (40 s OSL, 4000 s pause, 40 s OSL; [7]). For the samples from Tapuama, these data indicate a low temperature D_e plateau region below 210 °C (Figure 4), and minimal thermal transfer in a similar temperature band (Figure 5). The samples from Cupari and Arapiuns showed similar characteristics. A plateau in D_e data is a possible indication of complete resetting but, given the medium aliquots utilized in this work, such an interpretation must be treated with some caution.

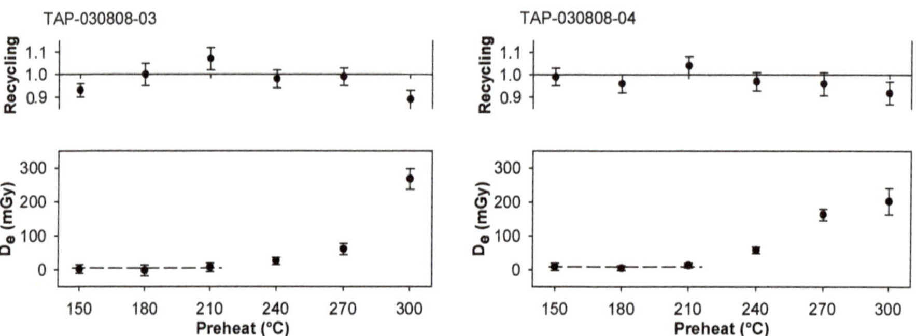

Figure 4. Determination of the equivalent dose (D_e) plateau test results on samples from Tapuama.

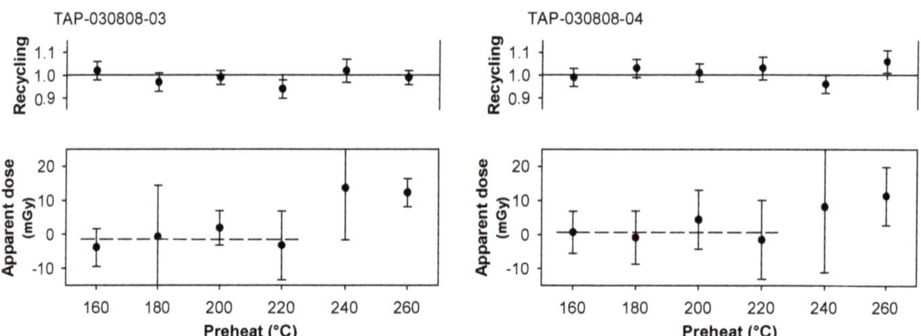

Figure 5. Thermal transfer test results on samples from Tapuama.

3.3.3. Dose Recovery Tests

Dose recovery tests with preheat variation [29] were carried out on all the samples investigated. Test results that typify the behavior of all samples are shown in Figure 6 for TAP-030808-03 and TAP-030808-04. The given dose was ~1.6 Gy (10 s beta exposure). This value was considerably larger than the D_e values indicated in Figures 4 and 7, and Table 1, but was considered prudent because the offset time [49] for our source had not been determined, and even doses administered for 1 s of source exposure would exceed D_e by a factor of ~40 for youngest samples. An implication of using a relatively large given dose is that evidence of the thermal transfer signal above ~220 °C is masked (Figure 6). Despite this, measured-to-given ratios for all samples were close to unity and imply the first sensitivity measurement is appropriate to the preceding natural OSL [32,33], and support choice of preheat (see Table 1) from D_e plateau tests and thermal transfer tests.

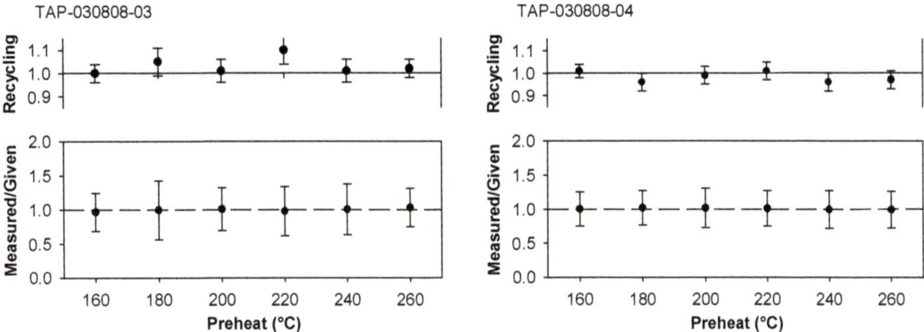

Figure 6. Dose recovery test data on samples from Tapuama.

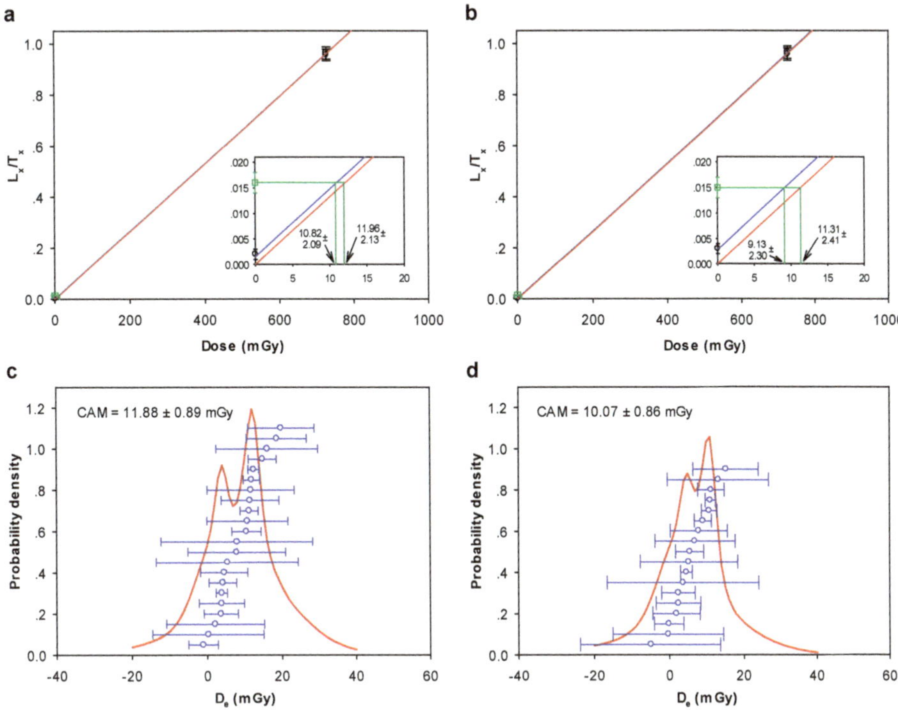

Figure 7. (**a**) and (**b**): Examples of growth curves of sensitivity-corrected OSL (L_x/T_x) with regenerative dose for two aliquots of Arapiuns (ARA)-040808-05. Only growth to first regenerative recycled point is shown, with linear fit corresponding to all regenerative dose points. Inset figures show details of interpolation with the natural OSL in the first 20 mGy of the growth curve. Blue line is fit to L_x/T_x value for zero regenerative dose, red line is fit forced through the origin, and green symbol and lines are interpolation of natural OSL to both growth curve fits. Corresponding D_e values for both fits are shown. Recuperation in (**a**) and (**b**) is 9.7 ± 5.4% and 19.5 ± 7.2%, respectively. (**c**) and (**d**): Distribution of D_e data when growth curves are forced through the origin (**c**), compared to growth curve fitting to zero regenerative dose (**d**). In (**d**), acceptance threshold for recuperation was set at 35%, with 18 of 22 aliquots accepted.

Table 1. Summary of OSL data.

Sample [a]	Lat., Long. (°S, °W)	n [b]	Preheat [c] (°C)	Recuperation threshold [d]	$\sigma_{b, De}$ [e] (%)	D_e [f] (mGy)	Dose Rate (mGya^{-1})	Age [g] (a)
ARA-040808-05	3°6′3″ 55°13′39″	18 (22)	220	35%	0	10.1 ± 0.86 [c]	0.42 ± 0.04	24 ± 3
TAP-030808-03	3°29′33″ 55°15′41″	12 (12)	200	Origin fit	0	4.10 ± 1.56 [c]	0.31 ± 0.02	13 ± 5
TAP-030808-04	3°29′31″ 55°15′39″	12 (12)	200	Origin fit	0	12.2 ± 2.83 [c]	0.36 ± 0.03	34 ± 8
CUP-030808-01	3°42′44″ 55°23′45″	10 (10)	200	5%	20.1	147 ± 9.80 [m]	0.46 ± 0.03	324 ± 29
CUP-030808-02	3°42′44″ 55°23′45″	23 (23)	220	5%	13.3	261 ± 7.48 [c]	0.47 ± 0.03	557 ± 35

[a] Samples are arranged from northerly-to-southerly sampling localities; further details in Section 2. [b] Number of aliquots accepted for D_e analysis (figures in parentheses are total number measured). Sample Cupari CUP-080308-01 had a lower quartz yield and only ten aliquots were measured for D_e analysis. [c] Ten-second preheat for D_e measurement chosen from a combination of preheat plateau, thermal transfer and dose recovery test data; cutheat was 160 °C for all measurements. [d] Individual D_e values were accepted if recuperation was below the specified threshold; for the Tapuama samples the growth curves were forced through the origin. [e] Over-dispersion in D_e data. [f] Superscript 'c' indicates central age model (CAM) result; superscript 'm' indicates minimum age model (MAM) result. [g] OSL ages quoted are in years (a) from 2009 with 1-sigma uncertainty.

3.4. Towards OSL Dating of Multi-Grain Quartz Aliquots

A summary of the OSL analysis is given in Table 1. D_e data, calculated using late background subtraction, were indistinguishable from those data analyzed using early background subtraction. An important observation is that recuperation measured during the D_e SAR cycle is significant for the quartz from Arapiuns (Figure 7a,b) and Tapuama. For the Arapiuns sample (ARA-040808-05), we calculate a similar D_e result when the growth curves are forced through the origin (Figure 7c; all aliquots accepted) compared to when the acceptance threshold for recuperation was set at 35% to achieve a satisfactory D_e dataset (Figure 7d; 18 of 22 aliquots accepted); for Tapuama (TAP-030808-03 and TAP-030808-04), recuperation was much more significant and as a consequence all growth curves were forced through the origin to obtain D_e values. High recuperation is somewhat surprising given the minimal thermal transfer for preheats <~220 °C (Figure 5), and the likelihood of numerous bleaching events occurring in these shoreline environments which have been linked to substantially reduced recuperation effect [50]. Given that the values measured are unusually high (e.g., for Tapuama aliquots, recuperation exceeds L_n/T_n by a factor ranging from ~1 to ~60), we suspect that the majority of the recuperation signal recorded could be a measurement artefact due to the comparatively large regenerative doses (lowest beta dose was ~800 mGy) used compared to the measured D_e values (Table 1). Furthermore, if for example the D_e values were ~400 mGy, then the majority would be accepted below the 5% threshold level. Future work will investigate whether there is a systematic dose-dependent effect on the size of the recuperation signal.

Over-dispersion (σ_b; [51]) values of 0% for the Arapiuns and Tapuama samples indicate the OSL signals were completely reset, although the true extent of resetting may not be revealed due to the medium-sized aliquots measured. For these samples the final D_e was calculated using the central age model (CAM) [52]. The second of the Cupari samples, CUP-030808-02, had a moderate over-dispersion value of 13.3%, but was also well suited to a CAM analysis. CUP-030808-01 was the only sample with a lower quartz yield and, subsequent to D_e plateau, thermal transfer and dose recovery tests, only 10 aliquots were available for D_e analysis. Over-dispersion for this sample was higher at 20.1% but, despite the low number of aliquots a minimum age model (MAM; [52]), analysis returned a result with a reasonable number of significant aliquots contributing to the MAM result (p-value = 0.333).

The OSL age for sample ARA-040808-05, from the shoreline dune feature at Arapiuns, is 24 ± 3 a from 2009. For Tapuama, the sample from the unvegetated distal end of the spit, TAP-030808-03, is 13 ± 5 a, and from the sands within the vegetated proximal end of the spit, TAP-030808-04, is 34 ±

8 a. Although we lack direct independent dating evidence, these are plausible ages for the Tapuama samples, with the sands from the unvegetated distal end of the spit of younger depositional age than the sands in the vegetated proximal end of the spit. Vegetation adds stability to sediments via root networks increasing cohesive strength, and by grasses, shrubs, and trees increasing surface roughness and dissipating some energy of wind or water; together, these lower the effectiveness of erosion by wind or water. The result for TAP-030808-04 is supportive of apparent decadal stability of other vegetated landforms (cf. Figure 2), whereas TAP-030808-03 suggests continual reworking and redeposition occurs in more active zones of the spit. The duplicated samples from the densely vegetated bar at Cupari have significantly older OSL ages of 324 ± 29 a and 557 ± 35 a; these ages in the 100s-of-years range are more consistent with youngest ages from other OSL studies of Tapajós sand bars [15]. Although the Cupari bar seems likely to be an older feature, we suspect that the discordance in the two ages may in part be related to poor resetting that is not apparent because of the large aliquots measured. For these samples the natural OSL is sufficiently large that smaller aliquots, or potentially single grains, could be measured to investigate this age discordance. ^{210}Pb data from a series of channel bottom cores from the Tapajós indicate sedimentation rates of 0.2–0.7 cmyr^{-1} in the upper stretch of the ria (consistent with the sampling localities in the work described here), and 0.2–1.9 cmyr^{-1} sampled across the entire ria [25]. If we make the assumption that these values represent sedimentation rates not only for clays, silts, and sands in the channel bottom but also for sands in shoreline features, we derive an age range of ~30–300 years (for 60 cm depth assuming uniform linear deposition) similar to the age range indicated from this OSL study.

This study demonstrates the potential of OSL to determine depositional age of very young fluvial landforms in the Rio Tapajós. Importantly it has revealed how future experimental approaches should be modified in the following ways: (1) use of ultra-low-dose beta source (and assessment of possible dose-dependency of recuperation); (2) optimization of aliquot size or single grain analyzes (including assessment of F-statistic [53] and un-logged age model approaches [6]); and (3) considering the seasonal river-level oscillation and extremely low external dose rates, careful assessment of water content fluctuation, accurate measurement of internal dose rates, and modelling of gamma and cosmic dose rates [46]. Given the nature of this study as one of feasibility of OSL approaches on a small selection of samples, it follows that geomorphological interpretation is somewhat speculative and should be limited. We propose future work with detailed stratigraphic and lateral sampling strategies which, in combination with differential remotely sensed imagery and hydrological data, will provide a sensitive monitor of how fluvial landforms change in response to land cover and land use change (including planned dam projects) and modern analogue data for depositional models of ancient tidal rhythmites.

The landforms investigated in this study were all quartz rich, but quartz yield was dependent on grain size distribution. For the ~5 cm core samples from Cupari, the majority of the sample was >310 μm, and quartz yields for sieve fractions <310 μm were correspondingly low. Conversely, ~5 cm core samples from Arapiuns and Tapuama had high quartz yields in the 250–310 μm and 212–250 μm sieve fractions, respectively. This indicates that in certain localities, a 1–2 cm sampling resolution may be possible.

4. Conclusions

In this study, we have investigated the feasibility of OSL dating of small samples of very young quartz collected from bar, spit and dune shoreline features along the Rio Tapajós, Brazilian Amazon. Five samples were collected from three study sites in ~65 cm by ~2 cm diameter vertical cores. Small subsamples from ~58–63 cm core depth were analyzed. The measured OSL signals exhibit very high specific luminescence sensitivity, have excellent dose recovery and recycling, essentially independent of preheat, show minimal thermal transfer below ~220 °C, and have a low temperature D_e plateau in a similar temperature band. Significant recuperation is observed for samples from two of the study sites but, given the minimal thermal transfer and likely numerous bleaching–burial cycles, we propose that the recuperation is possibly a measurement artifact due to the relatively high regenerative and test doses compared to the natural dose. Preliminary ages of features so far investigated range from 13 to 34 to several 100 years. Sampled sands are rich in quartz, and yields of 212–250 μm and 250–310 μm grains indicate high-resolution sampling is possible. These results demonstrate the potential of OSL for studying very young active fluvial processes in these settings.

Author Contributions: Conceptualization, J.Q.G.S., A.W.A. and M.M.C.; methodology, J.Q.G.S.; formal analysis, J.Q.G.S. and S.H.; investigation, J.Q.G.S., A.W.A. and M.M.C.; writing—original draft preparation, J.Q.G.S.; writing—review and editing, J.Q.G.S., A.W.A. and M.M.C.; visualization, J.Q.G.S.; supervision, J.Q.G.S.; project administration, J.Q.G.S.; funding acquisition, J.Q.G.S.

Funding: This work was supported by the Kansas State University Luminescence Research and Dating Laboratories, and the Department of Geology, Kansas State University.

Acknowledgments: We thank Jennifer Roozeboom (née Boswell) for assistance in the luminescence laboratories at Kansas State University, and the captain and crew of the Marco André, Santarém. We thank two anonymous referees and the academic editor for their careful and thoughtful reviews which have considerably improved this paper.

Conflicts of Interest: The authors declare no conflict of interest.

References

1. Thomsen, K.J.; Murray, A.S.; Jain, M.; Bøtter-Jensen, L. Laboratory fading rates of various luminescence signals from feldspar-rich sediment extracts. *Radiat. Meas.* **2008**, *43*, 1474–1486. [CrossRef]

2. Thomsen, K.J.; Murray, A.S.; Jain, M. Stability of IRSL signals from sedimentary K-feldspar samples. *Geochronometria* **2011**, *38*, 1–13. [CrossRef]

3. Olley, J.M.; Caitcheon, G.G.; Murray, A.S. The distribution of apparent dose as determined by optically stimulated luminescence in small aliquots of fluvial quartz: Implications for dating young sediments. *Quat. Sci. Rev.* **1998**, *17*, 1033–1040. [CrossRef]

4. Pietsch, T.J. Optically stimulated luminescence dating of young (<500 years old) sediments: Testing estimates of burial dose. *Quat. Geochronol.* **2009**, *4*, 406–422. [CrossRef]

5. Wallinga, J.; Hobo, N.; Cunningham, A.C.; Versendaal, A.J.; Makaske, B.; Middelkoop, H. Sedimentation rates on embanked floodplains determined through quartz optical dating. *Quat. Geochronol.* **2010**, *5*, 170–175. [CrossRef]

6. Arnold, L.J.; Roberts, R.G.; Galbraith, R.F.; DeLong, S.B. A revised burial dose estimation procedure for optical dating of young and modern-age sediments. *Quat. Geochronol.* **2009**, *4*, 306–325. [CrossRef]

7. Ballarini, M.; Wallinga, J.; Murray, A.S.; van Heteren, S.; Oost, A.P.; Bos, A.J.J.; van Eijk, C.W.E. Optical dating of young coastal dunes on a decadal time scale. *Quat. Sci. Rev.* **2003**, *22*, 1011–1017. [CrossRef]

8. Ballarini, M.; Wallinga, J.; Wintle, A.G.; Bos, A.J.J. Analysis of equivalent-dose distributions for single grains of quartz from modern deposits. *Quat. Geochronol.* **2007**, *2*, 77–82. [CrossRef]

9. Ballarini, M.; Wallinga, J.; Wintle, A.G.; Bos, A.J.J. A modified SAR protocol for optical dating of individual grains from young quartz samples. *Radiat. Meas.* **2007**, *42*, 360–369. [CrossRef]

10. Nielsen, A.; Murray, A.S.; Pejrup, M.; Elberling, B. Optically stimulated luminescence dating of a beach ridge plain in Northern Jutland, Denmark. *Quat. Geochronol.* **2006**, *1*, 305–312. [CrossRef]

11. Buckland, C.E.; Bailey, R.M.; Thomas, D.S.G. Using post-IR IRSL and OSL to date young (<200 yrs) dryland aeolian dune deposits. *Radiat. Meas.* **2019**, *126*, 106131. [CrossRef]

12. Soares, E.A.A.; Tatumi, S.H.; Riccomini, C. OSL age determinations of Pleistocene fluvial deposits in Central Amazonia. *Ann. Braz. Acad. Sci.* **2010**, *82*, 691–699. [CrossRef]

13. Fiore, M.; Soares, E.A.A.; Mittani, J.C.R.; Yee, M.; Tatumi, S.H. OSL dating of sediments from Negro and Solimões rivers – Amazon, Brazil. *Radiat. Phys. Chem.* **2014**, *95*, 113–115. [CrossRef]

14. Rossetti, D.F.; Cohen, M.C.; Tatumi, S.H.; Sawakuchi, A.O.; Cremon, É.H.; Mittani, J.C.; Bertani, T.C.; Munita, C.J.A.S.; Tudela, D.R.G.; Yee, M.; et al. Mid-Late Pleistocene OSL chronology in western Amazonia and implications for the transcontinental Amazon pathway. *Sediment. Geol.* **2015**, *330*, 1–15. [CrossRef]

15. Araki, L.A.; Sawakuchi, A.O.; Nogueira, L.; Turra, B.B. OSL dating of wave-generated sand bars in the lower Tapajós River, eastern Amazônia. In Proceedings of the 9th New World Luminescence Dating Workshop, Logan, UT, USA, 15–18 August 2013.

16. Caldas, M.M.; Walker, R.T.; Arima, E.; Perz, S.; Aldrich, S.; Simmons, C.; Wood, C. Theorizing Land Cover and Land Use Change: The Peasant Economy of Amazonian Deforestation. *Ann. Assoc. Am. Geogr.* **2007**, *97*, 86–110. [CrossRef]

17. Nepstad, D.C.; Stickler, C.M.; Soares-Filho, B.; Merry, F. Interactions among Amazon land use, forests and climate: Prospects for a near-term forest tipping point. *Philos. Trans. R. Soc. B* **2008**, *363*, 1737–1746. [CrossRef] [PubMed]

18. Arima, E.; Richards, P.; Walker, R.T.; Caldas, M.M. Statistical confirmation of indirect land use change in the Brazilian Amazon. *Environ. Res. Lett.* **2011**, *6*, 024010. [CrossRef]

19. Houghton, R.A. Land-use change and the carbon cycle. *Glob. Chang. Biol.* **1995**, *1*, 275–287. [CrossRef]

20. Dale, V.H.; Pearson, S.M.; Offerman, H.L.; O'Neill, R.V. Relating patterns of land-use change to faunal biodiversity in the Central Amazon. *Conserv. Biol.* **1994**, *8*, 1027–1036. [CrossRef]

21. Farella, N.; Lucotte, M.; Louchouarn, P.; Roulet, M. Deforestation modifying terrestrial organic transport in the Rio Tapajós, Brazilian Amazon. *Org. Geochem.* **2001**, *32*, 1443–1458. [CrossRef]

22. Archer, A.W. Review of Amazonian depositional systems. In *Fluvial Sedimentology VII*; Blum, M., Marriott, S., Leclair, S.F., Eds.; Blackwell: Oxford, UK, 2005; pp. 17–39.

23. Hoorn, C.; Roddaz, M.; Dino, R.; Soares, E.; Uba, C.; Mapes, R. The Amazonian Craton and its Influence on Past Fluvial Systems (Mesozoic-Cenozoic, Amazonia). In *Amazonia, Landscape and Species Evolution*, 1st ed.; Chapter 7; Hoorn, C., Wesselingh, F.P., Eds.; Blackwell Publishing Ltd.: Hoboken, NJ, USA, 2009; pp. 101–122.

24. Freitas, P.T.A.; Asp, N.E.; Souza-Filho, P.W.M.; Nittrouer, C.A.; Ogston, A.S.; Da Silva, M.S. Tidal influence on the hydrodynamics and sediment entrapment in a major Amazon River tributary—Lower Tapajós River. *J. S. Am. Earth Sci.* **2017**, *79*, 189–201. [CrossRef]

25. Fricke, A.T.; Nittrouer, C.A.; Ogston, A.S.; Nowacki, D.J.; Asp, N.E.; Souza Filho, P.W.M.; Da Silva, M.S.; Jalowska, A.M. River tributaries as sediment sinks: Processes operating where the Tapajós and Xingu rivers meet the Amazon tidal river. *Sedimentology* **2017**, *64*, 1731–1753. [CrossRef]

26. NASA. Shuttle Radar Topography Mission. Available online: http://www2.jpl.nasa.gov/srtm/ (accessed on 6 December 2019).

27. Earth Resources Observation and Science Center/U.S. Geological Survey/U.S. Department of the Interior. USGS 30 ARC-Second Global Elevation Data, GTOPO30. Research Data Archive at the National Center for Atmospheric Research, Computational and Information Systems Laboratory. 1997. Available online: https://doi.org/10.5065/A1Z4-EE71 (accessed on 6 December 2019).

28. Shen, R.; Pennell, K.; Suuberg, E.M. Influence of Soil Moisture on Soil Gas Vapor Concentration for Vapor Intrusion. *Environ. Eng. Sci.* **2013**, *30*, 628–637. [CrossRef] [PubMed]

29. Spencer, J.Q.G.; Robinson, R.A.J. Dating intramontane alluvial deposits from NW Argentina using luminescence techniques: Problems and potential. *Geomorphology* **2008**, *93*, 144–156. [CrossRef]

30. Bøtter-Jensen, L.; Andersen, C.E.; Duller, G.A.T.; Murray, A.S. Developments in radiation, stimulation and observation facilities in luminescence measurements. *Radiat. Meas.* **2003**, *37*, 535–541. [CrossRef]

31. Murray, A.S.; Wintle, A.G. Luminescence dating of quartz using an improved single-aliquot regenerative-dose protocol. *Radiat. Meas.* **2000**, *32*, 57–73. [CrossRef]

32. Murray, A.S.; Wintle, A.G. The single-aliquot regenerative-dose protocol: Potential for improvements in reliability. *Radiat. Meas.* **2003**, *37*, 377–381. [CrossRef]

33. Wintle, A.G.; Murray, A.S. A review of quartz optically stimulated luminescence characteristics and their relevance in single-aliquot regeneration dating protocols. *Radiat. Meas.* **2006**, *41*, 369–391. [CrossRef]

34. Wallinga, J.; Murray, A.S.; Bøtter-Jensen, L. Measurement of the dose in quartz in the presence of feldspar contamination. *Radiat. Prot. Dosim.* **2002**, *101*, 367–370. [CrossRef]

35. Lukas, S.; Spencer, J.Q.G.; Robinson, R.A.J.; Benn, D.I. Problems associated with luminescence dating of Late Quaternary glacial sediments in the NW Scottish Highlands. *Quat. Geochronol.* **2007**, *2*, 243–248. [CrossRef]

36. Morrocco, S.M.; Ballantyne, C.K.; Spencer, J.Q.G.; Robinson, R.A.J. Age and significance of aeolian sediment reworking on high plateaux in the Scottish Highlands. *Holocene* **2007**, *17*, 349–360. [CrossRef]

37. Ashley, G.M.; Ndiema, E.; Spencer, J.Q.G.; Harris, J.W.K.; Kiura, P.W. Paleoenvironmental context of archaeological sites, implications of subsistence strategies under Holocene climate change, northern Kenya. *Geoarchaeology* **2011**, *26*, 809–837. [CrossRef]

38. Ashley, G.M.; Ndiema, E.; Spencer, J.Q.G.; Du, A.; Lordan, P.T.; Kiura, P.W.; Dibble, L.; Harris, J.W.K. Paleoenvironmental Reconstruction of Dongodien, Lake Turkana, Kenya and OSL Dating of Site Occupation during Late Holocene Climate Change. *Afr. Archaeol. Rev.* **2017**, *34*, 345–362. [CrossRef]

39. Rittase, W.M.; Kirby, E.; McDonald, E.; Walker, J.D.; Gosse, J.; Spencer, J.Q.G.; Herrs, A.J. Temporal variations in Holocene slip rate along the central Garlock fault, Pilot Knob Valley, California. *Lithosphere* **2014**, *6*, 48–58. [CrossRef]

40. Spencer, J.Q.G.; Oviatt, C.G.; Pathak, M.; Fan, Y. Testing and refining the timing of hydrologic evolution during the latest Pleistocene regressive phase of Lake Bonneville. *Quat. Int.* **2015**, *362*, 139–145. [CrossRef]

41. Cunningham, A.C.; Wallinga, J. Selection of integration time intervals for quartz OSL decay curves. *Quat. Geochronol.* **2010**, *5*, 657–666. [CrossRef]

42. Duller, G.A.T. Assessing the error on equivalent dose estimates derived from single aliquot regenerative dose measurements. *Anc. TL* **2007**, *25*, 15–24.

43. Adamiec, G.; Aitken, M.J. Dose-rate conversion factors: Update. *Anc. TL* **1998**, *16*, 37–50.

44. Spencer, J.Q.; Owen, L.A. Optically stimulated luminescence dating of Late Quaternary glaciogenic sediments in the upper Hunza valley: Validating the timing of glaciation and assessing dating methods. *Quat. Sci. Rev.* **2004**, *23*, 175–191. [CrossRef]

45. Prescott, J.R.; Hutton, J.T. Cosmic ray contributions to dose rates for luminescence and ESR dating: Large depths and long-term time variations. *Radiat. Meas.* **1994**, *23*, 497–500. [CrossRef]

46. Spencer, J.Q.G.; Alghamdi, A.G.; Presley, D.; Huot, S.; Martin, L.; Mercier, N. Attic dust from historic buildings: Progress dating young mm-thick layers. In Proceedings of the 15th International Conference on Luminescence and Electron Spin Resonance Dating, Capetown, South Africa, 11–15 September 2017.

47. Vandenberghe, D.; De Corte, F.; Buylaert, J.P.; Kučera, J.; Van den haute, P. On the internal radioactivity in quartz. *Radiat. Meas.* **2008**, *43*, 771–775. [CrossRef]

48. Sawakuchi, A.O.; Mineli, T.D.; Nogueira, L.; Henrique Grohmann, C.; Guedes, C.C.F. OSL and IRSL as a proxy for the origin and transport of sands in the Amazon river. In Proceedings of the 14th International Conference on Luminescence and Electron Spin Resonance Dating, Montréal, QC, Canada, 7–11 July 2014.

49. Markey, B.; Bøtter-Jensen, L.; Duller, G.A.T. A new flexible system for measuring thermally and optically stimulated luminescence. *Radiat. Meas.* **2007**, *27*, 83–89. [CrossRef]

50. Aitken, M.J.; Smith, B.W.; Rhodes, E.J. Optical dating: Recapitulation on recuperation. In Proceedings of the a Workshop on Long and Short Range Limits in Luminescence Dating, Oxford, UK, 11–13 April 1989.

51. Galbraith, R.F.; Roberts, R.G.; Yoshida, H. Error variation in OSL palaeodose estimates from single aliquots of quartz: A factorial experiment. *Radiat. Meas.* **2005**, *39*, 289–307. [CrossRef]

52. Galbraith, R.F.; Roberts, R.G.; Laslett, G.M.; Yoshida, H.; Olley, J.M. Optical dating of single and multiple grains of quartz from Jinmium rock shelter, northern Australia: Part I, experimental design and statistical models. *Archaeometry* **1999**, *41*, 339–364. [CrossRef]

53. Spencer, J.Q.; Sanderson, D.C.W.; Deckers, K.; Sommerville, A.A. Assessing mixed dose distributions in young sediments identified using small aliquots and a simple two-step SAR procedure: The F-statistic as a diagnostic tool. *Radiat. Meas.* **2003**, *37*, 425–431. [CrossRef]

 *methods and protocols*

Article

Single Aliquot Regeneration (SAR) Optically Stimulated Luminescence Dating Protocols Using Different Grain-Sizes of Quartz: Revisiting the Chronology of Mircea Voda Loess-Paleosol Master Section (Romania)

Ștefana-M. Groza-Săcaciu [1,2], Cristian Panaiotu [3] and Alida Timar-Gabor [1,2,*]

1   Interdisciplinary Research Institute on Bio-Nano-Science, Babes-Bolyai University,
    400000 400271 Cluj-Napoca, Romania; smgroza@yahoo.com
2   Faculty of Environmental Science and Engineering, Babes-Bolyai University,
    400000 400294 Cluj-Napoca, Romania
3   Faculty of Physics, University of Bucharest, 030018 077125 Magurele, Romania; cristian.panaiotu@gmail.com
*   Correspondence: alida.timar@ubbcluj.ro

Received: 6 November 2019; Accepted: 20 February 2020; Published: 27 February 2020

Abstract: The loess-paleosol archive from Mircea Vodă (Romania) represents one of the most studied sections in Europe. We are applying here the current state of the art luminescence dating protocols for revisiting the chronology of this section. Analysis were performed on fine (4–11 µm) and coarse (63–90 µm) quartz extracts using the single aliquot regenerative (SAR) optically stimulated luminescence (OSL) dating protocol. Laboratory generated SAR dose response curves in the high dose range (5 kGy for fine quartz and 2 kGy for coarse quartz) were investigated by employing a test dose of either 17 or 170 Gy. The results confirm the previously reported different saturation characteristics of the two quartz fractions, with no evident dependency of the equivalent dose (D_e) on the size of the test dose. The OSL SAR ages are discussed and compared to the previously obtained results on quartz and feldspars. The previous reports regarding the chronological discrepancy between the two quartz fractions are confirmed. However, while previous investigations on other sites concluded that this discrepancy appears only above equivalent doses of about 100 Gy, here fine grain quartz ages underestimate coarse quartz ages starting with equivalent doses as low as around 50 Gy.

Keywords: luminescence dating; loess; optically stimulated luminescence; single aliquot regeneration protocol; quartz; grain size

---

1. Introduction

The development of the single-aliquot regenerative-dose (SAR) protocol [1] for optically stimulated luminescence (OSL) dating of quartz has revolutionized the luminescence dating method by giving rise to high precision equivalent dose estimates. Loess-paleosol sequences are important archives of the climatic changes that took place during the Pleistocene, but their significance can only be fully understood once a reliable and absolute chronology is available. Due to its quartz rich and windblown nature, loess is generally considered an ideal material for the application of OSL. However, although more precise ages can be obtained by SAR-OSL, the validation of the accuracy of these OSL ages by independent age control is hindered by the lack of methods which can directly date the depositional time of the sediments. In this context, the identification of the paleosol associated with marine isotope stage (MIS) 5 is known to yield valuable time control as the identification of this paleosol provides a minimum age threshold for the sediments underlying it, which should be no younger than ~130 ka. However, it is well known that the results of luminescence dating methods applied on quartz underestimate the expected ages for samples collected below this soil. For example, an age of 106 ± 16 ka (equivalent dose of 310 ± 9 Gy) was obtained for quartz grains of 4–11 μm from one sample taken immediately below the S_1 paleosol (associated with MIS 5) at Mircea Vodă loess paleosol site, Romania, while an increasing degree of age underestimation with depth was observed for samples taken from below S_1, S_2 and S_3 paleosols at the same location [2]. The same trend in age underestimates was reported in China. At Luochuan, Buylaert et al. [3] obtained an age on coarse (63–90 μm) quartz of 81± 7 ka (D_e= 229 ± 16 Gy) for the loess beneath the S_1 paleosol while Lai [4] reported lower ages than expected for samples older than 70 ka on 45–63 μm quartz. In the case of another site on the Chinese Loess Plateau (Zhongjiacai), Buylaert et al. [5] obtained for a sample taken from the lower part of the last interglacial paleosol an age of 69.8 ± 3.8 ka (D_e= 216 ± 6 Gy) on coarse (63–90 μm) quartz.

Another important issue which was raised relates to the choice of the quartz grain size. The use of coarse grains (so called inclusion dating) or fine grains (4–11 μm) has been proposed five decades ago for thermoluminescence dating of pottery, based on the different penetration powers of nuclear radiations in minerals by Fleming [6] and Zimmerman [7], respectively. However, in later geological applications in what regards OSL dating technique, the choice between these protocols was dictated by the dominant grain size within the investigated sedimentary unit. Consequently, it is common practice to use only one grain size fraction. A series of investigations carried out by our group during the last decade on quartz of different grain sizes extracted from loess yielded intriguing and concerning results. While ages obtained on fine (4–11 μm) and coarse (>63 μm) quartz samples were in good agreement up until ~40 ka, after this age the optical ages obtained on coarse (63–90 μm) quartz were reported to be systematically higher than those on fine (4–11 μm) quartz [8–13].

In the light of these findings, we are applying here the single aliquot regeneration dating protocol on quartz of different grain sizes for revisiting the chronology of Mircea-Vodă loess paleosol sequence in Romania. This is the site where we have reported for the first time various problems when investigating different quartz grain sizes [8] and we have subsequently applied alternative luminescence dating protocols on feldspars [14,15]. Additionally, there is a limited practice of performing interlaboratory comparison exercises in the field of luminescence dating. This represents an interesting endeavor since this new investigation takes place a decade later and in different laboratories (Ghent, Belgium [2,8] and Cluj-Napoca, Romania—current paper), using different samples from the same site. We are testing the robustness of the protocol by performing intrinsic rigor tests and we are discussing the accuracy of the obtained ages, also in the light of the results of the previous studies.

2. Optically Stimulated Luminescence Dating Methodology

2.1. Principles of Luminescence Dating

Optically stimulated luminescence was developed by Huntley et al. [16] and was aimed for establishing the chronology of sediments, especially those where the luminescent signal can be zeroed by exposure to sunlight before deposition such as loess, desert sands and coastal dunes [17]. This dating technique makes use of natural dosimeters (primarily quartz and feldspar grains) that have thermally stable traps capable of storing electrons that arise from the interaction of the environmental ionizing radiation during burial (these radiations coming from the decay of uranium, thorium and potassium in the sediment and from cosmic radiation) with the crystal lattice [18]. This trapped charge population builds up since the time of deposition. As such there is a functionality between the dose received by the crystal (hence the time as the dose rate is assumed to be constant) and the amount of trapped charge. Under controlled laboratory conditions this charge can be quantified in the form of a luminescence signal. The assumption on which the method is based on is that the growth of the luminescence signal in nature can be reproduced by performing controlled laboratory irradiations. Consequently, a dose response curve is constructed and the natural luminescent signal measured in the laboratory is expressed as an equivalent dose by interpolating the natural signal on this dose response curve. The luminescence age equation is shown below. The age is obtained by dividing the equivalent dose value (expressed in Gy) by the dose rate (expressed as Gy/ky).

$$\text{Age (ky)} = \text{Equivalent dose (Gy)/Dose rate (Gy/ky)}$$

2.2. Fine (4–11 μm) Versus Coarse (>63 μm) Quartz Grains Dating of Loess

Besides the type of mineral used for OSL measurements, the grain size also plays an important role. Commonly, the size is chosen depending on the dominant grain size of the investigated sedimentary unit. However, it is mandatory to take into consideration the depth at which the alpha, beta and gamma radiation penetrate the grain when age calculation is performed. The silt-sized (4–11 μm) fraction is fully penetrated by all three types of radiations (alpha, beta and gamma, respectively). The sand-sized (>63 μm) grain, on the other hand, receives less of the external beta dose rate due to the attenuation of these radiations in the grain [19,20] and the alpha dose is not homogenously delivered to the grain, the latter being concentrated in an exterior layer which is usually removed by hydrofluoric acid treatment. As a result, the alpha contribution can determine a dose rate for fine grains of even 40% of the total in comparison to coarse grains where the contribution is almost zero [20]. Additionally, an α-efficiency (a-value) must be incorporated when calculating the dose rate for fine grains (~4–11 μm) due to the different efficiency of α-particles compared to β- and γ-radiation in producing luminescence [21]. However, the need for using these different correction factors when dose rates are calculated is well known for decades [20].

It is generally believed that relying only on one fraction for OSL dating should lead to obtaining reliable chronologies. Dating studies on relatively young samples (D_e < 100 Gy) using multiple different grain sizes of quartz yielded accurate ages as confirmed by comparison with independent age control provided through tephrochronology [13,22] or radiocarbon dating [23,24]. In this dose range a good agreement has been reported when both fine and coarse quartz were used in order to obtain a chronology of the investigated loess sites [25–27]. On the other hand, for older samples the optical ages obtained on coarse quartz (>63–90 μm) were reported to be systematically higher than those on fine quartz (4–11 μm), resulting in a significant difference between the ages obtained on the two grain sizes and raising significant doubts on previously obtained chronologies for ages older than about 50 ka [8,11,12,28]. For large doses (>~500 Gy) the laboratory dose response can be well fitted only by a sum of two single saturating exponential functions [9]. Different saturation characteristics between the fine and coarse quartz fractions extracted from loess were noted, with the fine grains showing higher saturation characteristics worldwide [11]. This is most intriguing when correlated to the fact that the fine fraction underestimates the true ages sooner than the coarse ones. At the moment, the source of the age discrepancy is not fully understood, but it is thought to reside, at least partly, in the different saturation characteristics of fine grains compared to the coarse grains, and in the differences reported between the laboratory and the natural dose response curves as reported by Timar-Gabor and Wintle [29] for Romanian loess as well as by Chapot et al. [30] for loess in China.

3. Studied Site

3.1. Location and Importance

The Middle and Lower Danube Basins contain the westernmost part of the Eurasian steppe belt, covering the Pannonian Basin and reaching up until the Danube flows into the Black Sea. Here, loess intercalated with paleosols plateaus developed on top of accumulations of fluvial deposits in subsiding areas during the Quaternary. These loess-paleosols sequences (LPSs) are considered to be important and continuous paleoclimatic archives, displaying similar sedimentological and pedological properties to deposits from China and Central Asia [31,32].

The Lower Danube Basin encompasses the area outlined by the Iron Gates gorges, the Black Sea, the Southern Carpathians and the Balkans. The basin is divided into three main regions—the Bulgarian Danube Plain, the Romanian Plain and the Dobrogea Plateau. For the Romanian part, the loess-like deposits are predominant [33]. They were first described in the works of Ana Conea and were given a proposed chronology based on pedostratigraphic methods [34,35]. Dating studies later focused on the Romanian Plain and Dobrogea loess, with a small number of Middle to late Pleistocene loess-paleosol sites being investigated using modern techniques—Mostiștea [14,36–38], Lunca [39], Mircea Vodă, Costinești, Tuzla [2,8,12,31,32,35,40–42] and Urluia [43] (Figure 1).

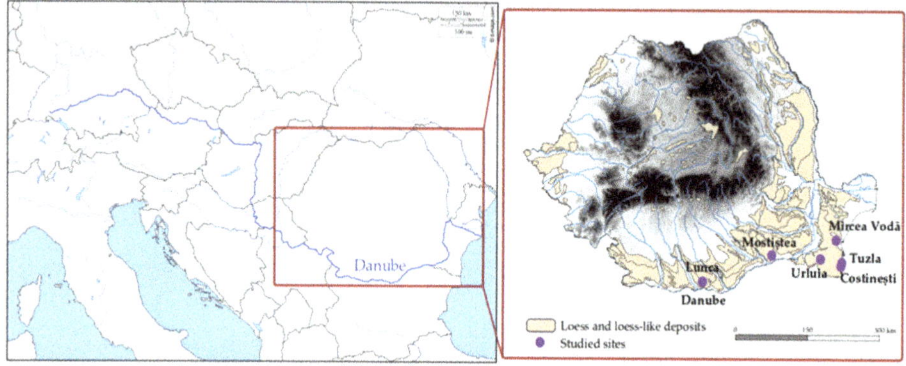

Figure 1. Showing the location of the loess and loess-like deposits in Romania alongside the previously investigated sites—Mircea Vodă, Tuzla, Costinești, Urluia, Mostiștea and Lunca.

The loess-paleosol archive from Mircea Vodă (48°19′15″ N, 28°11′21″ E) is situated in the Dobrogea region, in the proximity of the Danube River, the Black Sea and the Karasu valley. It is considered to be a key section, being one of the most studied sections in Eastern Europe. Six well developed pedocomplexes (covering the last 17 Marine Isotope Stages (MIS)) are comprised in the approximately 26 m thick eolian deposit with no visible hiatuses, overlaying Tertiary and Mesozoic sediments [41].

Previous sedimentological, geochemical and environmental magnetic results showed that the loess from Mircea Vodă displays similarities with the loess from Serbia (Vojvodina) and China (Chinese Loess Plateau) [31,41,44–46]. More precisely, the site displays similar major element ratios and geochemical fingerprint as the Serbian loess from the Vojvodina region, with Danube alluvial sediments being also the main loess source [31]. Furthermore, there is a resemblance in the concentration related magnetic parameters, diffuse reflectance spectroscopy results and soil color proxies for hematite and goethite, with the magnetic grain size and mineralogy being also similar to that of Chinese LPSs [44–46]. It was observed that the background susceptibilities for Mircea Vodă and Serbian loess sites are in the same range (21×10^{-8} -22×10^{-8} $m^3 \cdot kg^{-1}$) [31] and that the characteristic magnetic susceptibility patterns of their paleosols can be correlated with corresponding patterns in the susceptibility record of Chinese LPSs [31,41].

Grain-size analysis concluded that throughout the section, silt and fine sand (>16 μm) dominate, while in the lower part of the section there is a larger amount of clay-sized material [8]. The section also exhibits overall pedogenic processes, thus suggesting that loess deposition took place at the same time as weak pedogenesis [8].

From a geochemical point of view, Mircea Vodă exhibits higher carbonate content than the Serbian sites, probably as a result of a more arid climate [31]. The loess units, formed during glacial periods, are dominated by windblown coarse ferromagnetic minerals and have a high quartz content alongside a trend to higher zirconium and hafnium content [31,46]. The paleosol layers are dominated by fine ferromagnetic minerals produced during interglacial pedogenesis, with small amounts of coarser eolian magnetic grains [46]. With the sediment budget being attributed to the Danube River, due to the origin of the quartz and zircon and the bi- and three-modal distribution of the grain-size in loess layers, an additional input from the Ukrainian glaciofluvial deposits and local sand dune fields was also proposed [8,31].

3.2. Stratigraphy

As previously mentioned, the Mircea Vodă section displays six pedocomplexes, comprising at least 700 ka of paleoclimate. The S_0 layer is a steppe soil which displays similarities with the L_3 unit in what regards magnetic granulometry [31,45]. An interstadial pedocomplex (L_1S_1) of the last glacial cycle is comprised in the L_1 unit [2,31].

The S_1 pedocomplex has been identified as a gray-brown fossil steppe soil [38,41]. It displays three magnetic susceptibility peaks—a dominating peak in the lower half of the unit which may represent MIS 5e and two additional weakly expressed susceptibility peaks, probably representing MIS 5a and MIS 5c [41].

The S_2 pedocomplex has also been identified as a gray-brown fossil steppe soil [38,44]. It comprises three clearly separated peaks and it is attributed to MIS 7. Moreover, due to its characteristic magnetic susceptibility pattern, it can be correlated with the Chinese Loess Plateau sections which show similar enhanced magnetism resulting from interglacial pedogenesis [47].

Due to its paleopedological characteristics, the S_3 unit can be identified as a fossil steppe or forest-steppe soil [44]. The S_3 does not show the characteristic double peak like other nearby sections does (Batajnica, Serbia), but it has the strongest magnetic enhancement and it corresponds to MIS 9 [42]. The S_4 paleosol is correlated with MIS 11 [44].

The S_5 paleosol is correlated with MIS 13-15 and has been classified as a fossil (chromic) Cambisol and Luvisol [48]. It is the best-developed soil of the Brunhes-chron therefore it may represent a marker horizon in this area [41]. The S_6 unit shows two susceptibility peaks, the latter most probably indicating the interglacial formation of MIS 17 or MIS 19 [41].

3.3. Previous Studies on Mircea Vodă Section

In order to obtain a continuum time-depth model for the Mircea Vodă section, modeling has been employed on the magnetic susceptibility data by Timar et al. [2] by using Match-2.3 software [49] and the stack of 57 globally distributed benthic $\delta^{18}O$ records as the target curve [50]. Two tie points have been used for the upper part (0 ka) and the bottom (626 ka) of the section [2]. The modeling results, similar to the previously obtained chronostratigraphy by Buggle et al. [41], show that paleosols and loess units correspond to interglacial and glacial periods, therefore being correlated with odd and even marine isotopes, respectively. Moreover, both models assign the weakly developed paleosol embedded in L_1 to the MIS 3 interstadial, thus disproving the chronology proposed by Conea [34,35].

Mircea Vodă was the first section in Romania to be dated using optically stimulated luminescence (OSL) methods based on fine quartz (4–11 μm) by Timar et al. [2]. At the same time, Bălescu et al. [38] investigated alkali feldspars extracted from three samples taken from L_1, L_2 and L_3 loess units, with the age results being in broad (stratigraphic interpretation) agreement with those obtained by Timar et al. [2]. The quartz luminescence study of Timar et al. [2] focused on the last four glacial periods, with 9 samples being taken from the uppermost loess layer (L_1) and three more from L_2, L_3 and L_4 loess units, respectively (Figure 2). Timar-Gabor et al. [8] later presented a comparison on ages obtained on coarse (63–90 μm) quartz. The two OSL datasets were not in agreement as one would generally expect. As a result, the same samples have been investigated by Vasiliniuc et al. [14,15,51] by using polymineral fine (4–11 μm) fraction extracted from the same material used by Timar et al. [2].

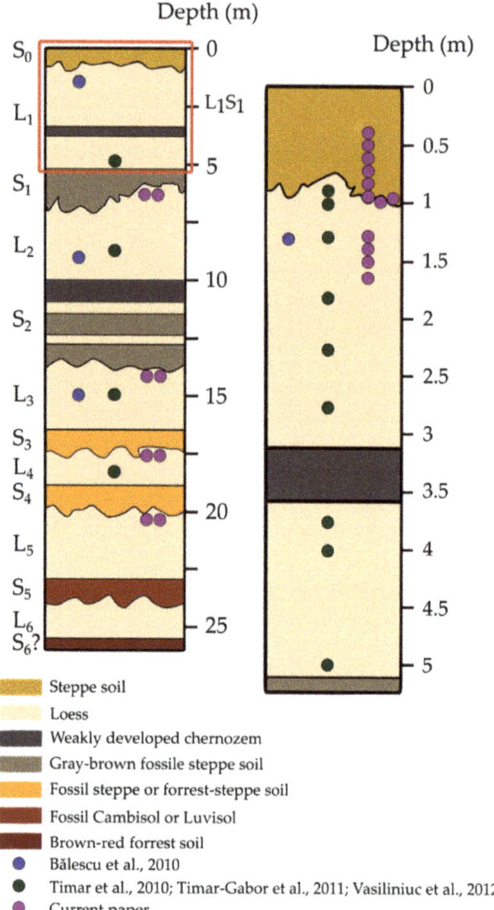

Figure 2. Stratigraphic column of the Mircea Vodă loess-paleosol section (drawn by the authors based on field observation and in accordance with previous stratigraphy presented by a Bălescuet al. [38]). The column on the right represents the first approximately 5 m at a higher resolution. The colored circles represent the position of the samples which were investigated in previous studies and current paper.

Luminescence Characteristics and Behavior

The first OSL chronology obtained for the Mircea Vodă section was reported by Timar et al. [2] on fine (4–11 μm) quartz fraction extracted from 12 samples (MV 01–13) (Figure 2). Later on, Timar-Gabor et al. [8] focused on the coarse (63–90 μm) quartz fraction obtained from the same samples. The luminescence characteristics were studied by applying the SAR protocol [1] (Table 1). The OSL signal for both fine (4–11 μm) and coarse (63–90 μm) quartz grains exhibited rapid decay during optical stimulation, with the natural, regenerated and calibration quartz signals being indistinguishable from one another [2,8]. In order to further assess whether the signal is dominated by the fast component, LM-OSL measurements have been performed (Table S1). Once more, the natural signal for both quartz fractions proved to match the signal from the calibration quartz, displaying no significant dependency on the different preheat temperatures used [2]. This was also confirmed by LM-OSL dose response curves (DRC) constructed up to 1 kGy [9].

Moreover, the SAR measurement sequence proved to be accurately corrected for sensitivity changes based on the results obtained for recycling and IR depletion tests (both ratios within 10% from unity). Recuperation tests shown that thermal transfer is not a significant issue either, with signals measured following a zero dose being <0.3% of the sensitivity corrected natural signal. The dose recovery tests (Table S1) also showed that known laboratory given doses can be successfully measured over the entire dose range (from ~28–~480 Gy) for both fine and coarse quartz [2,8].

However, the equivalent doses obtained on fine (4–11 μm) quartz were lower than those obtained on coarse (63–90 μm) quartz, in contradiction with what is presumed considering the expectations based on dose rates [2,8]. In other words, due to the fact that the fine grains have received alpha dose, they should display higher equivalent doses compared to coarse grains. This raised one significant issue—why are different results obtained despite the similar OSL characteristics and behavior? In order to further investigate this issue, pulse annealing measurements on both quartz fractions have been employed in order to assess the potential contamination of the OSL dosimetric trap with an unstable component. The results disproved the contamination scenario, with the signal being confirmed to be thermally stable [8].

The only noticeable difference between the two grain sizes was seen in the response of the signal as function of dose. Sensitivity corrected dose response curves built up to ~700 Gy either fitted with single saturating exponential or a sum of a single saturating exponential and a linear component showed different growth patterns for the two quartz fractions [8] (Table S1). Later on, Timar-Gabor et al. [9] further investigated dose response curves up to 10 kGy for both quartz grain sizes and Timar-Gabor et al. [28] investigated the reproducibility of the dose response curves up to 15 kGy on coarse (63–90 μm) quartz (Table S1). In the first experiment the curves were fitted better with a sum of two saturating exponentials function and the coarse grains saturated much earlier [9]. For the latter experiment the dose response constructed up to 15 kGy could be well reproduced following repeated light exposure and irradiation cycles [28].

The site has also been investigated by Bălescu et al. [38], alongside two more loess sites from Eastern Romania—Tuzla and Mostiștea. At Mircea Vodă, samples have been taken from the L_1, L_2 and L_3 units, from which 60–80 μm alkali feldspars fraction was extracted. The measurement protocol used was the multiple aliquot additive dose method (MAAD) [52] (Table 1).

Bearing in mind the issues rose by the quartz results, Vasiliniuc et al. [14,15,51] tried a different approach. Their studies focused on luminescence properties and ages obtained for polymineral fine (4–11 μm) material extracted from previously investigated samples by Timar et al. [2].

Feldspar dating was carried out by Vasiliniuc et al. [14] who used IRSL signals by employing the post-IR IRSL SAR protocol [5,53] (Table 1). Two preheat post-IR IR stimulation temperature combinations were used. In the first, a 60 s preheat treatment at 250 °C was followed by 100 s IR stimulation at 50 °C (IR_{50}) and a second 100 s stimulation at 225 °C (post-IR_{50} IR_{225}). The second choice of measurement parameters involved a 60 s preheat treatment at 325 °C was followed by 100 s IR stimulation at 50 °C (IR_{50}) and a second 100 s stimulation at 300 °C (post-IR_{50} IR_{300}) (Table S1). Residual doses obtained were between 3.8 ± 0.3 Gy and 17.0 ± 0.4 Gy for the post-IR IR_{225} and between 11.3 ± 0.3 Gy and 33.2 ± 1.1 Gy for the post-IR IR_{300}, the latter results being similar to those obtained by Thiel et al. [54] and Stevens et al. [55].

The observation of both natural and laboratory induced (during dose recovery tests) signals above the saturation level of the dose response curve, in the case of post-IR IR_{300} signals lead to the conclusion that these signals suffer from dose dependent initial sensitivity changes. On the other hand, the post-IR IR_{225} signals were observed to successfully pass the SAR performance tests in terms of recycling ratio, recuperation and dose recovery. For old samples both natural and regenerated signals (measured during dose recovery tests) were observed to correspond to the saturating region of the dose-response curve, indicating that the small fading rate determined for this signal is probably an artefact of the measurement procedure. The uncorrected ages obtained using the post-IR IR_{225} signals for samples taken from L_2, L_3 and L_4 were found in good agreement with the results of time-depth modelling based on magnetic susceptibility data.

Table 1. Previous studies and protocols used for the chronology of Mircea Vodă section.

Authors/Year	Mineral	Stratigraphical Units Investigated	Measurement Protocol
Bălescu, S.; Lamothe, M.; Panaiotu, C.; Panaiotu, C. (2010) [38]	alkali feldspars (60–80 μm)	• 1 sample from L_2 • 1 sample from L_3 • 1 sample from L_4	Multiple aliquot additive dose method (MAAD)
Timar, A.; Vandenberghe, D.; Panaiotu, E.C.; Panaiotu, C.G.; Necula, C.; Cosma, C.; van den haute, P. (2010) [2]	quartz (4–11 μm)	• 9 samples from L_1 • 1 sample from L_2 • 1 sample from L_3 • 1 sample from L_4	SAR (CW-OSL)
Timar-Gabor, A.; Vandenberghe, D.A.G.; Vasiliniuc, Ş.; Panaiotu, E.C.; Panaiotu, C.G.; Dimofte, D.; Cosma, C. (2011) [8]	quartz (63–90 μm)	• 9 samples from L_1 • 1 sample from L_2 • 1 sample from L_3 • 1 sample from L_4	SAR (CW-OSL)
Timar-Gabor, A.; Vasiliniuc, Ş.; Vandenberghe, D.A.G.; Cosma, C.; Wintle, A.G.; (2012) [14]	quartz (4–11 and 63–90 μm)	• 2 samples from L_1 • 1 sample from L_2 • 1 sample from L_3	SAR (CW-OSL) Dose response curves constructed up to 1200 Gy using LM-OSL signals
Vasiliniuc, Ş.; Vandenberghe, D.A.G.; Timar-Gabor, A.; Panaiotu, C.; Cosma, C.; van den Haute, P. (2012) [14]	polymineral grains (4–11 μm)	• 5 samples from L_1 • 1 sample from L_2 • 1 sample from L_3 • 1 sample from L_4	Post- IR IR$_{225, 300}$
Vasiliniuc, Ş.; Vandenberghe, D.A.G.; Timar-Gabor, A.; Cosma, C.; Van Den haute, P. (2013) [51]	polymineral grains (4–11 μm)	• 9 samples from L_1 • 1 sample from L_2 • 1 sample from L_3 • 1 sample from L_4	Double SAR (CW-OSL)
Vasiliniuc, Ş.; Vandenberghe, D.A.G.; Timar-Gabor, A.; van den Haute, P. (2013) [15]	polymineral grains (4–11 μm)	• 9 samples from L_1 • 1 sample from L_2 • 1 sample from L_3 • 1 sample from L_4	Modified SAR -IRSL at 115°C and 250°C
Timar-Gabor, A.; Constantin, D.; Buylaert, J.P.; Jain, M.; Murray, A.S.; Wintle, A.G. (2015) [28]	quartz (63–90 μm)	• 2 samples from L_1	SAR (CW-OSL) Dose response curves constructed up to 15 kGy

3.4. Current Study on Mircea Vodă

3.4.1. Sampling, Preparation and Analytical Facilities

For the current paper, investigations were performed on 20 new samples from Mircea Vodă section. The first 12 samples (2MV 40–MV 2.6) were taken from the Pleistocene/Holocene transition (Figure 3a), while doublet samples (2MV 570, L3, L4 and L5) were taken directly beneath the S_1–S_4 units, respectively (Figure 3b). The sampling procedure was carried out by using stainless steel tubes inserted horizontally in the freshly cleaned profile.

Standard laboratory sample preparation was then performed under red light conditions. The bulk material was first treated with HCl (35% concentration) for carbonate removal and H_2O_2 (30% concentration) for organic matter removal. After each of these steps the samples were rinsed 3 times. The coarse fraction (63–90 μm) was extracted by wet and dry sieving, after which the material was treated with 40% HF for 60 min and a 60 min bath in 10% HCl. Attenberg cylinders were used for obtaining the fine fraction (<11 μm). The material was afterwards etched with 35% hexafluorosilicic acid (H_2SiF_6) for 10 days and centrifuged with distilled water in order to attain the 4–11 μm quartz grains [56,57]

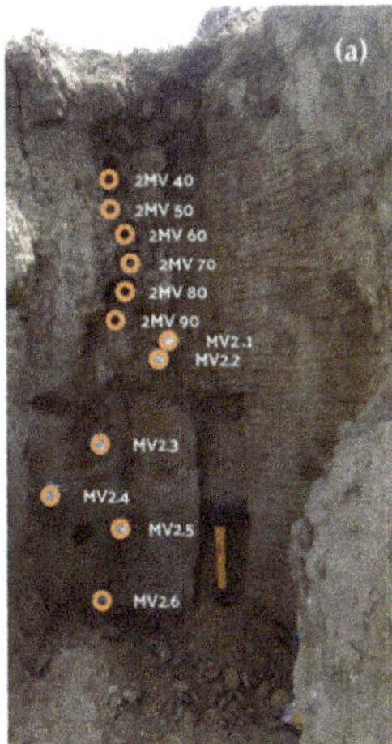
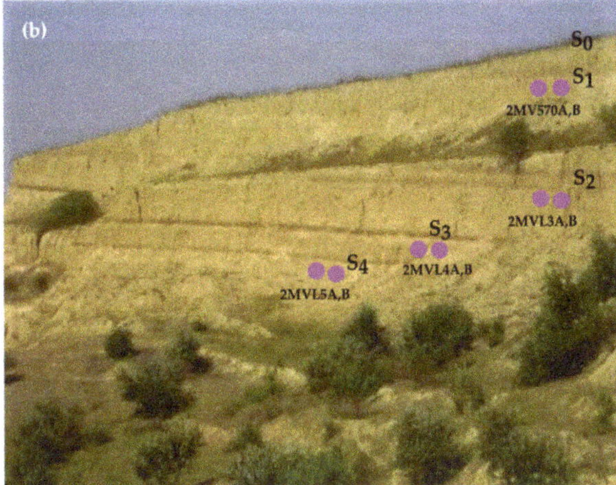

Figure 3. (**a**) Sample positions on the field in the Holocene soil (S_0) and L_1 loess unit; (**b**) relative positions for the collected doublet samples beneath the S_1–S_4.

For measurement purposes the coarse (63–90 μm) quartz grains were mounted on stainless steel disks using silicone oil as adhesive. The fine (4–11 μm) quartz grains were settled on aluminum disks from a 2 mg/mL suspension in acetone.

The samples were measured on Risø TL/OSL-DA-20 readers [58] with the stimulation being performed by blue light emitting diodes (470 ± 30 nm) and IR light emitting diodes (875 ± 80 nm). The luminescence emissions were detected by an incorporated bialkaline EMI 9235QA photomultiplier (maximum detection efficiency ~400 nm) through a 7.5 mm thick Hoya U-340 UV filter. Irradiations were carried out using a ^{90}Sr-^{90}Y radioactive source which was calibrated using gamma irradiated fine and coarse calibration quartz [59]. Radionuclide specific activities were measured though high-resolution gamma spectrometry using a coaxial detector with high purity germanium well detector (120 cm^3 volume), full width at half maximum (FWHM) of 1.40 keV at 122 keV and a full width at half maximum (FWHM) of 2.30 keV at 1332 keV. IAEA 312 and IAEA 327 standards have been used for relative calibration.

3.4.2. Luminescence Measurements

For $\mathbf{D_e}$ determination for the fine (4–11 μm) and coarse (63–90 μm) fractions the single-aliquot regenerative dose (SAR) protocol was used [1,60]. To corroborate the previous quartz studies on Mircea Vodă, the $\mathbf{D_e}$ dependency on the preheat treatment was assessed for one doublet sample from the L_4 unit. The test concluded that there is no systematic variation for the 200–280 °C temperature range (Figure 4). Thus, for consistency reasons with our previous studies a preheat temperature of 220 °C for 10 s and a cutheat of 180 °C have been further used, alongside a test dose of 17 Gy. A high-temperature bleach by stimulation with blue LEDs for 40 s at 280 °C at the end of each test dose signal measurement was employed (Table 1). TL was recorded during the preheat procedure. The OSL signal used was recorded during the first 0.308 s of stimulation and an early background subtraction has been applied from the 1.69–2.31 s interval [61].

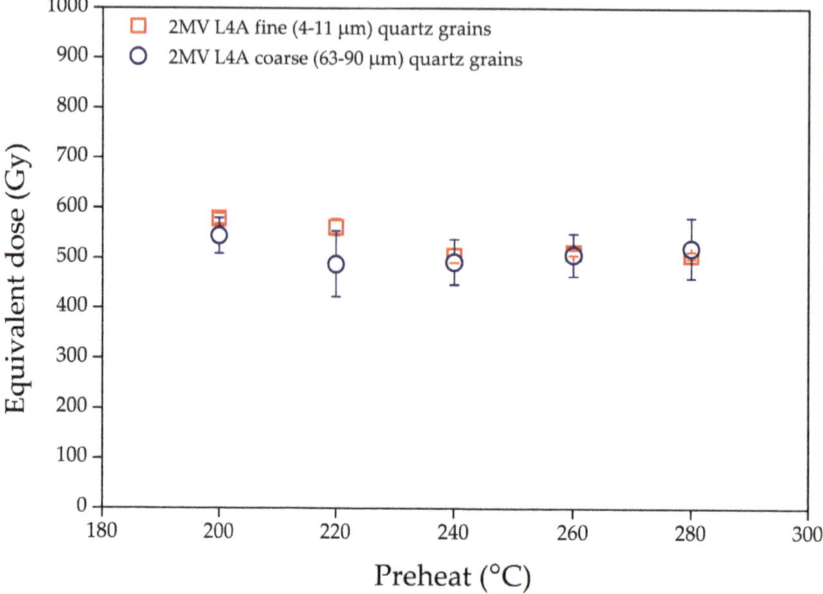

Figure 4. Equivalent dose dependence on preheat temperature for sample 2MV L4A for fine (4–11 μm) quartz fraction (open red squares) and coarse (63–90 μm) quartz fraction (open blue circles).

For determining equivalent doses at least 8 aliquots have been measured per sample per quartz fraction. The accepted aliquots exhibited good recycling and IR depletion ratios, though the average values in the case of coarse (63–90 μm) quartz were 2% and 4%, respectively, lower than those obtained by Timar-Gabor et al. [8]. In the case of the 63–90 μm quartz extracts, 22% of the measured aliquots were rejected due to poor IR and recycling, while for the 4–11 μm only 1% of the aliquots were rejected and only due to poor IR depletion values. In Figure 5 representative CW-OSL decay and dose response curves for two samples are presented. By interpolating the sensitivity corrected natural OSL signals onto the dose response curve constructed the equivalent doses were determined (Table 2).

Table 2. Summary of the equivalent doses, radionuclide activities, calculated dose rates and optical ages. The luminescence and dosimetry data are indicated alongside the random uncertainties and the optical ages are indicated alongside the overall uncertainties. All uncertainties are standard uncertainties. Specific activities were measured on a well detector by high resolution gamma spectrometry. The ages were calculated assuming water content of 20%. The total dose rate includes the contribution from cosmic rays [62], gamma, beta and alpha (for 4–11 μm quartz grains) radiations. An internal dose rate of 0.01 ± 0.002 Gy/ka [63] alongside a beta attenuation and etching factor of 0.94 ± 0.05 [19] were taken into consideration for the coarse (63–90 μm) quartz fraction. The alpha efficiency factor for the 4–11 μm quartz grains was that of 0.04 ± 0.02 [64]. The optical ages marked with asterisk (*) were obtained for samples which were found to be close to saturation levels (between 71% and 87%, Figure 6).

Unit Code	Sampling Depth (m)	Laboratory Code	Grain Size (μm)	Equivalent Dose (Gy)	Recycling Ratio	Recuperation (%)	IR depletion Ratio	Total dose Rate (Gy/ka)	Cosmic dose Rate (Gy/ka)	Age (ka)	Random Error (%)	Systematic Error (%)
S_0/L_1	0.4	2MV 40	4–11	15.4 ± 0.2	1.02 ± 0.01	0.11 ± 0.03	0.99 ± 0.01	2.91 ± 0.05	0.22 ± 0.03	5.3 ± 0.5	2.3	9.9
			63–90	12.9 ± 1.5	1.03 ± 0.01	0.15 ± 0.07	0.98 ± 0.01	3.44 ± 0.05		5.3 ± 0.7	11.8	7.6
S_0/L_1	0.5	2MV 50	4–11	20.3 ± 0.3	1.02 ± 0.01	0.07 ± 0.03	0.95 ± 0.01	3.05 ± 0.07	0.21 ± 0.03	6.7 ± 0.7	2.7	9.9
			63–90	17.3 ± 1.9	1.05 ± 0.01	0.14 ± 0.04	0.98 ± 0.01	2.55 ± 0.06		6.8 ± 0.9	11.2	7.6
S_0/L_1	0.6	2MV 60	4–11	23.6 ± 0.5	1.04 ± 0.01	0.05 ± 0.05	0.99 ± 0.01	3.15 ± 0.05	0.21 ± 0.03	7.5 ± 0.8	2.7	9.9
			63–90	22.1 ± 1.6	1.02 ± 0.01	0.21 ± 0.10	0.98 ± 0.01	2.64 ± 0.05		8.4 ± 0.9	7.4	7.6
S_0/L_1	0.7	2MV 70	4–11	25.6 ± 0.3	1.03 ± 0.01	0.07 ± 0.03	0.98 ± 0.01	2.88 ± 0.05	0.20 ± 0.03	8.9 ± 0.9	2.1	9.9
			63–90	26.6 ± 2.1	1.03 ± 0.01	0.09 ± 0.02	0.98 ± 0.01	2.41 ± 0.05		11.0 ± 1.2	8.1	7.6
S_0/L_1	0.8	2MV 80	4–11	31.2 ± 0.4	1.00 ± 0.01	0.07 ± 0.03	0.95 ± 0.01	2.81 ± 0.05	0.20 ± 0.03	11.1 ± 1.1	2.0	10.0
			63–90	36.4 ± 2.9	1.02 ± 0.01	0.09 ± 0.03	0.95 ± 0.01	2.35 ± 0.05		15.5 ± 1.7	8.2	7.6
S_0/L_1	0.9	2MV 90	4–11	35.9 ± 0.6	0.99 ± 0.01	0.19 ± 0.04	0.98 ± 0.01	2.77 ± 0.05	0.19 ± 0.03	12.9 ± 1.3	2.5	10.0
			63–90	36.0 ± 2.5	1.03 ± 0.01	0.05 ± 0.02	0.97 ± 0.01	2.32 ± 0.05		15.5 ± 1.6	7.2	7.6
S_0/L_1	0.93	MV 2.1	4–11	28.3 ± 0.6	1.04 ± 0.01	0.03 ± 0.03	0.94 ± 0.01	2.75 ± 0.04	0.19 ± 0.03	14.0 ± 1.4	2.3	9.7
			63–90	36.1 ± 2.0	1.02 ± 0.01	0.11 ± 0.03	0.98 ± 0.01	2.31 ± 0.04		15.6 ± 1.5	5.8	7.6
S_0/L_1	0.99	MV 2.2	4–11	51.7 ± 0.6	0.99 ± 0.01	0.04 ± 0.01	0.98 ± 0.01	2.88 ± 0.05	0.19 ± 0.03	18.0 ± 1.8	2.2	9.7
			63–90	53.6 ± 2.6	1.01 ± 0.01	0.04 ± 0.02	0.97 ± 0.01	2.42 ± 0.05		22.1 ± 2.1	5.2	7.7
S_0/L_1	1.23	MV 2.3	4–11	51.2 ± 0.9	1.02 ± 0.02	0.04 ± 0.02	0.90 ± 0.01	2.81 ± 0.06	0.18 ± 0.03	18.2 ± 1.9	2.8	9.9
			63–90	62.1 ± 3.0	1.01 ± 0.01	0.15 ± 0.07	0.97 ± 0.01	2.53 ± 0.05		26.4 ± 2.5	5.3	7.6
S_0/L_1	1.35	MV 2.4	4–11	46.8 ± 0.8	1.03 ± 0.02	0.04 ± 0.02	0.98 ± 0.01	2.92 ± 0.04	0.18 ± 0.03	16.1 ± 1.6	2.2	9.9
			63–90	60.8 ± 3.8	1.02 ± 0.01	0.09 ± 0.03	0.97 ± 0.01	2.44 ± 0.04		24.9 ± 2.5	6.4	7.7
S_0/L_1	1.47	MV 2.5	4–11	59.7 ± 1.0	1.03 ± 0.01	0.04 ± 0.02	0.92 ± 0.01	2.85 ± 0.05	0.18 ± 0.03	20.9 ± 2.1	2.3	9.9
			63–90	79.0 ± 4.1	1.00 ± 0.01	0.06 ± 0.02	0.97 ± 0.01	2.39 ± 0.04		33.1 ± 3.1	5.5	7.7

Table 2. *Cont.*

Unit Code	Sampling Depth (m)	Laboratory Code	Grain Size (µm)	Equivalent Dose (Gy)	Recycling Ratio	Recuperation (%)	IR depletion Ratio	Total dose Rate (Gy/ka)	Cosmic dose Rate (Gy/ka)	Age (ka)	Random Error (%)	Systematic Error (%)
S_0/L_1	1.67	MV 2.6	4–11	67.4 ± 0.6	1.00 ± 0.01	0.05 ± 0.01	0.98 ± 0.01	2.91 ± 0.05	0.17 ± 0.03	22.8 ± 2.3	2.0	10.1
			63–90	86.9 ± 3.5	1.00 ± 0.01	0.06 ± 0.03	0.98 ± 0.01	2.42 ± 0.04		35.9 ± 3.2	4.4	7.7
L_2	5.70	2MV 570A	4–11	331 ± 6	0.96 ± 0.01	0.10 ± 0.01	0.95 ± 0.01	3.03 ± 0.06	0.11 ± 0.02	109 ± 11	2.6	10.0
			63–90	303 ± 13	0.97 ± 0.01	0.06 ± 0.02	0.96 ± 0.01	2.54 ± 0.05		120 ± 11*	4.7	7.8
L_2	5.70	2MV 570B	4–11	331 ± 3	0.96 ± 0.01	0.09 ± 0.01	0.96 ± 0.01	2.95 ± 0.06	0.11 ± 0.02	112 ± 12	2.2	10.1
			63–90	353 ± 13	0.97 ± 0.01	0.05 ± 0.02	0.95 ± 0.01	2.46 ± 0.05		143 ± 13*	4.2	7.8
L_3	13.70	2MV L3A	4–11	514 ± 16	0.98 ± 0.01	0.06 ± 0.01	0.98 ± 0.01	2.88 ± 0.05	0.06 ± 0.01	179 ± 19	2.4	10.3
			63–90	475 ± 19	0.96 ± 0.01	0.010 ± 0.02	0.94 ± 0.01	2.39 ± 0.04		199 ± 18*	4.3	8.0
L_3	13.70	2MV L3B	4–11	501 ± 11	0.98 ± 0.01	0.05 ± 0.01	0.99 ± 0.01	2.86 ± 0.06	0.06 ± 0.01	175 ± 18	2.2	10.2
			63–90	501 ± 23	0.95 ± 0.01	0.11 ± 0.02	0.94 ± 0.01	2.38 ± 0.05		210 ± 20*	5.0	8.0
L_4	17.70	2MV L4A	4–11	567 ± 9	0.99 ± 0.01	0.05 ± 0.002	1.00 ± 0.01	3.14 ± 0.05	0.04 ± 0.01	180 ± 19	2.3	10.4
			63–90	577 ± 23	0.97 ± 0.01	0.13 ± 0.02	0.97 ± 0.01	2.60 ± 0.05		222 ± 20*	-	-
L_4	17.70	2MV L4B	4–11	477 ± 6	1.00 ± 0.01	0.07 ± 0.004	0.91 ± 0.01	3.41 ± 0.06	0.04 ± 0.01	140 ± 13	2.2	8.8
			63–90	555 ± 26	0.98 ± 0.01	0.17 ± 0.03	0.95 ± 0.01	2.31 ± 0.05		240 ± 23*	-	-
L_5	20.50	2MV L5A	4–11	566 ± 8	0.98 ± 0.01	0.07 ± 0.003	0.98 ± 0.01	2.66 ± 0.06	0.04 ± 0.01	213 ± 22	2.5	10.2
			63–90	425 ± 39	0.98 ± 0.01	0.21 ± 0.09	0.98 ± 0.01	2.22 ± 0.05		192 ± 24*	-	8.0
L_5	20.50	2MV L5B	4–11	556 ± 13	0.99 ± 0.01	0.06 ± 0.01	0.98 ± 0.01	2.98 ± 0.06	0.04 ± 0.01	187 ± 20	2.9	-
			63–90	443 ± 35	0.94 ± 0.02	0.43 ± 0.10	0.97 ± 0.03	2.48 ± 0.05		179 ± 20*	8.1	8.0

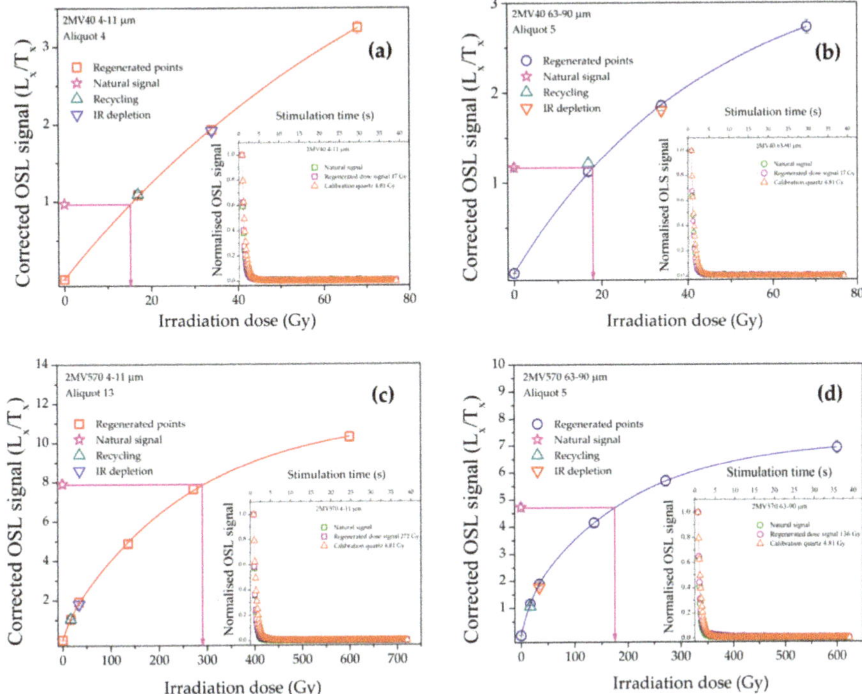

Figure 5. SAR dose response curves for accepted aliquots from samples 2MV40 (**a**,**b**) and 2MV570 (**c**,**d**) for both quartz fractions. Natural signals are represented as stars. Error bars are smaller than the symbols. The comparison between the normalized decay curves (the number of counts in each data channel divided by the number of counts measured in the first channel of stimulation) of the natural OSL signals, the regenerated signals and the decay of the calibration quartz is represented in the insets.

Based on previous reports, the CW-OSL growth curves up to high doses for both fine and coarse quartz is best described by a sum of two saturating exponential functions [65–68] of the form:

$$I(D) = I_0 + A_1*(1 - \exp(-D/D_{01})) + A_2*(1 - \exp(-D/D_{02}))$$

where the parameters are: I—intensity of the signal for a given dose D; I_0—intercept; A_1, A_2—saturation amplitudes of the two exponential components; D_{01}, D_{02} -doses which represent the onset of saturation of each exponential function.

In order to assess the closeness to saturation of the coarse quartz (63–90 μm) natural signal, CW-OSL growth curves were constructed up to 1 kGy for the old samples taken from L_3, L_4 and L_5 loess units. The ratio between the average sensitivity corrected signal (L_{nat}/T_{nat}) and the corrected luminescence signals measured for the 1000 Gy regenerative dose ($L_x/T_{x1000Gy}$) was calculated since for this dose it was observed that the dose response curve is very close to saturation. The coarse (63–90 μm) quartz natural signal for these older samples reached between 71% and 87% of the laboratory saturation level. All equivalent doses are presented in Table 2.

For both quartz fractions, the average sensitivity-corrected natural signals (L_{nat}/T_{nat}) for the 2MV570, 2MV3, 2MVL4 and 2MVL5 samples were plotted as a function of the expected D_e (Figure 6). Bearing in mind the fact that these samples were taken directly under the paleosol units, the expected ages were found based on the climatic records of benthic δ^{18}O [50]. The expected D_e was thus calculated by multiplying these ages with the annual dose values obtained in Table 2. A relative uncertainty of

10% was considered. It can be noted that the samples are in field saturation, meaning that the natural signals are no longer increasing with depth. The averaged natural signal for fine quartz is 59% of the average sensitivity corrected signal for the 5000 Gy regenerative dose (a dose high enough for the laboratory dose response to approach saturation—see Figure 7) and the coarse quartz natural signal is 75% of the average sensitivity corrected signal for the 2000 Gy regenerative dose (a dose high enough for the laboratory dose response to approach saturation—see Figure 7). These ratios of the natural signals to laboratory saturation levels for the two grain sizes are in agreement with previous reports from Timar-Gabor et al. [9] for an infinitely old sample. Differences of 60% (4–11 µm quartz) and 80% (63–90 µm quartz) were reported between the natural and laboratory dose response curves constructed for the nearby loess-paleosol site of Costinești [28]. Similar discrepancies between the natural and the laboratory dose response curves were reported by Chapot et al. [30] on samples taken from Luochuan, China.

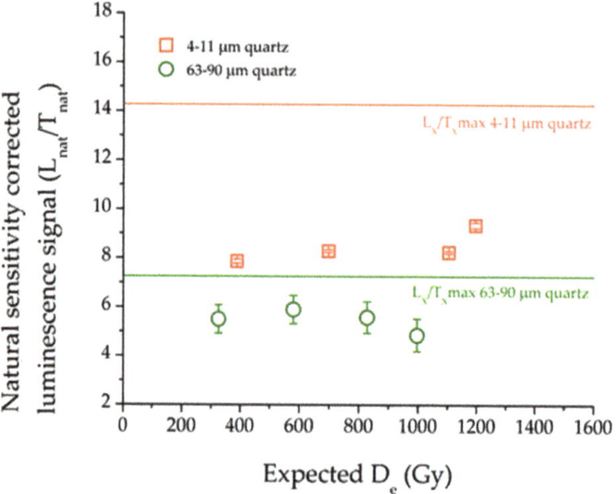

Figure 6. The natural sensitivity corrected luminescence signal (L_x/T_x) for samples 2MV570, 2MV3, 2MVL4 and 2MVL5 for 4–11 µm quartz (open squares) and 63–90 µm quartz (open circles) plotted against the expected D_e. The average maximum sensitivity corrected signal (L_x/T_x max) for a regenerative dose of 2000 Gy in the case of coarse grains and 5000 Gy in the case of fine grains (taken from the data presented in Figure 7) is shown as a reference for both quartz fractions.

Extended CW-OSL growth curves were subsequently constructed for samples 2MV L3A and 2MV L4A up to 5 kGy (for 4–11 µm quartz grains) and 2 kGy (for 63–90 µm quartz grains) using at least 6 regenerative points and a test dose of 17 Gy as well as a test dose of 170 Gy (Figure 7a–d). The data was fitted using a sum of two exponential functions (Table 3). Results obtained using a test dose of 17 Gy confirm the different saturation characteristics between the two quartz fractions [9,11,22,28], with average value of the two samples obtained for the fine (4–11 µm) fraction ($D_{01} = 88 \pm 22$ Gy; $D_{02} = 1194 \pm 142$ Gy) similar to those reported by Timar-Gabor et al. [11] ($D_{01} = 151 \pm 5$ Gy; $D_{02} = 1411 \pm 64$ Gy) and the coarse (63–90 µm) fraction ($D_{01} = 50 \pm 8$ Gy; $D_{02} = 427 \pm 54$ Gy) close to values from Timar-Gabor et al. [11] ($D_{01} = 44$ Gy; $D_{02} = 452$ Gy), Murray et al. [65] ($D_{01} = 44$ Gy; $D_{02} = 450$ Gy for 180–250 µm quartz) and Pawley et al. [66] ($D_{01} = 51$ Gy; $D_{02} = 320$ Gy for 125–180 µm quartz). For a test dose of 170 Gy, there is a general trend for the D_{01} and D_{02} values to decrease (Figure 7a–d). For the 2MV L3A and 2MV L4A fine (4–11 µm) quartz the D_{01} values decreased by 34% and 18%, while the D_{02} values decreased by 31% and 19%, respectively. The D_{01} values for coarse (63–90 µm) quartz were lower than those using a test dose of 17 Gy by 24% and 22% and the D_{02} values were lower by 18% and 10%, respectively (Table 3). As a result, the use of the 170 Gy test dose increased the closeness of the natural corrected luminescence signals to the saturation levels for both quartz fractions by 3.5% up to 16%, with the coarse (63–90 μm) quartz for the two oldest samples reaching ≥86% of saturation level. The equivalent doses were obtained by measuring at least 3 aliquots (Table 3). Even though there is a slight increase in the values obtained the results remain consistent within uncertainties. It should be noted that poor recycling ratios have been observed for fine (4–11 μm) quartz when the 170 Gy test dose was employed (4 aliquots measured), with values lower than unity by 27%. However, this effect was not significant for the coarse (63–90 μm) quartz measured with a test dose of 170 Gy with recycling values contained in the 0.9–1.1 interval.

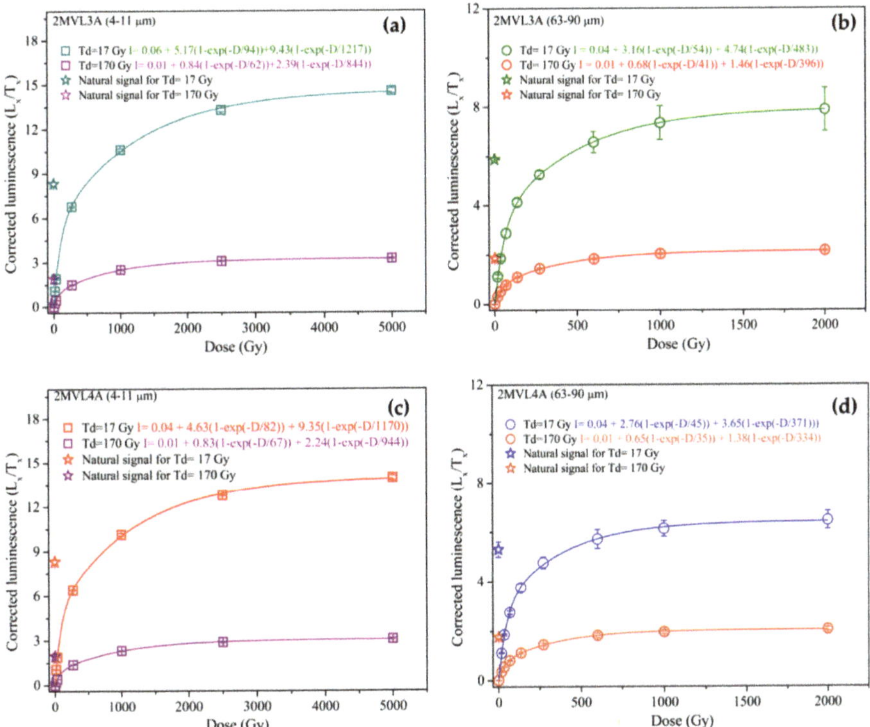

Figure 7. Comparison of growth curves for samples 2MVL3A (**a**,**b**) and 2MVL4A (**c**,**d**) for both quartz fractions. The curves were best described by a sum of two exponential exponentials function. At least three aliquots have been used in order to obtain the average corrected luminescence signals used to construct de growth curves. A preheat temperature of 220 °C for 10 s and a cutheat of 180 °C have been employed.

Table 3. Fitting parameters for OSL dose response curves constructed up to 5000 Gy (4–11 μm quartz) and 2000 Gy (63–90 μm quartz) for samples 2MV L3A and 2MV L4A.

Sample	Test Dose (Gy)	y_0	y_0 Error	A_1	A_1 Error	D_{01}	D_{01} Error	A_2	A_2 Error	D_{02}	D_{02} Error	Reduced χ^2	R^2	D_e (Gy)	Closeness of Natural Signal to Laboratory Saturation Level (%)
4–11 μm	17	0.06	0.14	5.2	0.4	94	18	9.4	0.4	1217	112	0.022	0.999	493 ± 28	57
2MV L3A	170	0.01	0.02	0.8	0.06	62	8	2.4	0.06	844	42	0.05×10^{-2}	0.999	505 ± 26	59
4–11 μm	17	0.04	0.12	4.6	0.3	82	13	9.4	0.3	1170	88	0.017	0.999	584 ± 29	60
2MV L4A	170	0.01	0.03	0.8	0.08	67	12	2.2	0.07	944	63	0.09×10^{-2}	0.999	604 ± 47	63

Sample	Test Dose (Gy)	y_0	y_0 Error	A_1	A_1 Error	D_{01}	D_{01} Error	A_2	A_2 Error	D_{02}	D_{02} Error	Reduced χ^2	R^2	D_e (Gy)	
63–90 μm	17	0.04	0.05	3.2	0.2	54	4	4.7	0.2	483	30	0.002	0.999	584 ± 29	74
2MV L3A	170	0.01	0.02	0.7	0.05	41	5	1.5	0.05	396	25	0.03×10^{-2}	0.999	560 ± 73	86
63–90 μm	17	0.04	0.07	2.8	0.3	45	6	3.7	0.2	371	41	0.005	0.999	440 ± 27	83
2MV L4A	170	0.01	0.02	0.7	0.06	35	5	1.4	0.05	334	24	0.3×10^{-2}	0.999	628 ±89	89

4. Ages and Discussion

Previous luminescence dating studies on Mircea Vodă site (Figure 8a) revealed an age discrepancy between the two quartz fractions investigated that is still not yet understood. The ages obtained ranged from 8.7 ± 1.3 to 159 ± 24 ka for fine silt-sized (4–11 µm) quartz and from 16 ± 2 ka to 230 ± 31 ka for fine sand-sized (63–90 µm) quartz, the difference varying between 20% to 70% [2,8]. Despite the fact that both datasets were consistent with the stratigraphic position of the samples, the fine (4–11 µm) quartz ages for the three samples taken from the L_2, L_3 and L_4 loess units were interpreted as underestimates. The post-IR IR$_{225}$ signal was considered more reliable than the previously obtained quartz ages for the L_2, L_3 and L_4 units. These ages are presented alongside the quartz ages in Figure 8a.

Due to the importance of Mircea Vodă loess-paleosol master section, the current study aimed to obtain a more detailed chronological framework. In this regard 13 samples have been collected from the Holocene soil (S_0) and L_1 loess unit and doublet samples have been collected directly beneath the S_1–S_4 units, respectively. The luminescence ages were found to be mostly in agreement with their stratigraphically corresponding results reported by Timar-Gabor et al. [8] and Vasiliniuc et al. [14]. The overall uncertainties associated with the new OSL ages are dominated by the systematic uncertainties caused by the time-averaged water content, a-value and beta attenuation factors. The overall contribution from random sources of uncertainties scatters around 2.5% for the fine (4–11 µm) quartz (generally 10 aliquots per sample) and ranges from 4.2% to 11.8% for the coarse (63–90 µm) quartz. It has been proposed that the source for such a large spread could be attributed to reduced number of coarse quartz grains on the disk compared to fine aliquots, thus reducing the variability in the luminescence properties. Other studies have also reported similar spread in coarse quartz data [23,25].

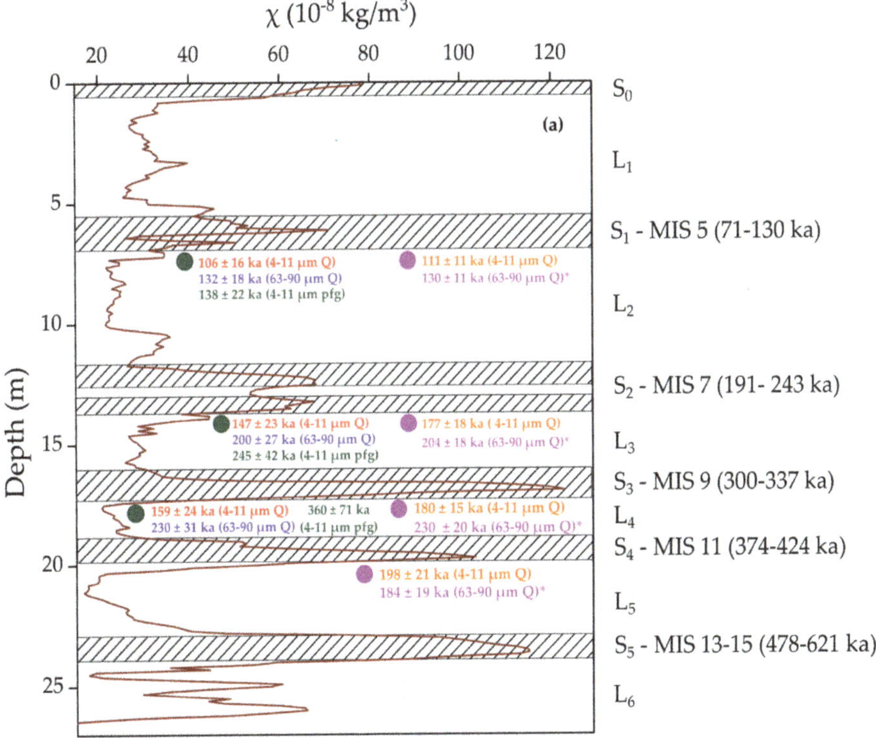

Figure 8. *Cont.*

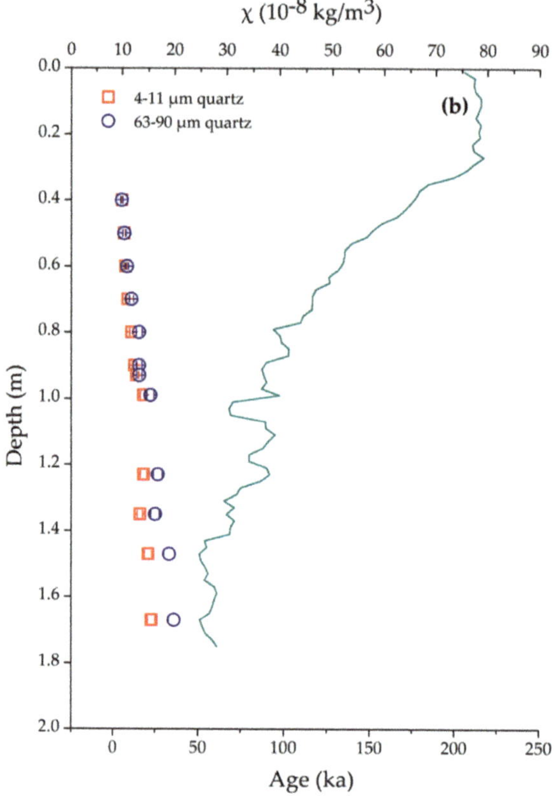

Figure 8. (**a**) Schematic representation of the loess (L) and paleosol units (S; hatched area) with magnetic susceptibility values (χ) from [2].The boundaries of paleosols developed during odd marine isotope stages (MIS) are after [49]. The ages for the old samples (green circles) and new samples (purple circles) are shown as follows: written in red—[2]; blue—[8]; green—[14] and written in orange—4–11 µm quartz current paper; purple—63–90 µm quartz current paper. The current ages represent the weighted results from the doublet samples. The optical ages marked with asterisk (*) were obtained for samples which were found to be close to saturation levels. (**b**) Plot of new optical ages as a function of depth alongside new magnetic susceptibility data. The fine (4–11 µm) quartz ages are represented as open squares and the coarse (63–90 µm) quartz ages are represented as open circles.

The top part of the profile that encompasses the Pleistocene/Holocene transition (L_1/S_0 unit) is characterized by increasing ages with profile depth, ranging from 5.3 ± 0.5 ka to 22.8 ± 2.3 ka for fine (4–11 µm) quartz and from 5.3 ± 0.7 ka to 35.9 ± 3.2 ka for coarse (63–90 µm) quartz (Figure 8b). In the Holocene soil, for ages up to about 11 ka OSL ages obtained for coarse and fine quartz agree, as expected and previously reported for such young samples [27]. The older samples taken from the L_1/S_0 transition and the L_1 loess unit yielded ages that no longer agree within uncertainties. The fine (4–11 µm) ages continue to be younger than coarse (63–90 µm) ages. As previously reported by Timar-Gabor et al. [8] on Mircea Vodă, Timar-Gabor et al. [9] on Mostiștea, Constantin et al. [12] on Costinești, Timar-Gabor et al. [10] on Orlovat in Serbia as well as by Timar-Gabor et al. [11] in China, the equivalent doses are higher for coarse (63–90 µm) quartz, which is unexpected considering the annual dose rate.

The age discrepancy between the two quartz fractions appears to begin sooner than previously reported. For Romanian, Serbian and Chinese loess samples, SAR-OSL ages divergence arose beyond

~40 ka (>~100 Gy) [11]. Only for samples collected from Lunca section (southern Wallachian Plain, see Figure 1) such a difference occurred starting with samples as young as <30 ka (~80 Gy) [39].

For the samples starting with L_2 downwards, the results obtained on doublet samples on each quartz fractions were found in agreement within uncertainties. Consequently, weighted ages have been calculated for each grain size fraction on the doublet samples.

The doublet samples collected from the L_2 loess unit yielded weighted ages on the two quartz fractions of 111 ± 11 ka for fine (4–11 μm) quartz and 130 ± 11 ka for coarse (63–90 μm) quartz. Fine quartz ages underestimate the expected age, while the coarse quartz age is in broad agreement with the expected age. These results are consistent with those reported by Timar-Gabor et al. [8] and Vasiliniuc et al. [14].

The weighted average ages recovered from the L_3 samples were that of 177 ± 18 ka for 4–11 μm quartz and 204 ± 18 ka for 63–90 μm quartz. The fine quartz age clearly underestimates the expected age, the offset being of 37%, while for the coarse quartz age the underestimation is of 19%. In the case of coarse quartz it was observed that the signal was close to laboratory saturation levels, with an average of 80%, while in the case of fine quartz, the natural signals were interpolated significantly below the saturation level of the laboratory dose response curve.

In what regards the samples from L_4 and L_5 loess units, the ages on fine (4–11 μm) quartz are 180 ± 15 ka and 198 ± 21 ka and the ages for coarse (63–90 μm) quartz are 230 ± 20 ka and 184 ± 19 ka, respectively. The two set of ages underestimate severely the expected values from stratigraphical boundaries considerations [50]. The coarse(63–90 μm) quartz signals were found to be close to laboratory saturation for both L_4 and L_5 samples (85% and 77%, respectively), with the closeness to laboratory saturation being more pronounced when a larger test dose was used. On the other hand, the same statement cannot be made for fine (4–11 μm) quartz (e.g., for 2MVL5 sample the natural signals were at 60% of the laboratory saturation level).

The reasons behind the age discrepancies between fine (4–11 μm) quartz and coarse (63–90 μm), especially in the case of the Mircea Vodă site have been proposed, discussed and investigated in previous studies. Microdosimetry could be a reason, but one should bear in mind the fact that such an age discrepancy could only arise from a dose rate difference of approximately 1 Gy/ka [8]. The purity of the quartz extracts was checked by comparing the natural and regenerated signals with the calibration quartz, by performing the IR depletion test [69] and by checking the 110 °C TL peak recorder during preheats. The results dismiss the possibility of a feldspathic component which would end in different age results between the two fractions. This also concurs with previous time resolved OSL (TR-OSL) results reported by Timar-Gabor et al. [28]. In what regards partial bleaching, residual doses of the order of at least tens or even hundreds of Gy would be needed in order to cause such an age offset for all samples investigated. This is not to be expected in the case of quartz.

5. Conclusions

Fine (4–11 μm) and coarse (63–90 μm) quartz have been investigated by applying SAR-OSL protocol in order to augment the existing chronological framework from Mircea Vodă loess-paleosol master section. The age results for the Pleistocene/Holocene transition have shown that fine and coarse fractions agree only up to ~20 ka. For samples older than this, fine grains quartz ages underestimate coarse quartz ages. Thus, the discrepancy between two datasets occurs sooner than previously shown for other sites [11]. The reason for this difference is yet not understood.

As previously reported the 63–90 μm quartz does not underestimate the expected geological ages and agrees with post-IR IR_{225} for samples collected just below S_1. For older samples coarse quartz SAR OSL signals approach (86%) laboratory saturation and also enter field saturation. For the counterpart fine grains quartz OSL natural signals are significantly below laboratory saturation levels. However, the fine quartz ages underestimate the expected ages. Therefore, these ages should be taken as minimum ages. Investigation on extended growth curves up to 5 kGy (for 4–11 μm quartz grains)

Methods Protoc. **2020**, *3*, 19

and 2 kGy (for 63–90 µm quartz grains) using test doses of a different order of magnitude (17 and 170 Gy) have concluded that the equivalent dose was insensitive to the size of the test dose.

Supplementary Materials: The following are available online at http://www.mdpi.com/2409-9279/3/1/19/s1.

Author Contributions: Project administration, conceptualization, methodology and funding acquisition was undergone by A.T.-G.; resources, writing—review and editing, supervision and validation was undergone by A.T.-G. and C.P.; investigation and was undergone by C.P. and Ș.-M.G.-S.; writing—original draft preparation, review and editing was undergone by Ș.-M.G.-S. All authors have read and agreed to the published version of the manuscript.

Funding: This research was funded the European Research Council (ERC) under the European Union Horizon 2020 research and innovation program ERC-2015-STG (Grant Agreement No. 678106).

Acknowledgments: Daniela Constantin, Dan Veres and Ramona Balc are highly acknowledged for their contribution in sampling campaigns.

Conflicts of Interest: The authors declare no conflict of interest.

References

1. Murray, A.S.; Wintle, A.G. Luminescence dating of quartz using an improved single-aliquot regenerative-dose protocol. *Radiat. Meas.* **2000**, *32*, 57–73. [CrossRef]

2. Timar, A.; Vandenberghe, D.; Panaiotu, E.C.; Panaiotu, C.G.; Necula, C.; Cosma, C.; van den haute, P. Optical dating of Romanian loess using fine-grained quartz. *Quat. Geochronol.* **2010**, *5*, 143–148. [CrossRef]

3. Buylaert, J.P.; Vandenberghe, D.; Murray, A.S.; Huot, S.; De Corte, F.; Van den Haute, P. Luminescence dating of old (>70 ka) Chinese loess: A comparison of single aliquot OSL and IRSL techniques. *Quat. Geochronol.* **2012**, *10*, 75–80. [CrossRef]

4. Lai, Z.P. Chronology and the upper dating limit for loess samples from Luochuan section in the Chinese Loess Plateau using quartz OSL SAR protocol. *J. Asian Earth Sci.* **2010**, *37*, 176–185. [CrossRef]

5. Buylaert, J.P.; Murray, A.S.; Vandenberghe, D.; Vriend, M.; De Corte, F.; Van den haute, P. Optical dating of Chinese loess using sand-sized quartz: Establishing a time frame for Late Pleistocene climate changes in the western part of the Chinese Loess Plateau. *Quat. Geochronol.* **2008**, *3*, 99–113. [CrossRef]

6. Fleming, S.J. Study of thermoluminescence of crystalline extracts from pottery. *Archaeometry* **1966**, *9*, 170–173. [CrossRef]

7. Zimmerman, D.W. Thermoluminescent dating using fine grains from pottery. *Archaeometry* **1971**, *13*, 29–52. [CrossRef]

8. Timar-Gabor, A.; Vandenberghe, D.A.G.; Vasiliniuc, S.; Panaiotu, C.E.; Panaiotu, C.G.; Dimofte, D.; Cosma, C. Optical dating of Romanian loess: A comparison between silt-sized and sand-sized quartz. *Quat. Int.* **2011**, *240*, 62–70. [CrossRef]

9. Timar-Gabor, A.; Vasiliniuc, S.; Vandenberghe, D.A.G.; Cosma, C.; Wintle, A.G. Investigations into the reliability of SAR-OSL equivalent doses obtained for quartz samples displaying dose response curves with more than one component. *Radiat. Meas.* **2012**, *47*, 740–745. [CrossRef]

10. Timar-Gabor, A.; Constantin, D.; Marković, S.B.; Jain, M. Extending the area of investigation of fine versus coarse quartz optical ages from the Lower Danube to the Carpathian Basin. *Quat. Int.* **2014**, *388*, 168–176. [CrossRef]

11. Timar-Gabor, A.; Buylaert, J.-P.; Guralnik, B.; Trandafir-Antohi, O.; Constantin, D.; Anechitei-Deacu, V.; Jain, M.; Murray, A.S.; Porat, N.; Hao, Q.; et al. On the importance of grain size in luminescence dating using quartz. *Radiat. Meas.* **2017**, *106*, 464–471. [CrossRef]

12. Constantin, D.; Begy, R.; Vasiliniuc, S.; Panaiotu, C.; Necula, C.; Codrea, V.; Timar-Gabor, A. High-resolution OSL dating of the Costinești section (Dobrogea, SE Romania) using fine and coarse quartz. *Quat. Int.* **2014**, *334*, 20–29. [CrossRef]

13. Anechitei-Deacu, V.; Timar-Gabor, A.; Fitzsimmons, K.E.; Veres, D.; Hambach, U. Multi-method investigations on quartz grains of different sizes extracted from a loess section in Southeast Romania interbedding the Campanian Ignimbrite ash layer. *Geochronometria* **2013**, *41*, 1–14. [CrossRef]

14. Vasiliniuc, Ș.; Vandenberghe, D.A.G.; Timar-Gabor, A.; Panaiotu, C.; Cosma, C.; van den Haute, P. Testing the potential of elevated temperature post-IR IRSL signals for dating Romanian loess. *Quat. Geochronol.* **2012**, *10*, 75–80. [CrossRef]

15. Vasiliniuc, S.; Vandenberghe, D.A.G.; Timar-Gabor, A.; Van Den haute, P. Conventional IRSL dating of Romanian loess using single aliquots of polymineral fine grains. *Radiat. Meas.* **2013**, *48*, 60–67. [CrossRef]

16. Huntley, D.J.; Godfrey-Smith, D.I.; Thewalt, M.L.W. Optical dating of sediments. *Nature* **1985**, *313*, 105–107. [CrossRef]

17. Singhvi, A.K.; Porat, N. Impact of luminescence dating on geomorphological and palaeoclimate research in drylands. *Boreas* **2008**, *37*, 536–558. [CrossRef]

18. Wintle, A.G. Luminescence dating: Where it has been and where it is going. *Boreas* **2008**, *37*, 471–482. [CrossRef]

19. Mejdahl, V. Thermoluminescence dating: Beta-dose attenuation in quartz grains. *Archaeometry* **1979**, *21*, 61–72. [CrossRef]

20. Aitken, M.J. *Thermoluminescence Dating*; Academic Press: London, UK, 1985; pp. 252–262.

21. Schmidt, C.; Bösken, J.; Kolb, T. Is there a common alpha-efficiency in polymineral samples measured by various infrared stimulated luminescence protocols? *Geochronometria* **2018**, *45*, 160–172. [CrossRef]

22. Constantin, D.; Timar-Gabor, A.; Veres, D.; Begy, R.; Cosma, C. SAR-OSL dating of different grain-size quartz from a sedimentary section in southern Romania interbedding the Campanian Ignimbrite/Y5 ash layer. *Quat. Geochronol.* **2012**, *10*, 81–86. [CrossRef]

23. Lomax, J.; Fuchs, M.; Preusser, F.; Fiebig, M. Luminescence based loess chronostratigraphy of the Upper Palaeolithic site Krems-Wachtberg, Austria. *Quat. Int.* **2014**, *351*, 88–97. [CrossRef]

24. Trandafir, O.; Timar-Gabor, A.; Schmidt, C.; Veres, D.; Anghelinu, N.; Hambach, U.; Simon, S. OSL dating of fine and coarse quartz from a Palaeolithic sequence on the Bistriţa Valley (Northeastern Romania). *Quat. Geochronol.* **2014**, *30*, 487–492. [CrossRef]

25. Fuchs, M.; Kreutzer, S.; Rousseau, D.-D.; Antoine, P.; Hatté, C.; Lagroix, F.; Moine, O.; Gauthier, C.; Svoboda, J.; Lisá, L. The loess sequence of Dolní Věstonice, Czech Republic: A new OSL-based chronology of the Last Climatic Cycle. *Boreas* **2013**, *42*, 664–677. [CrossRef]

26. Kreutzer, S.; Fuchs, M.; Meszner, S.; Faust, D. OSL chronostratigraphy of a loess-palaeosol sequence in Saxony/Germany using quartz of different grain sizes. *Quat. Geochronol.* **2012**, *10*, 102–109. [CrossRef]

27. Constantin, D.; Veres, D.; Panaiotu, C.; Anechitei-Deacu, V.; Groza, S.M.; Begy, R.; Kelemen, S.; Buylaert, J.P.; Hambach, U.; Marković, S.B.; et al. Luminescence age constraints on the Pleistocene-Holocene transition recorded in loess sequences across SE Europe. *Quat. Geochronol.* **2019**, *49*, 71–77. [CrossRef]

28. Timar-Gabor, A.; Constantin, D.; Buylaert, J.P.; Jain, M.; Murray, A.S.; Wintle, A.G. Fundamental investigations of natural and laboratory generated SAR dose response curves for quartz OSL in the high dose range. *Radiat. Meas.* **2015**, *81*, 150–156. [CrossRef]

29. Timar-Gabor, A.; Wintle, A.G. On natural and laboratory generated dose response curves for quartz of different grain sizes from Romanian loess. *Quat. Geochronol.* **2013**, *18*, 34–40. [CrossRef]

30. Chapot, M.S.; Roberts, H.M.; Duller, G.A.T.; Lai, Z.P. A comparison of natural-and laboratory-generated dose response curves for quartz optically stimulated luminescence signals from Chinese Loess. *Radiat. Meas.* **2012**, *47*, 1045–1052. [CrossRef]

31. Buggle, B.; Glaser, B.; Zöller, L.; Hambach, U.; Marković, S.; Glaser, I.; Gerasimenko, N. Geochemical characterization and origin of Southeastern and Eastern European loesses (Serbia, Romania, Ukraine). *Quat. Sci. Rev.* **2008**, *27*, 1058–1075. [CrossRef]

32. Fitzsimmons, K.E.; Marković, S.B.; Hambach, U. Pleistocene environmental dynamics recorded in the loess of the middle and lower Danube basin. *Quat. Sci. Rev.* **2012**, *41*, 104–118. [CrossRef]

33. Codrea, V. *Geologia Cuaternarului: Notiuni de Baza; (Quaternary Geology. Basic Notions)*; Litographed course; Babes-Bolyai University: Cluj Napoca, Romania, 1998.

34. Conea, A. Profils de loess en Roumanie. La stratigraphy des loess d'Europe. In *Bulletin de l'Association Française Pour L'étude du Quaternaire*; Fink, J., Ed.; Suppl. INQUA: Budapest, Hungary, 1986; pp. 127–134.

35. Conea, A. *Formatiuni Cuaternareîn Dobrogea (Loessuri si Paleosoluri) (Quaternary Units in Dobrogea)*, 1st ed.; Editura Academiei RSR: Bucuresti, Romania, 1970; p. 234.

36. Panaiotu, C.G.; Panaiotu, E.C.; Grama, A.; Necula, C. Paleoclimatic Record from a Loess-Paleosol Profile in Southeastern Romania. *Phys. Chem. Earth A* **2001**, *26*, 893–989. [CrossRef]

37. Necula, C.; Panaiotu, C. Application of dynamic programming to the dating of a loess-paleosol sequences. *Rom. Rep. Phys.* **2008**, *60*, 157–171.

38. Bălescu, S.; Lamothe, M.; Panaiotu, C.; Panaiotu, C. La chronologie IRSL des séquences loessiques de l'est de la Roumanie. *Quaternaire* **2010**, *21*, 115–126. [CrossRef]
39. Constantin, D.; Cameniţă, A.; Panaiotu, C.; Necula, C.; Codrea, V.; Timar-Gabor, A. Fine and coarse-quartz SAR-OSL dating of Last Glacial loess in Southern Romania. *Quat. Int.* **2015**, *357*, 33–43. [CrossRef]
40. Rădan, S.C.; Rădan, M. Study of the geomagnetic field structure in Tertiary in the context of magnetostratigraphic scale elaboration. I—The Pliocene. *Anu. Inst. Geol. României* **1998**, *70*, 215–231.
41. Buggle, B.; Hambach, U.; Glaser, B.; Gerasimenko, N.; Marković, S.; Glaser, I.; Zöller, L. Stratigraphy, and spatial and temporal paleoclimatic trends in Southeastern/Eastern European loess-paleosol sequences. *Quat. Int.* **2009**, *196*, 86–106. [CrossRef]
42. Bălescu, S.; Lamothe, M.; Mercier, N.; Huot, S.; Bălteanu, D.; Billardd, A.; Hus, J. Luminescence chronology of Pleistocene loess deposits from Romania: Testing methods of age correction for anomalous fading in alkali feldspars. *Quat. Sci. Rev.* **2003**, *22*, 967–973. [CrossRef]
43. Fitzsimmons, K.E.; Hambach, U. Loess accumulation during the last glacial maximum: Evidence from Urluia, southeastern Romania. *Quat. Int.* **2014**, *334*, 74–85. [CrossRef]
44. Buggle, B.; Hambach, U.; Müller, K.; Zöller, L.; Marković, S.B.; Glaser, B. Iron mineralogical proxies and Quaternary climate change in SE-European loess-paleosol sequences. *Catena* **2014**, *117*, 4–22. [CrossRef]
45. Necula, C.; Panaiotu, C. Rock magnetic properties of a loess-paleosols complex from Mircea Vodă (Romania). *Rom. Rep. Phys.* **2012**, *64*, 516–527.
46. Necula, C.; Panaiotu, C.; Heslop, D.; Dimofte, D. Climatic control of magnetic granulometry in the Mircea Vodă loess/paleosol sequence (Dobrogea, Romania). *Quat. Int* **2013**, *293*, 5–14. [CrossRef]
47. Heslop, D.; Langereis, C.G.; Dekkers, M.J. A new astronomical timescale for the loess deposits of Northern China. *Earth Planet. Sci. Lett.* **2000**, *184*, 125–139. [CrossRef]
48. Buggle, B.; Hambach, U.; Kehl, M.; Marković, S.B.; Zöller, L.; Glaser, B. The progressive evolution of a continental climate in southeast-central European lowlands during the Middle Pleistocene recorded in loess paleosol sequences. *Geology* **2013**, *41*, 771–774. [CrossRef]
49. Lisiecki, L.E.; Lisiecki, P.A. Application of dynamic programming to the correlation of paleoclimate records. *Paleoceanography* **2002**, *17*, 1049. [CrossRef]
50. Lisiecki, L.E.; Raymo, M.E. A Pliocene-Pleistocene stack of 57 globally distributed benthic $\delta^{18}O$ records. *Paleoceanography* **2005**, *20*, PA1003. [CrossRef]
51. Vasiliniuc, Ş.; Vandenberghe, D.A.G.; Timar-Gabor, A.; Cosma, C.; van den Haute, P. Combined IRSL and post-IR OSL dating of Romanian loess using single aliquots of polymineral fine grains. *Quat. Int.* **2013**, *293*, 15–21. [CrossRef]
52. Wallinga, J.; Murray, A.; Duller, G. Underestimation of equivalent dose in single-aliquot optical dating of feldspars caused by preheating. *Radiat. Meas.* **2000**, *32*, 691–695. [CrossRef]
53. Wacha, L.; Frechen, M. The geochronology of the "Gorjanović loess section" in Vukovar, Croatia. *Quat. Int.* **2011**, *240*, 87–99. [CrossRef]
54. Thiel, C.; Buylaert, J.P.; Murray, A.; Terhorst, B.; Hofer, I.; Tsukamoto, S.; Frechen, M. Luminescence dating of the Stratzing loess profile (Austria)—Testing the potential of an elevated temperature post-IR IRSL protocol. *Quat. Int.* **2011**, *234*, 23–31. [CrossRef]
55. Stevens, T.; Marković, S.B.; Zech, M.; Hambach, U.; Sümegi, P. Dust deposition and climate in the Carpathian Basin over an independently dated last glacial-interglacial cycle. *Quat. Sci. Rev.* **2011**, *30*, 662–681. [CrossRef]
56. Lang, A.; Lindauer, S.; Kuhn, R.; Wagner, G.A. Procedures used in optically and infrared stimulated luminescence dating of sediments in Heidelberg. *Anc. TL* **1996**, *14*, 7–11.
57. Frechen, M.; Schweitzer, U.; Zander, A. Improvements in sample preparation for the fine grain technique. *Anc. TL* **1996**, *14*, 15–17.
58. Thomsen, K.J.; Bøtter-Jensen, L.; Jain, M.; Denby, P.M.; Murray, A.S. Recent instrumental developments for trapped electron dosimetry. *Radiat. Meas.* **2008**, *43*, 414–521. [CrossRef]
59. Hansen, V.; Murray, A.S.; Buylaert, J.-P.; Yeo, E.-Y.; Thomsen, K. A new irradiated quartz for beta source calibration. *Radiat. Meas.* **2015**, *81*, 123–127. [CrossRef]
60. Murray, A.S.; Wintle, A.G. The single aliquot regenerative dose protocol: Potential for improvements in reliability. *Radiat. Meas.* **2003**, *37*, 377–381. [CrossRef]
61. Cunningham, A.C.; Wallinga, J. Selection of integration intervals for quartz OSL decay curves. *Quat. Geochronol.* **2010**, *5*, 657–666. [CrossRef]

62. Prescott, J.R.; Hutton, J.T. Cosmic ray contributions to dose rates for luminescence and ESR dating: Large depths and long-term time variations. *Radiat. Meas.* **1994**, *23*, 497–500. [CrossRef]

63. Vandenberghe, D.; De Corte, F.; Buylaert, J.-P.; Kučera, J.; Van den haute, P. On the internal radioactivity of quartz. *Radiat. Meas.* **2008**, *43*, 771–775. [CrossRef]

64. Rees-Jones, J. Optical dating of young sediments using fine-grain quartz. *Anc. TL* **1995**, *13*, 9–14.

65. Murray, A.S.; Svendsen, J.I.; Mangerund, J.I.; Astakhov, V.I. Testing the accuracy of quartz OSL dating using a known-age Eemian site on the river Sula, northern Russia. *Quat. Geochronol.* **2007**, *2*, 102–109. [CrossRef]

66. Pawley, S.M.; Toms, P.; Armitage, S.J.; Rose, J. Quartz luminescence dating of Anglian Stage (MIS 12) fluvial sediments: Comparison of SAR age estimates to the terrace chronology of the Middle Thames valley, UK. *Quat. Geochronol.* **2010**, *5*, 569–582. [CrossRef]

67. Lowick, S.E.; Preusser, F. Investigating age underestimation in the high dose region of optically stimulated luminescence using fine grain quartz. *Quat. Geochronol.* **2011**, *6*, 33–41. [CrossRef]

68. Lowick, S.E.; Preusser, F.; Wintle, A.G. Investigating quartz optically stimulated luminescence dose-response curves at high doses. *Radiat. Meas.* **2010**, *45*, 975–984. [CrossRef]

69. Duller, G.A.T. Distinguishing quartz and feldspar in single grain luminescence measurements. *Radiat. Meas.* **2003**, *37*, 161–165. [CrossRef]

Review

Review of the Post-IR IRSL Dating Protocols of K-Feldspar

Junjie Zhang and Sheng-Hua Li *

Department of Earth Sciences, The University of Hong Kong, Hong Kong 999077, China; junjie@connect.hku.hk
* Correspondence: shli@hku.hk

Received: 17 October 2019; Accepted: 10 January 2020; Published: 14 January 2020

Abstract: Compared to quartz, the infrared stimulated luminescence (IRSL) of K-feldspar saturates at higher dose, which has great potential for extending the dating limit. However, dating applications with K-feldspar has been hampered due to anomalous fading of the IRSL signal. The post-IR IRSL (pIRIR) signal of K-feldspar stimulated at a higher temperature after a prior low-temperature IR stimulation has significantly lower fading rate. Different dating protocols have been proposed with the pIRIR signals and successful dating applications have been made. In this study, we review the development of various pIRIR dating protocols, and compare their performance in estimating the equivalent dose (D_e). Standard growth curves (SGCs) of the pIRIR signals of K-feldspar are introduced. Single-grain K-feldspar pIRIR dating is presented and the existing problems are discussed.

Keywords: OSL dating; K-feldspar; post-IR IRSL; standard growth curve (SGC); single-grain

1. Introduction

Optically stimulated luminescence (OSL) dating is applied to date the last exposure event of a mineral grain to sunlight [1]. The OSL ages are obtained by the equation: Age = D_e/D_r, where D_e is the equivalent dose, representing the radiation dose that mineral grains have received after being blocked from sunlight, and the D_r is the environmental dose rate mainly dominated by the radioactive elements U, Th, K. Quartz has long been used as the ideal mineral for OSL dating, because of its abundance in nature, its signal stability, and its susceptibility to bleaching, especially after the single-aliquot regenerative-dose (SAR) protocol was established [2–4]. However, the OSL signal of quartz saturates with relatively low radiation dose, which has limited its application in dating older samples. Studies have shown that the quartz OSL ages were underestimated when the samples were older than ~100 ka [5–7], a situation usually attributed to the early saturation behavior of quartz OSL signal [8]. In some case, such as in the aeolian deposits from the Chinese Loess Plateau, age underestimation began at ~70 ka [9,10], which may be related to abnormally short lifetime of the OSL signal [11,12].

Compared to quartz, the infrared stimulated luminescence (IRSL) of K-feldspar has several advantages: the IRSL signal of K-feldspar saturates at higher dose, which has the potential to date older samples; the IRSL signal intensity of K-feldspar is much greater than quartz OSL signal, providing higher data precision; and the K element within K-feldspar grains provides an internal component for the dose rate, which is not affected by variations of the external environment [13–15]. However, the application of K-feldspar IRSL dating has long been hampered by a phenomenon called anomalous fading [16–18]. The thermally stable luminescence signal of K-feldspar fades under ambient temperature, which is suggested to be a result of quantum tunneling between the electron traps and holes [16–19]. Fading correction method has been proposed to address this problem [14,20,21]. With this solution, the fading rate of IRSL signal for each individual sample needs to be measured, which requires a large amount of laboratory work. In addition, fading correction for older samples would become more complicated because of the dose-dependent behavior of fading rate [21–24], and age overestimation after fading correction has been widely reported [25–29]. An isochron dating method has also been proposed [30,31]. In this method, it is assumed that the IRSL signal induced by internal dose does not fade, and D_e values of K-feldspar fractions with different grain sizes need to be measured, which is also laborious.

It would be highly advantageous if a non-fading signal of K-feldspar could be identified. Studies that aimed to isolate the non-fading IRSL signals have driven the development of post-IR IRSL (pIRIR) dating protocols, and a detailed review has been published in 2014 [32]. Here, we briefly introduce the pIRIR developments before 2014, and focus on the improvements since 2014.

2. Development of pIRIR Dating Protocols

2.1. Two-Step IR Stimulation

Thomsen et al. [33] found that the IRSL signal stimulated at 225 °C after a preceding IR stimulation at 50 °C (post-IR IRSL) faded much more slowly. The 50 °C IR stimulation (IR_{50}) is supposed to remove the close trap-hole pairs which are prone to tunneling; while the left distant trap-hole pairs stimulated at the subsequent high temperature experience less fading [33–36]. Following the observation of Thomsen et al. [33], Buylaert et al. [37] tested the two-step post-IR IRSL ($pIRIR_{50, 225}$) dating protocol (Table 1) on K-feldspar extracts from various depositional environments. The fading rate (g_{2days} value) of the $pIRIR_{50, 225}$ signal (mean value, 1.62 ± 0.06%/decade) was significantly lower than the fading rate of the IR_{50} signal (mean value, 3.23 ± 0.13%/decade) [37]. After fading correction, the $pIRIR_{50, 225}$ and IR_{50} signals had similar ages; however, the smaller fading rate made the $pIRIR_{50, 225}$ ages less dependent on the assumptions behind the fading correction model [37].

In order to isolate a more stable pIRIR signal, Thiel et al. [38] increased the temperature of the second-step stimulation from 225 °C to 290 °C (Table 1). The measured fading rate of the $pIRIR_{50, 290}$ signal of loess samples from Austria was 1.0–1.5%/decade, compared to 2.1–3.6%/decade of the IR_{50} signal [38]. However, based on observations that the natural $pIRIR_{50, 290}$ signal of a sample from below the Brunhes/Matuyama geomagnetic reversal boundary was saturated and that a quartz sample also showed a measured fading rate of 1.3 ± 0.3%/decade, Thiel et al. [38] suggested that the fading rate of the $pIRIR_{50, 290}$ signal was a laboratory artifact, and fading uncorrected $pIRIR_{50, 290}$ ages were the most appropriate. Buylaert et al. [39] systematically tested the $pIRIR_{50, 290}$ protocol and concluded that it was a robust dating method for Middle and Late Pleistocene sediments without fading correction.

The pIRIR signals stimulated at higher temperatures (e.g., $pIRIR_{50, 225}$, $pIRIR_{50, 290}$, $pIRIR_{200, 290}$) are more stable than the IR_{50} signal; however, they are also more difficult to bleach [33,37,38]. Thus, the residual doses of the high-temperature pIRIR signals are a concern in dating. Buylaert et al. [40] reported that the residual doses of the $pIRIR_{50, 225}$ and $pIRIR_{50, 290}$ signals from modern aeolian samples of the Chinese Loess Plateau were in the range of 2–8 Gy and 5–19 Gy, respectively. The $pIRIR_{50, 290}$ residual doses of three modern samples from an alluvial bar in Murray et al. [41] were measured to be 5–10 Gy. While quite high $pIRIR_{50, 290}$ residual doses (20–55 Gy) were reported for modern glaciofluvial sediments from a glaciated bay [42]. In various solar bleaching experiments, the measured residual doses of the pIRIR signals varied within a wide range, from less than 2 Gy to >40 Gy [38,39,43–45], and the residual doses increased systematically with higher preheat temperature and stimulation temperature [46–48]. A positive relationship between the residual doses and D_e values has been documented in numerous studies [39,49–54]. With a fixed bleaching time, a linear function between the residual doses and D_e values can be determined, and a minimal residual dose can be obtained by extrapolating to $D_e = 0$ Gy. The minimal residual dose of the $pIRIR_{50, 225/290}$ signal is mostly in the range of 4–7 Gy, which represents the contribution of an un-bleachable component that ubiquitously exists in fully-bleached samples [39,49,50,52–54]. In addition, some studies showed that with extremely long bleaching durations, the residual doses were also reduced to <10 Gy [29,51,52]. In summary, the residual doses of $pIRIR_{290}$ signal for partially-bleached sediments (e.g., glaciofluvial deposits) could be high (up to ~50 Gy), but the residual doses for well-bleached sediments are, in most cases, smaller than 10 Gy. Some studies argued that bleaching under different natural conditions (e.g., sub-aqueous and sub-aerial) had different efficiency due to different spectrum, which questioned the strategy of residual dose subtraction for D_e measurements [51,55].

For well-bleached old samples, a D_e overestimation of <10 Gy would result in an age overestimation of <~3 ka (e.g., for typical loess sediments), which has small effect. However, it becomes significant for young samples (e.g., of Holocene ages). In order to reduce the effect of residual dose, Reimann et al. [27] proposed a modified two-step pIRIR protocol with the first IR stimulation at 50 °C and the second IR stimulation at 180 °C, together with a preheat temperature of 200 °C (Table 1). The residual dose of the $\text{pIRIR}_{50, 180}$ signal was reduced to ~1 Gy [27]. Reimann et al. [27] showed that the $\text{pIRIR}_{50, 180}$ signal was sufficiently stable to date well-bleached middle Holocene and late Pleistocene samples, as the $\text{pIRIR}_{50, 180}$ ages agreed very well with quartz OSL ages and radiocarbon ages. To make it suitable to date even younger samples (e.g., of hundreds of years), the second IR stimulation was decreased to 150 °C, with a preheat treatment at 180 °C [28,56]. In several studies, the second IR stimulation temperature was changed to 170 °C with the preheat temperature at 200 °C to date Holocene aeolian deposits [57–59].

Table 1. Different pIRIR protocols: pIRIR$_{50, 225}$; pIRIR$_{50, 180}$; pIRIR$_{50, 290}$; pIRIR$_{200, 290}$; MET-pIRIR; 'SAR with solar'; 'MAR with heat'.

| Step | pIRIR50,225 | pIRIR50, 180 | pIRIR50, 290 | pIRIR200, 290 | MET-pIRIR | 'SAR with Solar' | 'MAR with Heat' |
	Buylaert et al. [37]	Reimann et al. [27]	Thiel et al. [38]	Li and Li [60]	Li and Li [26]	Li et al. [61]	Li et al. [62], Modified
1 *	Regenerative dose, D$_i$	Regenerative dose, D$_i$	Regenerative dose, D$_i$	Regenerative dose, D$_i$	Regenerative dose, D$_i$	Regenerative dose, D$_i$	Regenerative dose, D$_i$
2	Preheat at 250 °C for 60 s	Preheat at 200 °C for 60 s	Preheat at 320 °C for 60 s	Preheat at 320 °C for 60 s	Preheat at 300 °C for 10 s	Preheat at 300 °C for 60 s	Preheat at 320 °C for 60 s
3	IR for 100 s at 50 °C	IR for 100 s at 50 °C	IR for 200 s at 50 °C	IR for 200 s at 200 °C	IR for 100 s at 50 °C	IR for 100 s at 50 °C	IR for 100 s at 50 °C
4	IR for 100 s at 225 °C	IR for 100 s at 180 °C	IR for 200 s at 290 °C	IR for 200 s at 290 °C	IR for 100 s at 100 °C	IR for 100 s at 100 °C	IR for 100 s at 100 °C
5	Test dose, D$_t$	Test dose, D$_t$	Test dose, D$_t$	Test dose, D$_t$	IR for 100 s at 150 °C	IR for 100 s at 150 °C	IR for 100 s at 150 °C
6	Preheat at 250 °C for 60 s	Preheat at 200 °C for 60 s	Preheat at 320 °C for 60 s	Preheat at 320 °C for 60 s	IR for 100 s at 200 °C	IR for 100 s at 200 °C	IR for 100 s at 200 °C
7	IR for 100 s at 50 °C	IR for 100 s at 50 °C	IR for 200 s at 50 °C	IR for 200 s at 200 °C	IR for 100 s at 250 °C	IR for 100 s at 250 °C	IR for 100 s at 250 °C
8	IR for 100 s at 225 °C	IR for 100 s at 180 °C	IR for 200 s at 290 °C	IR for 200 s at 290 °C	Test dose, D$_t$	Test dose, D$_t$	IR for 100 s at 300 °C
9	IR at 290 °C for 40 s	Return to step 1	IR at 325 °C for 100 s	IR at 325 °C for 100 s	Preheat at 300 °C for 10 s	Preheat at 300 °C for 60 s	Cutheat to 500 °C
10	Return to step 1		Return to step 1	Return to step 1	IR for 100 s at 50 °C	IR for 100 s at 50 °C	Test dose, D$_t$
11					IR for 100 s at 100 °C	IR for 100 s at 100 °C	Preheat at 320 °C for 60 s
12					IR for 100 s at 150 °C	IR for 100 s at 150 °C	IR for 100 s at 50 °C
13					IR for 100 s at 200 °C	IR for 100 s at 200 °C	IR for 100 s at 100 °C
14					IR for 100 s at 250 °C	IR for 100 s at 250 °C	IR for 100 s at 150 °C
15					IR at 320 °C for 100 s	Solar simulator for 2 h	IR for 100 s at 200 °C
16					Return to step 1	Return to step 1	IR for 100 s at 250 °C
17							IR for 100 s at 300 °C

* For SAR protocols, in the first cycle, i = 0 and D$_0$ = 0, and the natural signal is measured. The sequence is run with several regenerative doses including a zero dose and a repeat dose, to build the growth curve.

2.2. Multi-Step IR Stimulation

Li and Li [26] proposed a multiple-elevated-temperature (MET) pIRIR dating protocol, with five-step IR stimulations from 50 °C to 250 °C with an increment of 50 °C (Table 1). An advantage of this protocol is that there are multiple D_e values corresponding to IR stimulations at different temperatures. Once the D_e reaches a plateau in the high temperature range, it would provide firm evidence that the high-temperature pIRIR signals are sufficiently stable and experience negligible anomalous fading (Figure 1) [26]. Later, the five-step MET-pIRIR protocol was modified to a six-step MET-pIRIR protocol, with the highest IR stimulation temperature increased from 250 °C to 300 °C, and the preheat temperature increased from 300 °C to 320 °C [63]. For the studied loess and desert samples from north China, the fading rates of the IR_{50} signal were 3–5%/decade, while the fading rates of MET-pIRIR$_{250}$ or MET-pIRIR$_{300}$ signals were very close to zero [26,63].

From the solar bleaching experiments, residual doses of the MET-pIRIR$_{250}$ signal were mostly in the range of 2–10 Gy, although in some cases they were up to ~20 Gy [26,64–68]. In order to make the MET-pIRIR protocol suitable for dating Holocene sediments, Fu and Li. [69] developed a modified low-temperature MET-pIRIR protocol. The preheat temperature was set to 200 °C and the five-step IR stimulation was performed from 50 °C to 170 °C with a step of 30 °C. The residual dose of the MET-pIRIR$_{170}$ signal was generally less than 1 Gy [69]. For Holocene samples, Fu and Li. [69] observed an age plateau between the stimulation temperatures of 110–170 °C. The five-step stimulation can be simplified to a three-step stimulation at temperatures of 110, 140, 170 °C. With the three-step protocol, the age plateau still existed at the stimulation temperatures of 140 °C and 170 °C, indicating that the pIRIR signals at 140 °C and 170 °C can be considered as sufficiently stable over the timescale of Holocene period [69]. Thus, Fu and Li. [69] further simplified the three-step protocol to a two-step protocol, with IR stimulations at 110 °C and 170 °C. Although no age plateau can be observed with the two-step IR stimulation, the pIRIR$_{110, 170}$ signal can still provide identical ages as the five-step pIRIR$_{170}$ signal, the three-step pIRIR$_{170}$ signal, and the quartz OSL signal. However, when the first IR stimulation temperature was decreased from 110 °C to 50 °C, the corresponded pIRIR$_{50, 170}$ ages were slightly underestimated [69].

As pIRIR signals stimulated at higher temperatures are more difficult to be bleached, several studies have proposed that the different bleaching rates of MET-pIRIR signals can be applied to infer the degree of bleaching, and thus to trace the transport history of sediments [70,71].

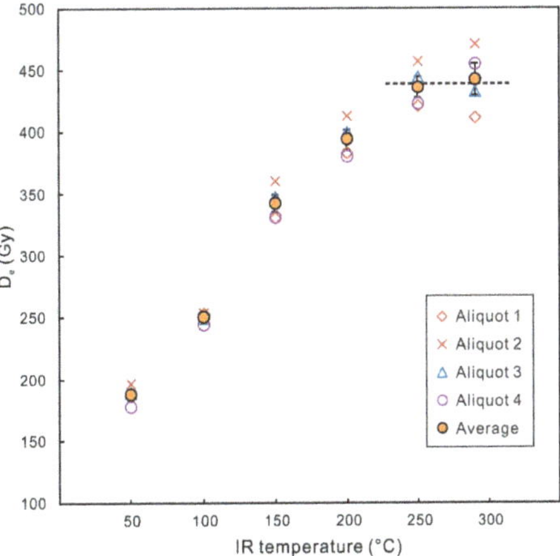

Figure 1. D_e versus IR stimulation temperature in the MET-pIRIR protocol. The sample used is 14LC-11.0, from the top of the L2 layer (second loess layer, corresponding to Marine Isotope Stage 6) in the Luochuan section, Chinese Loess Plateau (Zhang et al. [72]). Four aliquots were measured. The error bar of the average is the standard error calculated from four aliquots.

2.3. Modified Protocols to Extend the Dating Limit

The two-step pIRIR or MET-pIRIR protocols discussed above are all based on the conventional SAR protocol, with a subsequent test dose signal (T_x) to correct for the change in the sensitivity of the regenerative dose signal (L_x). Each aliquot is measured repeatedly with several cycles to build its individual growth curve [8]. The growth curve of the luminescence signal with the radiation dose can be fitted with a single saturating exponential function:

$$I = I_0 + I_{max} * \left(1 - e^{-D/D_0}\right), \tag{1}$$

where I is the luminescence signal, D is the radiation dose, and D_0 is called the characteristic saturation dose which quantifies the saturation behavior of the signal.

For quartz, D_0 varies significantly between different grains and samples, but in most cases it is smaller than 200 Gy [73–76]. It is suggested that for reliable dating with quartz, the D_e should not exceed the $2D_0$ limit [8]. Although D_0 of the K-feldspar pIRIR signal varies with different samples and experimental parameters (e.g., test dose, IR stimulation temperature), it is usually around 200–500 Gy with the conventional SAR protocol (Table 2). Applying the $2D_0$ limit, the dating range of the K-feldspar pIRIR signals would not exceed 1000 Gy. To extend the dating limit, modified protocols have been proposed based on the dose-dependent sensitivity of the MET-pIRIR signals from K-feldspar [61,62,77]. Li et al. [62] applied the multiple-aliquot regenerative-dose (MAR) protocol, but added a second test dose (T_2) after a 'cutheat to 600 °C' treatment behind the first test dose (T_1). The T_1/T_2 signal was applied to represent the dose-dependent sensitivity. For the MET-pIRIR$_{250}$ signal, the D_0 of the T_1/T_2 signal was ~740 Gy, which was significantly larger than that of the L_x/T_1 signal (~340 Gy). In addition, the L_x/T_2 signal also had a D_0 of ~770 Gy, close to that of the T_1/T_2 signal. Chen et al. [77] presented that with the L_x/T_2 signal, the D_e of a sample from the fifth paleosol layer (S5) of the Chinese loess sequence, corresponding to Marine Isotope Stage (MIS) 13–15, was estimated to be 1360 ± 200 Gy, which was broadly consistent with the expected D_e of 1550 ± 72 Gy. The modified MAR protocol is termed 'MAR with heat' in the text below (Table 1). Li et al. [61] modified the conventional SAR protocol, by adding a solar beaching treatment behind each cycle—'SAR with solar' (Table 1). It was found that solar bleaching was able to reset the luminescence sensitivity. Hence, the regenerative dose signal (L_x) or the test dose signal (T_x) can be used alone to estimate the D_e. Here, the T_x is used to represent the dose-dependent sensitivity. While the D_0 of the L_x/T_x signal was ~400 Gy, the D_0 of both the L_x and T_x signals was ~800 Gy [61].

These modified MAR and SAR protocols have greatly increased the D_0 of the K-feldspar pIRIR signal, and have extended the dating limit of K-feldspar to ~1500 Gy. Zhang and Li [78] proposed that a D_0 of ~800 Gy was very likely to be the intrinsic property of the pIRIR signals. In the conventional SAR protocols, the signals of the test dose (D_t) would be overestimated due to the effect of the preceding regenerative dose [43,78–82]. Multiple hypotheses have been proposed to account for the overestimation, such as the thermally transferred signal [43,79], signal inheritance [80,81], and dose dependent sensitivity change [62,82]. The test dose signal following a larger regenerative dose would be overestimated in a higher degree, and the corresponding L_x/T_x would be underestimated more significantly; hence the fitted growth curves have apparently lower D_0 values (200–500 Gy) compared to the intrinsic D_0 (~800 Gy). A larger test dose would reduce such an effect, thus a positive relationship between D_0 and D_t has been observed in numerous studies (Figure 2) [78–81,83,84].

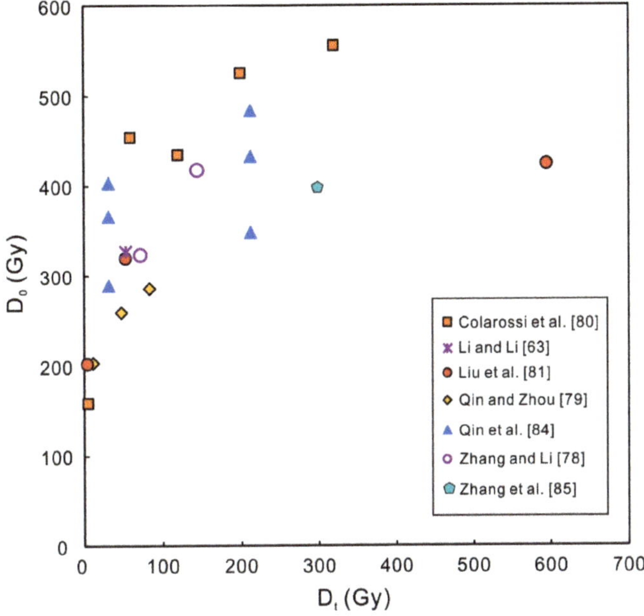

Figure 2. Characteristic saturation dose (D_0) of the pIRIR signals versus the test dose (D_t), with the SAR protocol. Data are from Table 2. A general positive relationship exists between D_0 and D_t. Please note that the pIRIR signals in this graph includes different kinds of signals, such as $\text{pIRIR}_{50,\,170}$, $\text{pIRIR}_{50,\,225}$, $\text{pIRIR}_{50,\,290}$, $\text{MET-pIRIR}_{250/300}$ signals. So the D_0 is scattered even with the same D_t.

Table 2. Characteristic saturation dose (D_0) of the growth curves from the conventional SAR protocol with different pIRIR signals.

Signal	D0 (Gy)	Test Dose (Gy)	Reference
$\text{pIRIR}_{50,\,295}$	204 ± 5	12	Qin and Zhou [79]
	286 ± 24	84	
$\text{pIRIR}_{50,\,290}$	203	4	Liu et al. [81]
	320	53	
	424	595	
$\text{pIRIR}_{50,\,225}$	159 ± 87	5	Colarossi et al. [80]
	455	60	
	435	120	
	526	200	
	556 ± 66	320	
$\text{pIRIR}_{50,\,290}$	290	32	Qin et al. [84]
	349	213	
$\text{pIRIR}_{50,\,225}$	367	32	
	433	213	
$\text{pIRIR}_{50,\,170}$	404	32	
	484	213	
MET-pIRIR_{250}	327 ± 16	54	Li and Li [63]
MET-pIRIR_{300}	250 ± 12	54	
MET-pIRIR_{250}	324 ± 5	72.5	Zhang and Li [78]
	417 ± 9	145	
MET-pIRIR_{300}	396 ± 13	300	Zhang et al. [85]

3. Comparison between Different pIRIR Protocols

Several studies have been performed to investigate the $pIRIR_{290}$ D_e dependence on the prior-IR stimulation temperature, and these studies suggested that the D_e values did not change significantly with the prior-IR stimulation temperature varied in the range of 50–260 °C [39,52–54,86]. From a comparison of the D_e values obtained with the $pIRIR_{50, 290}$, $pIRIR_{200, 290}$, MET-$pIRIR_{250}$ signals, Li and Li [60] showed that the estimated D_e values were consistent between the three signals when D_e was less than ~400 Gy; however, when the expected D_e exceeded ~400 Gy, the $pIRIR_{50, 290}$ signal had underestimated D_e results compared to the other two signals. Li and Li [60] suggested that the 50 °C prior-IR stimulation was too weak to completely remove the prone-to-fade signal. Qiu and Zhou. [87] compared the performance of four signals, which were $pIRIR_{50, 290}$, $pIRIR_{200, 290}$, three-step $pIRIR_{200, 290}$ signal with first stimulation at 50 °C, second stimulation at 200 °C and last stimulation at 290 °C, IRoff-$pIRIR_{200, 290}$ signal with isothermal holding (IR-off) for 200 s before the 290 °C IR stimulation, respectively. The D_e values of their tested samples were within the range of 400–900 Gy. Only the $pIRIR_{50, 290}$ signal had underestimated the D_e values, whereas the other three signals provided consistent D_e results [87]. Buylaert et al. [39] revealed that when the prior-IR stimulation temperature was increased from 50 °C to 200 °C, the intensity of corresponded $pIRIR_{200, 290}$ signal was only 7% of the $pIRIR_{50, 290}$ signal. However, it was later found that the $pIRIR_{290}$ signal intensity was still sufficiently high to guarantee precise measurements even when the prior-IR stimulation temperature was as high as 260 °C, and thus Buylaert et al. [88] applied the $pIRIR_{200, 290}$ protocol to date the last interglacial paleosol (S1) in the Chinese Loess Plateau. Stevens et al. [89] reported that D_e values were underestimated when the prior-IR stimulation temperature was below 140 °C, but a D_e plateau had been reached when the prior-IR stimulation temperature was ≥170 °C; and the $pIRIR_{200, 290}$ protocol was adopted to date the Chinese loess back to the S2 layer (second paleosol layer, corresponding to MIS 7). Ito et al. [29] carried out prior-IR stimulation temperature test on marine terrace deposits from Japan, and the results showed that the $pIRIR_{290}$ D_e plateau existed when the prior-IR temperature was within the range of 100–200 °C; while the $pIRIR_{290}$ D_e was underestimated with a prior-IR temperature of 50 °C and overestimated with a prior-IR temperature of 250 °C. A study performed on rock slices showed that the $pIRIR_{290}$ signal from naturally saturated slices was close to the laboratory saturation level only when the first-IR stimulation temperature was high (e.g., 200 °C or 250 °C) [81]. These studies suggest that the first-IR stimulation is better to be performed at a higher temperature (e.g., 200 °C) when dating older samples with the $pIRIR_{290}$ signal.

All of the foregoing comparisons in this section are based on the SAR protocol. However, the upper dating limit of these pIRIR SAR protocols has seldom been studied. A sample from the fifth loess layer (L5) of the Mangshan loess-palaeosol sequence of China was dated to be 401 ± 35 ka with the $pIRIR_{200, 290}$ protocol, which was much younger than the expected age within MIS 12 [87]. Zhang et al. [85] showed that with the SAR protocols, irrespective of the $pIRIR_{200, 290}$ signal or the MET-$pIRIR_{250/300}$ signal, the D_e values were underestimated when the expected D_e exceeded ~800 Gy, whereas the 'MAR with heat' protocol could provide reliable D_e estimates. The SAR D_e underestimation began to occur at the D_e value of ~800 Gy, which was close to twice the D_0 (~400 Gy with SAR protocol) for samples in Zhang et al. [85]. This indicates that the empirical $2D_0$ limit is also applicable to D_e measurements with K-feldspar.

Figure 3A shows that for a loess sample with the D_e of ~640 Gy, the SAR D_e values are smaller compared to the MAR D_e values for the low-temperature signals, but they become similar for the high-temperature (250 °C and 300 °C) signals. Several studies have illustrated that sensitivity correction in the first cycle (natural signal measurement) of the SAR protocol is unsuccessful for the low-temperature IRSL signals (e.g., IR_{50}) of K-feldspar when a high preheat temperature (e.g., >200 °C) is applied, and D_e underestimation exists for those low-temperature signals [67,84,90–92]. Age underestimation of the low-temperature IRSL signals has two sources with the SAR protocol—the anomalous fading and the failure of sensitivity correction. Some studies proposed that the failure of sensitivity correction was related to the increased electron trapping probability caused by the first preheat treatment [67,90,91]; while a recent study suggested that it was a combined effect of the decrease in the electron trapping probability and the increase in recombination probability [84]. The sensitivity correction of the high-temperature pIRIR signals is generally acceptable [67,84,91,92], which is why consistent SAR D_e and MAR D_e values can be obtained at the temperatures of 250 °C and 300 °C for relatively young samples (Figure 3A). However, the SAR D_e value becomes smaller than the MAR D_e value for a sample with D_e of ~900 Gy (Figure 3B) [85]. Despite of the generally successful sensitivity correction of the high-temperature pIRIR signals, both slight SAR D_e underestimation or overestimation can still occur, depending on the stimulation temperature of the pIRIR signals, the size of the D_e and the test dose, as well as sample origins [52,67,82,84,89]. Zhang et al. [85] performed dose recovery tests on the samples from their study, and observed a 10% overestimation for the dose recovery ratios with a recovery dose of 900 Gy, which cannot explain the SAR D_e underestimation for these old samples.

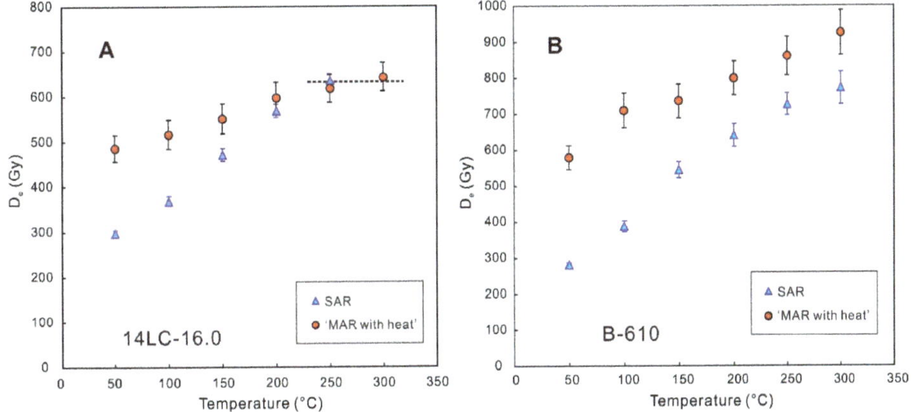

Figure 3. Comparison of D_e values obtained by the conventional SAR protocol (Li and Li [63]) and the modified 'MAR with heat' protocol (Li et al. [62]), with a six-step IR stimulation from 50 °C to 300 °C. (**A**) Sample 14LC-16.0 is from the base of the L2 layer (second loess layer, MIS 6) in the Luochuan section, Chinese Loess Plateau. The higher degree of SAR D_e underestimation compared to MAR D_e at low IR stimulation temperatures is due to the failure of sensitivity correction. Note that the D_e values are still consistent between the SAR and MAR protocols at higher IR stimulation temperatures (250 °C and 300 °C). (**B**) Sample B-610 is from the Jingbian section of Chinese Loess Plateau. Note that the SAR D_e values are still underestimated at high IR stimulation temperatures compared to the MAR D_e values, because D_e is already larger than 800 Gy. Figure 3B is modified from Zhang et al. [85].

The empirical $2D_0$ limit was initially proposed for quartz [8]. A recent study argued that the D_e underestimation beyond the $2D_0$ limit of quartz was caused by the rejection of the 'saturated' aliquots or grains which resulted in a truncated D_e distribution [93]. Instead of the conventional 'mean D_e' method, Li et al. [93] proposed a 'mean L_n/T_n' method to overcome this problem. The 'mean L_n/T_n' method applies the mean of re-normalized natural signal (L_n/T_n) to calculate the final D_e, thus no grains or aliquots would be abandoned when their natural signals lie above the saturation level of the growth curve. This method has been successfully applied in dating quartz of archeological cave sediments with D_e values up to ~300 Gy [94]. However, for the K-feldspar samples whose SAR D_e values were underestimated in Zhang et al. [85], no 'saturated' aliquots were observed and discarded. Applying the 'mean L_n/T_n' method cannot overcome the D_e underestimation of K-feldspar in Zhang et al. [85]. Several studies revealed that the natural dose response curve of quartz saturated at lower dose than the laboratory dose response curve, and they suggested it was the reason for quartz D_e underestimation in the high dose range [74,95–97]. Whether the natural dose response curve of K-feldspar also saturates at lower dose than the laboratory dose response curve needs further research.

4. Standard Growth Curves

The standard growth curve (SGC) was initially proposed for quartz to simplify the measurement procedure of D_e estimation [98]. Test dose standardized OSL signal ($L_x/T_x \times D_t$) was applied to construct the SGC [98]. With the SGC, only the sensitivity corrected natural signal (L_n/T_n) needs to be measured to estimate the D_e, which has greatly improved the efficiency of OSL dating [98]. However, some studies reported that if different D_t values were used, the $L_x/T_x \times D_t$ signal were still deviated from each other, with larger $L_x/T_x \times D_t$ for larger D_t [99,100]. Later, a regenerative dose normalization (re-normalization) procedure was proposed [101,102]. For each individual aliquot, the sensitivity corrected signals (L_x/T_x) are firstly re-normalized by a signal (L_{r1}/T_{r1}) from the aliquot itself. The re-normalized signal (I) is obtained by the following equation:

$$I = \frac{L_x/T_x}{L_{r1}/T_{r1}}, \tag{2}$$

where L_x/T_x are the sensitivity corrected signals of all the regenerative doses and L_{r1}/T_{r1} is the sensitivity corrected signal of a specific regenerative dose (D_{r1}). The I values of different aliquots and samples are plotted together, against the regenerative doses, to fit the SGC with appropriate functions such as the single saturating exponential (SSE) function, double saturating exponential (DSE) function, and general-order kinetic (GOK) function [103]. To estimate D_e, two cycles of the conventional SAR protocol need to be performed. The first cycle is to measure the natural signal (L_n/T_n) and the second cycle to measure the signal (L_c/T_c) of a regenerative dose, which is here termed the calibration dose (D_c). It is better to use a D_c that is close to the expected D_e value [78,102,104]. The L_c/T_c is used to calibrate the L_n/T_n:

$$f(D_e) = f(D_c)\frac{L_n/T_n}{L_c/T_c}, \tag{3}$$

where the $f(D_c)$ is the corresponded functional value of D_c on the SGC. The SGC D_e can be estimated from the $f(D_e)$ according to the SGC function.

The D_t applied in the SGC construction of K-feldspar was 24–66 Gy in Li et al. [102]. Due to the small size of the test dose, no dependence had been observed between the shape of the SGC and the D_t [102]. Zhang and Li [78] applied two larger D_t (72.5 Gy, 145 Gy) for SGC constructions, and reported that the SGC with a D_t of 145 Gy had a larger D_0 than the SGC with a D_t of 72.5 Gy. As illustrated above, D_0 generally increases with D_t (Figure 2). Zhang and Li [78] suggested that the D_t used for SGC D_e estimation should be close to the D_t used for SGC construction. Identical D_t for SGC construction and D_e estimation would always be the best choice.

A least-squares normalization procedure ('LS-normalization') was proposed to establish the SGCs for quartz from Haua Fteah cave, Libya [76]. Later, this LS-normalization procedure has also been applied in constructing SGCs for K-feldspar [92,105]. The 'LS-normalization' can further reduce the inter-grain and inter-aliquot variation of individual growth curves. It involves an iterative re-scaling and fitting process [76]. First, a starting curve is chosen. Then, individual growth curve of each aliquot or grain is re-scaled by multiplying a factor to make the sum of squared residuals—the difference between the observed values and the fitted values—is the smallest. All the rescaled-data are fitted again with a certain function. Iterate the re-scaling and fitting process until there is negligible change (<1%) in the results. The procedure can now be easily performed with the lsNORM function in the R package 'numOSL' [106].

For quartz, different grains or aliquots have quite different D_0 values, and different SGCs need to be built for quartz groups with different saturation behaviors [76,94]. In Figure 2, the D_0 values of K-feldspar are scatted even when the D_t is similar. That is because the figure includes different kinds of pIRIR signals (pIRIR$_{50, 170}$, pIRIR$_{50, 225}$, pIRIR$_{50, 290}$, MET-pIRIR$_{250}$, etc.). Also, the maximum regenerative doses used to build the growth curves in different studies are also different. Usually, the D_0 would increase when the maximum regenerative dose is larger [107]. When using the same D_t, the identical pIRIR signal from K-feldspar has quite similar saturation behavior between different samples [78,102,105]. Individual growth curves of different K-feldspar grains or aliquots from different continents become very close to each other after re-normalization or re-scaling, which indicates the existence of a global SGC of K-feldspar [102,105].

With a SGC, the machine time needs to estimate the D_e is only 1/3 of the time with a standard SAR approach, but the D_e values obtained by the SGC method are almost identical to the SAR D_e values [78,102]. Similar to quartz, the SGC also provides the 'mean L_n/T_n' solution to date K-feldspar samples with natural signals close to saturation [108], which cannot be accomplished with the conventional SAR approach.

5. Single-Grain Dating

The conventionally used aliquots in OSL dating usually contain hundreds of grains. The D_e values determined by aliquots are the mean of multiple grains. For partially-bleached samples, it would result in age overestimation. Single-grain dating measures the D_e values of individual grains, thus it has great advantages in dealing with partially-bleached sediments. By applying certain age models (e.g., [109–113]), the portion of fully-bleached grains can be distinguished and the last exposure event can be dated. Single-grain dating is also applicable to sediments that were well-bleached at deposition, but suffered disturbance after burial which resulted in mixing between different-aged grains (e.g., [114–118]). Several studies have applied single-grain dating of K-feldspar with the low-temperature IRSL signal (e.g., at 50 °C) [119–125]. Different methods were applied to overcome the fading problem, such as the *fadia* method [119–121], the isochron method [122], isolating 'zero'-fading grains [123], and fading correction [124,125]. After the pIRIR dating protocol was established, single-grain pIRIR dating with K-feldspar has been reported in numerous studies [43,108,126–137]. The machine time can be saved by performing prior-IR stimulations on all the grains simultaneously [108,129,137].

However, a general trend has been observed that brighter K-feldspar grains (higher signal sensitivity) yield higher D_e values [108,126,129,130,137]. Reimann et al. [126] applied pIRIR$_{50,\,180}$ dating protocol on K-feldspar single grains from southern Baltic Sea coast, NE Germany. A dependence of D_e values on the brightness of individual grains was observed—with higher D_e values for brighter grains, and only the brightest 30% of the K-feldspar grains yielded the mean single-grain pIRIR$_{50,\,180}$ ages that agreed with the age control [126]. As Reimann et al. [126] found that the fading rates of grains with different brightness were still close to each other, they suggested that the single-grain D_e dependence on brightness might be due to the different K contents in the grains (dimmer grains may contain less K). Brown et al. [129] performed K-feldspar single-grain pIRIR$_{50,\,225}$ dating for alluvial fan deposits in Baja California Sur, Mexico. In their study, the grains with brighter signal were found to have smaller fading rates. Rhodes [130] also reported a positive relationship between the grain brightness and D_e values, when dating individual K-feldspar grains from different locations with the pIRIR$_{50,\,225}$ signal. Rhodes [130] proposed an improved separation method to select the K-feldspar grains with density smaller than 2.565 g/cm^3—the 'Super-K' grains with higher brightness, rather than the regular K-feldspar fraction (density < 2.58 g/cm^3).

Jacobs et al. [108] carried out K-feldspar single-grain pIRIR$_{200,\,275}$ dating on sediments of the Denisova Cave in southern Siberia. For more than half of their samples, the weighted mean D_e values increased with a higher 'brightness threshold' of the grains (T_n), and similar pattern was also observed in the dose recovery ratios [108]. To make accurate D_e estimation, Jacobs et al. [108] determined a 'threshold' T_n for each sample, above which a mean D_e 'plateau' could be reached. The measured K contents of 60 individual K-feldspar grains (density < 2.58 g/cm^3) from Jacobs et al. [108] were mostly in the range of 12–14%, and no dependence of brightness on the K contents was observed. Guo et al. [137] compared the single-aliquot and single-grain MET-pIRIR$_{170}$ D_e results of K-feldspar samples from the Nihewan Basin, northern China. The mean D_e values of single-grain results were smaller than the those of the single-aliquot results (single-aliquot D_e was dominated by brighter grains inside the aliquot). Applying the 'brightness threshold' method, the mean single-grain D_e values corresponding to brighter grains became close to the single-aliquot D_e values. Fading rate tests showed that dimmer grains had higher fading rates, and Guo et al. [137] suggested that the discrepancy between the single-grain and single-aliquot D_e values were mainly due to the different fading rates of grains. However, a systematic increase in the fading-corrected D_e values with the 'brightness threshold' still existed and they proposed that brighter grains might still have slightly higher internal K contents than dimmer grains [137]. Guo et al. [137] have not observed a dependence of dose recovery ratios on the grain brightness. All the recovery ratios were close to unity irrespective of grain brightness, indicating that the sensitivity correction of natural signal (the first cycle in SAR protocol) is still successful for both bright and dim grains [137].

These studies presented above indicate that it is a ubiquitous phenomenon that brighter K-feldspar grains have higher D_e estimates, and can provide more reliable ages. By now, at least three factors may contribute to the lower D_e values for dim grains. One is that the dim grains have higher fading rates [129,137]. The second is that dim grains contain less internal K (less than 12–14 %), which corresponds to lower environmental dose rate [126,137]. The third is that sensitivity correction of natural signal is not successful for dim grains [108]. The true reason may be a combination of several factors and may also be sample-dependent. Therefore, in single-grain dating of K-feldspar, it is essential to exclude the dim grains with a 'brightness threshold', as age underestimation would be brought in if all grains are included to calculate the mean D_e value. A suitable 'brightness threshold' might be determined by the relationship between the mean D_e values and the 'brightness threshold'—the 'plateau' method [108,137].

6. Conclusions

The pIRIR signals of K-feldspar can be sufficiently stable to provide accurate age estimations without fading correction. However, care should be taken to choose the suitable pIRIR signal to date samples with different ages. For young samples (e.g., <10 ka), the low-temperature signals (e.g., $pIRIR_{50,180}$, MET-pIRIR$_{170}$) can be applied with a low preheat temperature (e.g., 200 °C) to avoid the age overestimation caused by residual doses. For intermediate-aged samples (e.g., 10–110 ka), the $pIRIR_{50,290}$, $pIRIR_{200,290}$, MET-pIRIR$_{250}$ signals are all suitable. For older samples, with D_e larger than 400 Gy (~110 ka with a typical dose rate of ~3.5 Gy/ka), the $pIRIR_{50,290}$ would underestimate the D_e values, while the $pIRIR_{200,290}$, MET-pIRIR$_{250}$ signals can still provide robust results. The empirical upper dating limit of $2D_0$ is also applicable to K-feldspar. With the modified protocols, such as 'MAR with heat' and 'SAR with solar', the dating limit can be increased to ~1500 Gy due to larger D_0.

The SGC of K-feldspar can greatly save the machine time needed for D_e measurement, while it provides D_e estimates almost identical to those of the standard SAR procedure. In addition, with the SGC, a 'mean L_n/T_n' approach for D_e estimation can be performed for samples whose natural signals are close to the saturation level of the growth curve. In single-grain dating of K-feldspar, bright grains usually have higher D_e values than the dim grains. Suitable 'brightness threshold' should be applied to exclude the dim grains to avoid age underestimation.

Author Contributions: Conceptualization, S.-H.L. and J.Z.; methodology, J.Z.; writing—original draft preparation, J.Z.; writing—review and editing, S.-H.L. All authors have read and agreed to the published version of the manuscript.

Funding: This research was founded by the grants to S.-H.L. from the Research Grant Council of the Hong Kong SAR, China (awards 17303014, 17307117).

Acknowledgments: We thank two anonymous reviewers and the editor—J.K. Feathers for their constructive comments.

References

1. Aitken, M.J. *An Introduction to Optical Dating*; Clarendon Press: Oxford, UK, 1998.
2. Murray, A.S.; Roberts, R.G. Measurement of the equivalent dose in quartz using a regenerative-dose single-aliquot protocol. *Radiat. Meas.* **1998**, *29*, 503–515. [CrossRef]
3. Murray, A.S.; Wintle, A.G. Luminescence dating of quartz using an improved single-aliquot regenerative-dose protocol. *Radiat. Meas.* **2000**, *32*, 57–73. [CrossRef]
4. Murray, A.S.; Wintle, A.G. The single aliquot regenerative dose protocol: Potential for improvements in reliability. *Radiat. Meas.* **2003**, *37*, 377–381. [CrossRef]
5. Murray, A.S.; Funder, S. Optically stimulated luminescence dating of a Danish Eemian coastal marine deposit: A test of accuracy. *Quat. Sci. Rev.* **2003**, *22*, 1177–1183. [CrossRef]
6. Murray, A.S.; Svendsen, J.I.; Mangerud, J.; Astakhov, V.I. Testing the accuracy of quartz OSL dating using a known-age Eemian site on the river Sula, northern Russia. *Quat. Geochronol.* **2007**, *2*, 102–109. [CrossRef]
7. Andreucci, S.; Sechi, D.; Buylaert, J.P.; Sanna, L.; Pascucci, V. Post-IR IRSL290 dating of K-rich feldspar sand grains in a wind-dominated system on Sardinia. *Mar. Pet. Geol.* **2017**, *87*, 91–98. [CrossRef]
8. Wintle, A.G.; Murray, A.S. A review of quartz optically stimulated luminescence characteristics and their relevance in single-aliquot regeneration dating protocols. *Radiat. Meas.* **2006**, *41*, 369–391. [CrossRef]
9. Lai, Z.P. Chronology and the upper dating limit for loess samples from Luochuan section in the Chinese Loess Plateau using quartz OSL SAR protocol. *J. Asian Earth Sci.* **2010**, *37*, 176–185. [CrossRef]
10. Buylaert, J.P.; Vandenberghe, D.; Murray, A.S.; Huot, S.; De Corte, F.; Van den Haute, P. Luminescence dating of old (> 70 ka) Chinese loess: A comparison of single-aliquot OSL and IRSL techniques. *Quat. Geochronol.* **2007**, *2*, 9–14. [CrossRef]
11. Fan, A.C.; Li, S.H.; Li, B. Observation of unstable fast component in OSL of quartz. *Radiat. Meas.* **2011**, *46*, 21–28. [CrossRef]
12. Lai, Z.P.; Fan, A.C. Examining Quartz Osl Age Underestimation for Loess Samples from Luochuan in the Chinese Loess Plateau. *Geochronometria* **2014**, *41*, 57–64. [CrossRef]
13. Duller, G.A.T. Behavioural studies of stimulated luminescence from feldspars. *Radiat. Meas.* **1997**, *27*, 663–694. [CrossRef]
14. Huntley, D.J.; Lamothe, M. Ubiquity of anomalous fading in K-feldspars and the measurement and correction for it in optical dating. *Can. J. Earth Sci.* **2001**, *38*, 1093–1106. [CrossRef]
15. Li, S.H.; Chen, Y.Y.; Li, B.; Sun, J.M.; Yang, L.R. OSL dating of sediments from deserts in northern China. *Quat. Geochronol.* **2007**, *2*, 23–28. [CrossRef]
16. Wintle, A.G. Anomalous Fading of Thermoluminescence in Mineral Samples. *Nature* **1973**, *245*, 143–144. [CrossRef]
17. Spooner, N.A. Optical Dating—Preliminary-Results on the Anomalous Fading of Luminescence from Feldspars. *Quat. Sci. Rev.* **1992**, *11*, 139–145. [CrossRef]
18. Spooner, N.A. The Anomalous Fading of Infrared-Stimulated Luminescence from Feldspars. *Radiat. Meas.* **1994**, *23*, 625–632. [CrossRef]
19. Visocekas, R. Tunneling Radiative Recombination in Labradorite—Its Association with Anomalous Fading of Thermo-Luminescence. *Nucl. Tracks Radiat. Meas.* **1985**, *10*, 521–529. [CrossRef]
20. Lamothe, M.; Auclair, M.; Hamzaoui, C.; Huot, S. Towards a prediction of long-term anomalous fading of feldspar IRSL. *Radiat. Meas.* **2003**, *37*, 493–498. [CrossRef]
21. Kars, R.H.; Wallinga, J.; Cohen, K.M. A new approach towards anomalous fading correction for feldspar IRSL dating—tests on samples in field saturation. *Radiat. Meas.* **2008**, *43*, 786–790. [CrossRef]
22. Li, B.; Li, S.-H. Investigations of the dose-dependent anomalous fading rate of feldspar from sediments. *J. Phys. D Appl. Phys.* **2008**, *41*, 225502. [CrossRef]
23. Morthekai, P.; Jain, M.; Murray, A.S.; Thomsen, K.J.; Botter-Jensen, L. Fading characteristics of martian analogue materials and the applicability of a correction procedure. *Radiat. Meas.* **2008**, *43*, 672–678. [CrossRef]
24. Wallinga, J.; Bos, A.J.J.; Dorenbos, P.; Murray, A.S.; Schokker, J. A test case for anomalous fading correction in IRSL dating. *Quat. Geochronol.* **2007**, *2*, 216–221. [CrossRef]
25. Li, G.-Q.; Zhao, H.; Chen, F.-H. Comparison of three K-feldspar luminescence dating methods for Holocene samples. *Geochronometria* **2011**, *38*, 14–22. [CrossRef]

26. Li, B.; Li, S.-H. Luminescence dating of K-feldspar from sediments: A protocol without anomalous fading correction. *Quat. Geochronol.* **2011**, *6*, 468–479. [CrossRef]
27. Reimann, T.; Tsukamoto, S.; Naumann, M.; Frechen, M. The potential of using K-rich feldspars for optical dating of young coastal sediments—A test case from Darss-Zingst peninsula (southern Baltic Sea coast). *Quat. Geochronol.* **2011**, *6*, 207–222. [CrossRef]
28. Madsen, A.T.; Buylaert, J.P.; Murray, A.S. Luminescence Dating of Young Coastal Deposits from New Zealand Using Feldspar. *Geochronometria* **2011**, *38*, 379–390. [CrossRef]
29. Ito, K.; Tamura, T.; Tsukamoto, S. Post-IR IRSL Dating of K-Feldspar from Last Interglacial Marine Terrace Deposits on the Kamikita Coastal Plain, Northeastern Japan. *Geochronometria* **2017**, *44*, 352–365. [CrossRef]
30. Li, B.; Li, S.-H.; Wintle, A.G.; Zhao, H. Isochron measurements of naturally irradiated K-feldspar grains. *Radiat. Meas.* **2007**, *42*, 1315–1327. [CrossRef]
31. Li, B.; Li, S.-H.; Wintle, A.G.; Zhao, H. Isochron dating of sediments using luminescence of K-feldspar grains. *J. Geophys. Res.-Earth Surf.* **2008**, *113*. [CrossRef]
32. Li, B.; Jacobs, Z.; Roberts, R.G.; Li, S.H. Review and assessment of the potential of post-IR IRSL dating methods to circumvent the problem of anomalous fading in feldspar luminescence. *Geochronometria* **2014**, *41*, 178–201. [CrossRef]
33. Thomsen, K.J.; Murray, A.S.; Jain, M.; Botter-Jensen, L. Laboratory fading rates of various luminescence signals from feldspar-rich sediment extracts. *Radiat. Meas.* **2008**, *43*, 1474–1486. [CrossRef]
34. Jain, M.; Ankjaergaard, C. Towards a non-fading signal in feldspar: Insight into charge transport and tunnelling from time-resolved optically stimulated luminescence. *Radiat. Meas.* **2011**, *46*, 292–309. [CrossRef]
35. Poolton, N.R.J.; Wallinga, J.; Murray, A.S.; Bulur, E.; Botter-Jensen, L. Electrons in feldspar I: On the wavefunction of electrons trapped at simple lattice defects. *Phys. Chem. Miner.* **2002**, *29*, 210–216. [CrossRef]
36. Poolton, N.R.J.; Ozanyan, K.B.; Wallinga, J.; Murray, A.S.; Botter-Jensen, L. Electrons in feldspar II: A consideration of the influence of conduction band-tail states on luminescence processes. *Phys. Chem. Miner.* **2002**, *29*, 217–225. [CrossRef]
37. Buylaert, J.P.; Murray, A.S.; Thomsen, K.J.; Jain, M. Testing the potential of an elevated temperature IRSL signal from K-feldspar. *Radiat. Meas.* **2009**, *44*, 560–565. [CrossRef]
38. Thiel, C.; Buylaert, J.P.; Murray, A.; Terhorst, B.; Hofer, I.; Tsukamoto, S.; Frechen, M. Luminescence dating of the Stratzing loess profile (Austria)—Testing the potential of an elevated temperature post-IR IRSL protocol. *Quat. Int.* **2011**, *234*, 23–31. [CrossRef]
39. Buylaert, J.P.; Jain, M.; Murray, A.S.; Thomsen, K.J.; Thiel, C.; Sohbati, R. A robust feldspar luminescence dating method for Middle and Late Pleistocene sediments. *Boreas* **2012**, *41*, 435–451. [CrossRef]
40. Buylaert, J.P.; Thiel, C.; Murray, A.S.; Vandenberghe, D.A.G.; Yi, S.W.; Lu, H.Y. IRSL and Post-IR IRSL Residual Doses Recorded in Modern Dust Samples from the Chinese Loess Plateau. *Geochronometria* **2011**, *38*, 432–440. [CrossRef]
41. Murray, A.S.; Thomsen, K.J.; Masuda, N.; Buylaert, J.P.; Jain, M. Identifying well-bleached quartz using the different bleaching rates of quartz and feldspar luminescence signals. *Radiat. Meas.* **2012**, *47*, 688–695. [CrossRef]
42. Alexanderson, H.; Murray, A.S. Luminescence signals from modern sediments in a glaciated bay, NW Svalbard. *Quat. Geochronol.* **2012**, *10*, 250–256. [CrossRef]
43. Nian, X.M.; Bailey, R.M.; Zhou, L.P. Investigations of the post-IR IRSL protocol applied to single K-feldspar grains from fluvial sediment samples. *Radiat. Meas.* **2012**, *47*, 703–709. [CrossRef]
44. Stevens, T.; Markovic, S.B.; Zech, M.; Hambach, U.; Sumegi, P. Dust deposition and climate in the Carpathian Basin over an independently dated last glacial-interglacial cycle. *Quat. Sci. Rev.* **2011**, *30*, 662–681. [CrossRef]
45. Li, Y.; Tsukamoto, S.; Frechen, M.; Gabriel, G. Timing of fluvial sedimentation in the Upper Rhine Graben since the Middle Pleistocene: Constraints from quartz and feldspar luminescence dating. *Boreas* **2018**, *47*, 256–270. [CrossRef]
46. Zhang, J.R.; Tsukamoto, S.; Nottebaum, V.; Lehmkuhl, F.; Frechen, M. D-e plateau and its implications for post-IR IRSL dating of polymineral fine grains. *Quat. Geochronol.* **2015**, *30*, 147–153. [CrossRef]
47. Wang, Y.X.; Chen, T.Y.; Chongyi, E.; An, F.Y.; Lai, Z.P.; Zhao, L.; Liu, X.J. Quartz OSL and K-feldspar post-IR IRSL dating of loess in the Huangshui river valley, northeastern Tibetan plateau. *Aeolian Res.* **2018**, *33*, 23–32. [CrossRef]

48. Roberts, H.M. Testing Post-IR IRSL protocols for minimising fading in feldspars, using Alaskan loess with independent chronological control. *Radiat. Meas.* **2012**, *47*, 716–724. [CrossRef]

49. Sohbati, R.; Murray, A.S.; Buylaert, J.P.; Ortuno, M.; Cunha, P.P.; Masana, E. Luminescence dating of Pleistocene alluvial sediments affected by the Alhama de Murcia fault (eastern Betics, Spain)—A comparison between OSL, IRSL and post-IR IRSL ages. *Boreas* **2012**, *41*, 250–262. [CrossRef]

50. Murray, A.S.; Schmidt, E.D.; Stevens, T.; Buylaert, J.P.; Markovic, S.B.; Tsukamoto, S.; Frechen, M. Dating Middle Pleistocene loess from Stari Slankamen (Vojvodina, Serbia)—Limitations imposed by the saturation behaviour of an elevated temperature IRSL signal. *Catena* **2014**, *117*, 34–42. [CrossRef]

51. Kars, R.H.; Reimann, T.; Ankjaergaard, C.; Wallinga, J. Bleaching of the post-IR IRSL signal: New insights for feldspar luminescence dating. *Boreas* **2014**, *43*, 780–791. [CrossRef]

52. Yi, S.W.; Buylaert, J.P.; Murray, A.S.; Lu, H.Y.; Thiel, C.; Zeng, L. A detailed post-IR IRSL dating study of the Niuyangzigou loess site in northeastern China. *Boreas* **2016**, *45*, 644–657. [CrossRef]

53. Li, G.Q.; Rao, Z.G.; Duan, Y.W.; Xia, D.S.; Wang, L.B.; Madsen, D.B.; Jia, J.; Wei, H.T.; Qiang, M.R.; Chen, J.H.; et al. Paleoenvironmental changes recorded in a luminescence dated loess/paleosol sequence from the Tianshan Mountains, arid central Asia, since the Penultimate Glaciation. *Earth Planet. Sci. Lett.* **2016**, *448*, 1–12. [CrossRef]

54. Li, G.Q.; Li, F.L.; Jin, M.; She, L.L.; Duan, Y.W.; Madsen, D.; Wang, L.B.; Chen, F. Late Quaternary lake evolution in the Gaxun Nur basin, central Gobi Desert, China, based on quartz OSL and K-feldspar pIRIR dating of paleoshorelines. *J. Quat. Sci.* **2017**, *32*, 347–361. [CrossRef]

55. Lowick, S.E.; Trauerstein, M.; Preusser, F. Testing the application of post IR-IRSL dating to fine grain waterlain sediments. *Quat. Geochronol.* **2012**, *8*, 33–40. [CrossRef]

56. Reimann, T.; Tsukamoto, S. Dating the recent past (<500 years) by pst-IR IRSL feldspar—Examples from the North Sea and Baltic Sea coast. *Quat. Geochronol.* **2012**, *10*, 180–187.

57. Li, G.Q.; Wen, L.J.; Xia, D.S.; Duan, Y.W.; Rao, Z.G.; Madsen, D.B.; Wei, H.T.; Li, F.L.; Jia, J.; Chen, F.H. Quartz OSL and K-feldspar pIRIR dating of a loess/paleosol sequence from arid central Asia, Tianshan Mountains, NW China. *Quat. Geochronol.* **2015**, *28*, 40–53. [CrossRef]

58. Wang, L.; Jia, J.; Zhao, H.; Liu, H.; Duan, Y.; Xie, H.; Zhang, D.D.; Chen, F. Optical dating of Holocene paleosol development and climate changes in the Yili Basin, arid central Asia. *Holocene* **2019**, *29*, 1068–1077. [CrossRef]

59. Buckland, C.E.; Bailey, R.M.; Thomas, D.S.G. Using post-IR IRSL and OSL to date young (<200 years) dryland aeolian dune deposits. *Radiat. Meas.* **2019**, *126*, 106131.

60. Li, B.; Li, S.H. A reply to the comments by Thomsen et al. on "Luminescence dating of K-feldspar from sediments: A protocol without anomalous fading correction". *Quat. Geochronol.* **2012**, *8*, 49–51. [CrossRef]

61. Li, B.; Roberts, R.G.; Jacobs, Z.; Li, S.H. A single-aliquot luminescence dating procedure for K-feldspar based on the dose-dependent MET-pIRIR signal sensitivity. *Quat. Geochronol.* **2014**, *20*, 51–64. [CrossRef]

62. Li, B.; Jacobs, Z.; Roberts, R.G.; Li, S.H. Extending the age limit of luminescence dating using the dose-dependent sensitivity of MET-pIRIR signals from K-feldspar. *Quat. Geochronol.* **2013**, *17*, 55–67. [CrossRef]

63. Li, B.; Li, S.H. Luminescence dating of Chinese loess beyond 130 ka using the non-fading signal from K-feldspar. *Quat. Geochronol.* **2012**, *10*, 24–31. [CrossRef]

64. Fu, X.; Li, B.; Li, S.H. Testing a multi-step post-IR IRSL dating method using polymineral fine grains from Chinese loess. *Quat. Geochronol.* **2012**, *10*, 8–15. [CrossRef]

65. Chen, Y.W.; Li, S.H.; Li, B. Residual doses and sensitivity change of post IR IRSL signals from potassium feldspar under different bleaching conditions. *Geochronometria* **2013**, *40*, 229–238. [CrossRef]

66. Li, B.; Roberts, R.G.; Jacobs, Z. On the dose dependency of the bleachable and non-bleachable components of IRSL from K-feldspar: Improved procedures for luminescence dating of Quaternary sediments. *Quat. Geochronol.* **2013**, *17*, 1–13. [CrossRef]

67. Zhang, J.J. Behavior of the electron trapping probability change in IRSL dating of K-feldspar: A dose recovery study. *Quat. Geochronol.* **2018**, *44*, 38–46. [CrossRef]

68. Rui, X.; Zhang, J.F.; Hou, Y.M.; Yang, Z.M.; Liu, Y.; Zhen, Z.M.; Zhou, L.P. Feldspar multi-elevated-temperature post-IR IRSL dating of the Wulanmulun Paleolithic site and its implication. *Quat. Geochronol.* **2015**, *30*, 438–444. [CrossRef]

69. Fu, X.; Li, S.-H. A modified multi-elevated-temperature post-IR IRSL protocol for dating Holocene sediments using K-feldspar. *Quat. Geochronol.* **2013**, *17*, 44–54. [CrossRef]

70. McGuire, C.; Rhodes, E.J. Downstream MET-IRSL single-grain distributions in the Mojave River, southern California: Testing assumptions of a virtual velocity model. *Quat. Geochronol.* **2015**, *30*, 239–244. [CrossRef]

71. Reimann, T.; Notenboom, P.D.; De Schipper, M.A.; Wallinga, J. Testing for sufficient signal resetting during sediment transport using a polymineral multiple-signal luminescence approach. *Quat. Geochronol.* **2015**, *25*, 26–36. [CrossRef]

72. Zhang, J.J.; Li, S.H.; Sun, J.M.; Hao, Q.Z. Fake age hiatus in a loess section revealed by OSL dating of calcrete nodules. *J. Asian Earth Sci.* **2018**, *155*, 139–145. [CrossRef]

73. Gong, Z.J.; Sun, J.M.; Lu, T.Y.; Tian, Z.H. Investigating the optically stimulated luminescence dose saturation behavior for quartz grains from dune sands in China. *Quat. Geochronol.* **2014**, *22*, 137–143. [CrossRef]

74. Timar-Gabor, A.; Wintle, A.G. On natural and laboratory generated dose response curves for quartz of different grain sizes from Romanian loess. *Quat. Geochronol.* **2013**, *18*, 34–40. [CrossRef]

75. Duller, G.A.T.; Botter-Jensen, L.; Murray, A.S. Optical dating of single sand-sized grains of quartz: Sources of variability. *Radiat. Meas.* **2000**, *32*, 453–457. [CrossRef]

76. Li, B.; Jacobs, Z.; Roberts, R.G. Investigation of the applicability of standardised growth curves for OSL dating of quartz from Haua Fteah cave, Libya. *Quat. Geochronol.* **2016**, *35*, 1–15. [CrossRef]

77. Chen, Y.W.; Li, S.H.; Li, B.; Hao, Q.Z.; Sun, J.M. Maximum age limitation in luminescence dating of Chinese loess using the multiple-aliquot MET-pIRIR signals from K-feldspar. *Quat. Geochronol.* **2015**, *30*, 207–212. [CrossRef]

78. Zhang, J.; Li, S.-H. Constructions of standardised growth curves (SGCs) for IRSL signals from K-feldspar, plagioclase and polymineral fractions. *Quat. Geochronol.* **2019**, *49*, 8–15. [CrossRef]

79. Qin, J.T.; Zhou, L.P. Effects of thermally transferred signals in the post-IR IRSL SAR protocol. *Radiat. Meas.* **2012**, *47*, 710–715. [CrossRef]

80. Colarossi, D.; Duller, G.A.T.; Roberts, H.M. Exploring the behaviour of luminescence signals from feldspars: Implications for the single aliquot regenerative dose protocol. *Radiat. Meas.* **2018**, *109*, 35–44. [CrossRef]

81. Liu, J.F.; Murray, A.; Sohbati, R.; Jain, M. The effect of test dose and first IR stimulation temperature on post-IR IRSL measurements on rock slices. *Geochronometria* **2016**, *43*, 179–187. [CrossRef]

82. Fu, X.; Li, S.H.; Li, B. Optical dating of aeolian and fluvial sediments in north Tian Shan range, China: Luminescence characteristics and methodological aspects. *Quat. Geochronol.* **2015**, *30*, 161–167. [CrossRef]

83. Carr, A.S.; Hay, A.S.; Powell, D.M.; Livingstone, I. Testing post-IR IRSL luminescence dating methods in the southwest Mojave Desert, California, USA. *Quat. Geochronol.* **2019**, *49*, 85–91. [CrossRef]

84. Qin, J.T.; Chen, J.; Li, Y.T.; Zhou, L.P. Initial sensitivity change of K-feldspar pIRIR signals due to uncompensated decrease in electron trapping probability: Evidence from radiofluorescence measurements. *Radiat. Meas.* **2018**, *120*, 131–136. [CrossRef]

85. Zhang, J.; Li, S.-H.; Wang, X.; Hao, Q.; Hu, G.; Chen, Y. Comparison of equivalent doses obtained with various post-IR IRSL dating protocols of K-feldspar. *Geochronometria* **2020**, in press.

86. Yi, S.; Buylaert, J.-P.; Murray, A.S.; Thiel, C.; Zeng, L.; Lu, H. High resolution OSL and post-IR IRSL dating of the last interglacial–glacial cycle at the Sanbahuo loess site (northeastern China). *Quat. Geochronol.* **2015**, *30*, 200–206. [CrossRef]

87. Qiu, F.Y.; Zhou, L.P. A new luminescence chronology for the Mangshan loess-palaeosol sequence on the southern bank of the Yellow River in Henan, central China. *Quat. Geochronol.* **2015**, *30*, 24–33. [CrossRef]

88. Buylaert, J.-P.; Yeo, E.-Y.; Thiel, C.; Yi, S.; Stevens, T.; Thompson, W.; Frechen, M.; Murray, A.; Lu, H. A detailed post-IR IRSL chronology for the last interglacial soil at the Jingbian loess site (northern China). *Quat. Geochronol.* **2015**, *30*, 194–199. [CrossRef]

89. Stevens, T.; Buylaert, J.P.; Thiel, C.; Ujvari, G.; Yi, S.; Murray, A.S.; Frechen, M.; Lu, H. Ice-volume-forced erosion of the Chinese Loess Plateau global Quaternary stratotype site. *Nat. Commun.* **2018**, *9*, 983. [CrossRef]

90. Wallinga, J.; Murray, A.; Duller, G. Underestimation of equivalent dose in single-aliquot optical dating of feldspars caused by preheating. *Radiat. Meas.* **2000**, *32*, 691–695. [CrossRef]

91. Kars, R.H.; Reimann, T.; Wallinga, J. Are feldspar SAR protocols appropriate for post-IR IRSL dating? *Quat. Geochronol.* **2014**, *22*, 126–136. [CrossRef]

92. Li, B.; Jacobs, Z.; Roberts, R.G. An improved multiple-aliquot regenerative-dose (MAR) procedure for post-IR IRSL dating of K-feldspar. *Anc. TL* **2017**, *35*, 1–10.

93. Li, B.; Jacobs, Z.; Roberts, R.G.; Galbraith, R.; Peng, J. Variability in quartz OSL signals caused by measurement uncertainties: Problems and solutions. *Quat. Geochronol.* **2017**, *41*, 11–25. [CrossRef]

94. Hu, Y.; Marwick, B.; Zhang, J.F.; Rui, X.; Hou, Y.M.; Yue, J.P.; Chen, W.R.; Huang, W.W.; Li, B. Late Middle Pleistocene Levallois stone-tool technology in southwest China. *Nature* **2019**, *565*, 82–85. [CrossRef] [PubMed]

95. Timar-Gabor, A.; Constantin, D.; Buylaert, J.P.; Jain, M.; Murray, A.S.; Wintle, A.G. Fundamental investigations of natural and laboratory generated SAR dose response curves for quartz OSL in the high dose range. *Radiat. Meas.* **2015**, *81*, 150–156. [CrossRef]

96. Anechitei-Deacu, V.; Timar-Gabor, A.; Thomsen, K.J.; Buylaert, J.P.; Jain, M.; Bailey, M.; Murray, A.S. Single and multi-grain OSL investigations in the high dose range using coarse quartz. *Radiat. Meas.* **2018**, *120*, 124–130. [CrossRef]

97. Anechitei-Deacu, V.; Timar-Gabor, A.; Constantin, D.; Trandafir-Antohi, O.; Del Valle, L.; Fornos, J.J.; Gomez-Pujol, L.; Wintle, A.G. Assessing the Maximum Limit of SAR-OSL Dating Using Quartz of Different Grain Sizes. *Geochronometria* **2018**, *45*, 146–159. [CrossRef]

98. Roberts, H.M.; Duller, G.A.T. Standardised growth curves for optical dating of sediment using multiple-grain aliquots. *Radiat. Meas.* **2004**, *38*, 241–252. [CrossRef]

99. Burbidge, C.I.; Duller, G.A.T.; Roberts, H.M. De determination for young samples using the standardised OSL response of coarse-grain quartz. *Radiat. Meas.* **2006**, *41*, 278–288. [CrossRef]

100. Shen, Z.X.; Mauz, B. Estimating the equivalent dose of late Pleistocene fine silt quartz from the Lower Mississippi Valley using a standardized OSL growth curve. *Radiat. Meas.* **2011**, *46*, 649–654. [CrossRef]

101. Li, B.; Roberts, R.G.; Jacobs, Z.; Li, S.H. Potential of establishing a 'global standardised growth curve' (gSGC) for optical dating of quartz from sediments. *Quat. Geochronol.* **2015**, *27*, 94–104. [CrossRef]

102. Li, B.; Roberts, R.G.; Jacobs, Z.; Li, S.H.; Guo, Y.J. Construction of a 'global standardised growth curve' (gSGC) for infrared stimulated luminescence dating of K-feldspar. *Quat. Geochronol.* **2015**, *27*, 119–130. [CrossRef]

103. Guralnik, B.; Li, B.; Jain, M.; Chen, R.; Paris, R.B.; Murray, A.S.; Li, S.H.; Pagonis, V.; Valla, P.G.; Herman, F. Radiation-induced growth and isothermal decay of infrared-stimulated luminescence from feldspar. *Radiat. Meas.* **2015**, *81*, 224–231. [CrossRef]

104. Peng, J.; Pagonis, V.; Li, B. On the intrinsic accuracy and precision of the standardised growth curve (SGC) and global-SGC (gSGC) methods for equivalent dose determination: A simulation study. *Radiat. Meas.* **2016**, *94*, 53–64. [CrossRef]

105. Li, B.; Jacobs, Z.; Roberts, R.G.; Li, S.H. Single-grain dating of potassium-rich feldspar grains: Towards a global standardised growth curve for the post-IR IRSL signal. *Quat. Geochronol.* **2018**, *45*, 23–36. [CrossRef]

106. Peng, J.; Li, B. Single-aliquot Regenerative-Dose (SAR) and Standardised Growth Curve (SGC) Equivalent Dose Determination in a Batch Model Using the Package 'numOSL'. *Anc. TL* **2017**, *35*, 32–53.

107. Timar-Gabor, A.; Buylaert, J.P.; Guralnik, B.; Trandafir-Antohi, O.; Constantin, D.; Anechitei-Deacu, V.; Jain, M.; Murray, A.S.; Porat, N.; Hao, Q.; et al. On the importance of grain size in luminescence dating using quartz. *Radiat. Meas.* **2017**, *106*, 464–471. [CrossRef]

108. Jacobs, Z.; Li, B.; Shunkov, M.V.; Kozlikin, M.B.; Bolikhovskaya, N.S.; Agadjanian, A.K.; Uliyanov, V.A.; Vasiliev, S.K.; O'Gorman, K.; Derevianko, A.P.; et al. Timing of archaic hominin occupation of Denisova Cave in southern Siberia. *Nature* **2019**, *565*, 594–599. [CrossRef]

109. Galbraith, R.F.; Roberts, R.G.; Laslett, G.M.; Yoshida, H.; Olley, J.M. Optical dating of single and multiple grains of quartz from jinmium rock shelter, northern Australia, part 1, Experimental design and statistical models. *Archaeometry* **1999**, *41*, 339–364. [CrossRef]

110. Thomsen, K.J.; Murray, A.S.; Botter-Jensen, L.; Kinahan, J. Determination of burial dose in incompletely bleached fluvial samples using single grains of quartz. *Radiat. Meas.* **2007**, *42*, 370–379. [CrossRef]

111. Arnold, L.J.; Roberts, R.G.; Galbraith, R.F.; DeLong, S.B. A revised burial dose estimation procedure for optical dating of young and modern-age sediments. *Quat. Geochronol.* **2009**, *4*, 306–325. [CrossRef]

112. Smedley, R.K. A new R function for the Internal External Uncertainty (IEU) model. *Anc. TL* **2015**, *33*, 16–21.

113. Hu, G.M.; Li, S.H. Simplified procedures for optical dating of young sediments using quartz. *Quat Geochronol.* **2019**, *49*, 31–38. [CrossRef]

114. Roberts, R.G.; Galbraith, R.F.; Yoshida, H.; Laslett, G.M.; Olley, J.M. Distinguishing dose populations in sediment mixtures: A test of single-grain optical dating procedures using mixtures of laboratory-dosed quartz. *Radiat. Meas.* **2000**, *32*, 459–465. [CrossRef]

115. Roberts, R.G.; Flannery, T.F.; Ayliffe, L.K.; Yoshida, H.; Olley, J.M.; Prideaux, G.J.; Laslett, G.M.; Baynes, A.; Smith, M.A.; Jones, R.; et al. New ages for the last Australian megafauna: Continent-wide extinction about 46,000 years ago. *Science* **2001**, *292*, 1888–1892. [CrossRef]

116. Jacobs, Z.; Duller, G.A.T.; Wintle, A.G.; Henshilwood, C.S. Extending the chronology of deposits at Blombos Cave, South Africa, back to 140 ka using optical dating of single and multiple grains of quartz. *J. Hum. Evol.* **2006**, *51*, 255–273. [CrossRef]

117. Feathers, J.; Kipnis, R.; Pilo, L.; Arroyo-Kalin, M.; Coblentz, D. How old is Luzia? Luminescence dating and stratigraphic integrity at Lapa Vermelha, Lagoa Santa, Brazil. *Geoarchaeology* **2010**, *25*, 395–436. [CrossRef]

118. Fu, X.; Cohen, T.J.; Fryirs, K. Single-grain OSL dating of fluvial terraces in the upper Hunter catchment, southeastern Australia. *Quat. Geochronol.* **2019**, *49*, 115–122. [CrossRef]

119. Lamothe, M.; Auclair, M. A solution to anomalous fading and age shortfalls in optical dating of feldspar minerals. *Earth Planet. Sci. Lett.* **1999**, *171*, 319–323. [CrossRef]

120. Lamothe, M.; Auclair, M. The fadia method: A new approach in luminescence dating using the analysis of single feldspar grains. *Radiat. Meas.* **2000**, *32*, 433–438. [CrossRef]

121. Balescu, S.; Lamothe, M.; Auclair, M.; Shilts, W.W. IRSL dating of Middle Pleistocene interglacial sediments from southern Quebec (Canada) using multiple and single grain aliquots. *Quat. Sci. Rev.* **2001**, *20*, 821–824. [CrossRef]

122. Li, B.; Li, S.H.; Duller, G.A.T.; Wintle, A.G. Infrared stimulated luminescence measurements of single grains of K-rich feldspar for isochron dating. *Quat. Geochronol.* **2011**, *6*, 71–81. [CrossRef]

123. Neudorf, C.M.; Roberts, R.G.; Jacobs, Z. Sources of overdispersion in a K-rich feldspar sample from north-central India: Insights from De, K content and IRSL age distributions for individual grains. *Radiat. Meas.* **2012**, *47*, 696–702. [CrossRef]

124. Trauerstein, M.; Lowick, S.; Preusser, F.; Rufer, D.; Schlunegger, F. Exploring fading in single grain feldspar IRSL measurements. *Quat. Geochronol.* **2012**, *10*, 327–333. [CrossRef]

125. Feathers, J.; Tunnicliffe, J. Effect of single-grain versus multi-grain aliquots in determining age for K-feldspars from southwestern British Columbia. *Anc. TL* **2011**, *29*, 53–58.

126. Reimann, T.; Thomsen, K.J.; Jain, M.; Murray, A.S.; Frechen, M. Single-grain dating of young sediments using the pIRIR signal from feldspar. *Quat. Geochronol.* **2012**, *11*, 28–41. [CrossRef]

127. Van Gorp, W.; Veldkamp, A.; Temme, A.J.A.M.; Maddy, D.; Demir, T.; van der Schriek, T.; Reimann, T.; Wallinga, J.; Wijbrans, J.; Schoorl, J.M. Fluvial response to Holocene volcanic damming and breaching in the Gediz and Geren rivers, western Turkey. *Geomorphology* **2013**, *201*, 430–448. [CrossRef]

128. Trauerstein, M.; Lowick, S.E.; Preusser, F.; Schlunegger, F. Small aliquot and single grain IRSL and post-IR IRSL dating of fluvial and alluvial sediments from the Pativilca valley, Peru. *Quat. Geochronol.* **2014**, *22*, 163–174. [CrossRef]

129. Brown, N.D.; Rhodes, E.J.; Antinao, J.L.; McDonald, E.V. Single-grain post-IR IRSL signals of K-feldspars from alluvial fan deposits in Baja California Sur, Mexico. *Quat. Int.* **2015**, *362*, 132–138. [CrossRef]

130. Rhodes, E.J. Dating sediments using potassium feldspar single-grain IRSL: Initial methodological considerations. *Quat. Int.* **2015**, *362*, 14–22. [CrossRef]

131. Gliganic, L.A.; Cohen, T.J.; Meyer, M.; Molenaar, A. Variations in luminescence properties of quartz and feldspar from modern fluvial sediments in three rivers. *Quat. Geochronol.* **2017**, *41*, 70–82. [CrossRef]

132. Reimann, T.; Roman-Sanchez, A.; Vanwalleghem, T.; Wallinga, J. Getting a grip on soil reworking—Single-grain feldspar luminescence as a novel tool to quantify soil reworking rates. *Quat. Geochronol.* **2017**, *42*, 1–14. [CrossRef]

133. Brill, D.; Reimann, T.; Wallinga, J.; May, S.M.; Engel, M.; Riedesel, S.; Bruckner, H. Testing the accuracy of feldspar single grains to date late Holocene cyclone and tsunami deposits. *Quat. Geochronol.* **2018**, *48*, 91–103. [CrossRef]

134. Riedesel, S.; Brill, D.; Roberts, H.M.; Duller, G.A.T.; Garrett, E.; Zander, A.M.; King, G.E.; Tamura, T.; Burow, C.; Cunningham, A.; et al. Single-grain feldspar luminescence chronology of historical extreme wave event deposits recorded in a coastal lowland, Pacific coast of central Japan. *Quat. Geochronol.* **2018**, *45*, 37–49. [CrossRef]

135. Schaarschmidt, M.; Fu, X.; Li, B.; Marwick, B.; Khaing, K.; Douka, K.; Roberts, R.G. pIRIR and IR-RF dating of archaeological deposits at Badahlin and Gu Myaung Caves—First luminescence ages for Myanmar. *Quat. Geochronol.* **2019**, *49*, 262–270. [CrossRef]

136. Smedley, R.K.; Buylaert, J.P.; Ujvari, G. Comparing the accuracy and precision of luminescence ages for partially-bleached sediments using single grains of K-feldspar and quartz. *Quat. Geochronol.* **2019**, *53*, 101007. [CrossRef]
137. Guo, Y.; Li, B.; Zhao, H. Comparison of single-aliquot and single-grain MET-pIRIR De results for potassium feldspar samples from the Nihewan Basin, northen China. *Quat. Geochronol.* **2020**, *56*, 101040. [CrossRef]

Protocol

Single-Grain Quartz OSL Characteristics: Testing for Correlations within and between Sites in Asia, Europe and Africa

Yue Hu [1,2,*], Bo Li [2,3] and Zenobia Jacobs [2,3]

[1]   Department of Archaeology, School of History and Culture, Sichuan University, Chengdu 610207, China
[2]   Centre for Archaeological Science, School of Earth and Environmental Science, University of Wollongong, Wollongong, NSW 2522, Australia; bli@uow.edu.au (B.L.); zenobia@uow.edu.au (Z.J.)
[3]   Australian Research Council (ARC) Centre of Excellence for Australian Biodiversity and Heritage, University of Wollongong, Wollongong, NSW 2522, Australia
*   Correspondence: yh280@scu.edu.cn

Received: 11 September 2019; Accepted: 16 December 2019; Published: 26 December 2019

Abstract: We studied the characteristics of the optically stimulated luminescence (OSL) signal of single-grain quartz from three sites in China, Italy, and Libya, including the brightness, decay curve and dose response curve (DRC) shapes, recuperation, and reproducibility. We demonstrate the large variation in OSL behaviors for individual quartz grains of different samples from different regions, and show that recuperation, sensitivity change, and reproducibility are independent of the brightness and decay curve shape of the OSL signals. The single-grain DRCs can be divided into at least eight groups with different characteristic saturation doses (D_0), and a standardized growth curve (SGC) can be established for each of the DRC groups. There is no distinctive difference in the shape of OSL decay curves among different DRC groups, but samples from different regions have a difference in the OSL sensitivities and decay shapes for different groups. Many of the quartz grains have low D_0 values (30–50 Gy), and more than 99% of the grains have D_0 values of <200 Gy. Our results raise caution against the dating of samples with equivalent dose values higher than 100 Gy, if there are many low-D_0 and 'saturated' grains.

Keywords: OSL; quartz; standardized growth curves; decay curve; saturation dose

---

1. Introduction

Single-grain optically stimulated luminescence (OSL) dating of quartz has been widely used for dating sediments, because of its inherent advantages over the single-aliquot method in identifying poorly behaved grains and dealing with insufficiently bleached samples and those affected by post-depositional mixing (e.g., [1–5]). It has previously been reported that different quartz grains may have substantially different OSL behaviors even for grains from the same sample or site (e.g., [6–8]). Understanding the intrinsic variability of OSL behaviors at the single-grain level is important for dating, since D_e estimation can be dependent on these behaviors, e.g., the shape of the dose response curve (DRC) (or characteristic saturation dose, D_0) [9–15], OSL sensitivity (or brightness) [16], and shape of OSL decay curves [17–19].

Previous studies have focused on the relationship between D_e estimates and the variability of a particular OSL characteristic for single grains, e.g., luminescence sensitivity [1,7,15,20], shape of OSL decay curve [1,6,7,15,19,20], and measurement uncertainties (including counting errors and instrument irreproducibility errors) [21]. However, none of these studies have systematically investigated correlations between multiple luminescence behaviors. In this paper, we studied quartz grains extracted from three sites in China, Italy, and Libya. We compared the OSL characteristics of different grains from the same and different samples from different sites. We investigated the relationship and correlation between the OSL characteristics, and discuss their implications for optical dating.

2. Sample Description, Preparation, and Measurements

Seven sediment samples were studied, including three from the Tianhuadong Cave (THD) [22], Yunnan Province, southwest China; one from the Visogliano (VISO) in Italy [23]; and three from the Haua Fteah Cave (HF) in Libya [14,24]. The samples from THD and HF have been described in detail in Hu et al. [25] and Jacobs et al. [14], respectively. These sites were chosen because their quartz grains generally have bright OSL signals, which allows us to study and compare their luminescence behavior in detail. Quartz grains of 150 to 180 or 180 to 212 µm in diameter were extracted using standard preparation procedures [26,27].

All OSL measurements were made on automated Risø TL-DA-20 luminescence readers equipped with a green laser (532 nm) [28]. Laboratory irradiations were carried out within the luminescence readers using calibrated $^{90}Sr/^{90}Y$ beta sources. For single-grain OSL measurements, standard Risø single grain discs were used (each disc contains 100 holes, each 300 µm in diameter and 300 µm deep) [29]. For the samples of 150 to 180 µm in diameter, extra care was taken to ensure that each hole contained only one grain (e.g., by picking up extra grains with a needle of static electricity). The discs were visually checked under a microscope to ensure that each hole contained only one grain. The spatial variation in the dose rate for individual grain positions was calibrated using gamma-irradiated calibration quartz standards. For the readers used in this study, the maximum difference of single-grain dose rates ranged from ~30% to ~60% (with relative standard deviation from ~5% to ~15%). The ultraviolet OSL emissions were detected by an Electron Tubes Ltd. 9235QA photomultiplier tube fitted with a 7.5-mm Hoya U-340 filter.

A single-aliquot regenerative-dose (SAR) procedure [30,31] was applied to measure the OSL signals and establish DRCs for individual grains. The SAR procedure (Table S1) involves measuring the OSL signals from the natural (burial) dose and from a series of regenerative doses, each of which was preheated at a specific temperature (240 °C) for 10 s prior to optical stimulation by the green laser beam for 2 s at 125 °C. A fixed test dose (D_t = ~10 Gy) was given after each natural and regenerative dose, with the induced test dose OSL signals used to monitor any sensitivity change that may have occurred during the SAR sequence. A cutheat to a temperature (180 °C) was applied to the test dose. From seven to nine regenerative doses (up to ~1200 Gy) were measured for each sample, including a duplicate regenerative dose to check on the validity of sensitivity correction and a 'zero dose' to monitor the extent of any 'recuperation' or 'thermal transfer' induced by the preheating. We also applied the OSL IR depletion-ratio test [32] at the end of the SAR sequence, using an infrared bleach of 40 s at 50 °C, to check for feldspar contamination.

3. Comparing OSL Characteristics

3.1. OSL Decay Curves and Signal Intensities

We first investigated the OSL decay signals from individual grains for different samples. Figure 1a–c shows the test-dose OSL decay curves of 10 grains for one sample (HF6031, THD-OSL4, VISO-OSL1) from each of the three sites. It is shown that most of the OSL signals decay to negligible levels after 0.5 s of green-laser stimulation. We calculated the net OSL intensity based on the integral from the first 0.1 s of OSL decay with subtraction of the background estimated from the last 0.3 s. The OSL intensity varies significantly from grain to grain. The net OSL signal intensities of the first test dose (T_n) range from a few tens to several tens of thousands of counts per 0.1 s of stimulation time (Figure 1d). The samples from HF have much brighter grains and a wider range of T_n intensities, followed by those from THD. In contrast, the samples from VISO are comparatively dimmer and have a much narrower range of T_n intensities. We calculated the cumulative percentage of luminescence intensities for all the samples, and it is observed that for most of the samples, 15% to 20% of the grains contribute ~80% of the total luminescence (Figure 1e).

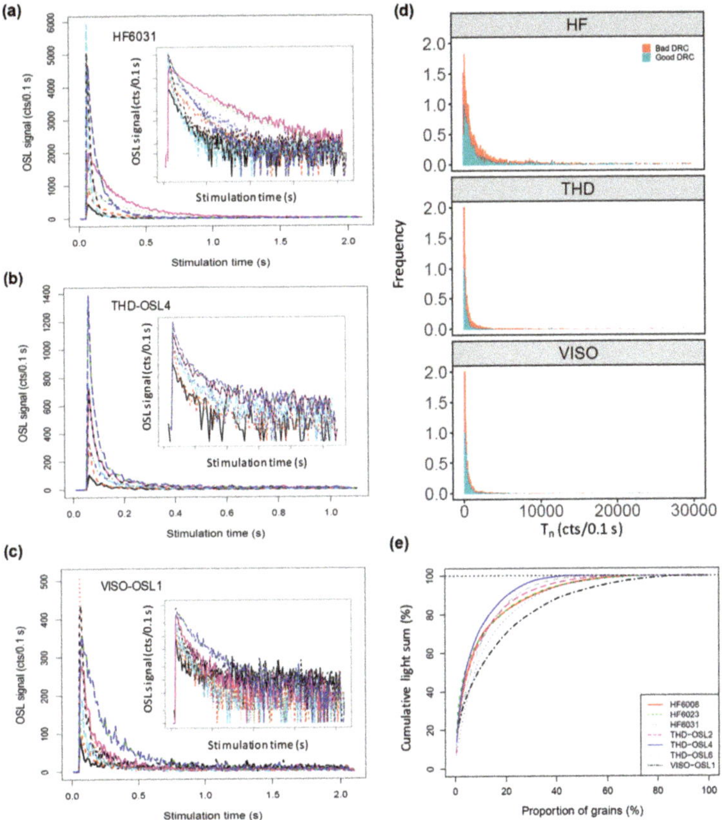

Figure 1. Ten representative OSL decay curves from three samples (**a**) HF6031, (**b**) THD-OSL4, and (**c**) VISO-OSL1. The inset panels show the same curves but with the *y*-axis on a logarithmic scale. (**d**) The density distribution of the test dose signal intensity (T_n) of individual grains from each site. Different colors represent the grains associated with poor or good DRCs, respectively (see Section 3.2). (**e**) Cumulative light sum plotted against the proportion of grains for each sample.

To investigate if there was any correlation between the OSL signal intensity and decay curve shape, we compared the fast ratios (FR) [33,34]—A quantitative measure of the dominance of the fast component in the initial OSL signal—of the T_n signal against its intensities. The fast ratios were calculated using the following equation:

$$\text{Fast Ratio} = \frac{F - S}{M - S} \tag{1}$$

where F is the initial intensity of the signal from the first 0.02 s of the decay curve, M is the signal intensity from 0.07 to 0.08 s, and S represents the contribution of slow components and background estimated from the signals from 0.5 to 0.6 s. Figure 2 shows the relationship between T_n and FR for samples from each of the three sites. It appears that brighter grains tend to have larger FR ratios, which is indicated by the positive correlation coefficients (R values) and is especially prominent for samples from VISO (R = 0.13) and THD (R = 0.18), although many dimmer grains also have larger FR ratios and uncertainties. We also tested using a later integral (0.9–1.0 s) for S, and we found no detectable change in the pattern of the relationship. This result suggests that selecting brighter grains may preferentially select grains dominated by the fast component.

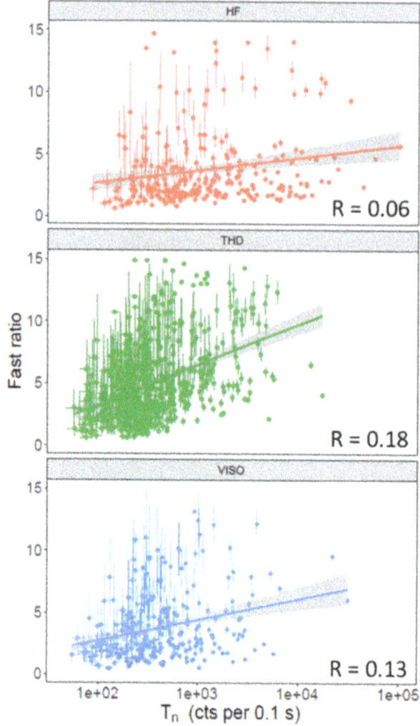

Figure 2. Relationship between the test dose signal intensity (T_n) and fast ratio of individual grains from each site.

3.2. Dose Response Curves

Apart from the variation in the OSL decay curve shapes and intensities, we also investigated the variability of DRCs among grains and samples. Grains were rejected if they had one or more of the following properties: (1) Test-dose signal following the natural dose (T_n) was too dim, i.e., the initial intensity was below the instrument detection limit (3σ below background intensity) and/or the relative standard error on the test dose measurement was more than 20%; (2) recuperation or thermal transfer was too high, i.e., the ratio between the sensitivity-corrected OSL signals for the zero dose and the largest regenerative dose was greater than 5%; and (3) the DRC data were too scattered to be fitted with suitable functions (e.g., a single saturating exponential function or a general-order kinetic (GOK) function [35]). We used a figure-of-merit (FOM) value of 10% and a reduced-chi-square (RCS) value of 5, as recommended by Peng and Li [36], as the upper limits for selecting satisfactory DRCs. The implementation of the rejection process was achieved using the functions provided in the R-package 'numOSL' [36]. The numbers of grains measured and rejected for each sample are summarized in Table 1. About 25%, 40%, and 65% of the grains from each of the three sites were rejected due to signals being too weak. Less than 5% of the grains in all samples had recuperation values greater than 5%. It is interesting to note that the proportions of grains with poor DRCs differed significantly from site to site, e.g., ~40% for HF, ~11% to 29% for THD, and 22% for VISO. Since the samples from HF are generally much brighter than those from the other sites, it is likely that the brighter grains may be preferentially rejected by the criterion of 'poor DRC'. In order to test this, we compared the signal intensity distributions of the T_n signal from the grains being associated with 'good' and 'bad' DRCs, respectively (Figure 1d). It shows that there was no distinctive difference between the T_n intensity distributions of the two groups, indicating that the different proportions of grains with poor DRCs reflects the variability of the grain's behavior among different sites.

Table 1. Number of single grains measured, rejected, and accepted for each sample, together with the reasons for their rejection. BG: background, RSE: relative standard error, FOM: figure of merit, RCS: reduced chi-squares.

Description		HF6008	HF6023	HF6031	THD-OSL2	THD-OSL4	THD-OSL6	VISO-OSL1
Total Measured		300	200	500	300	3000	300	2500
1. Weak signal	$T_n < 3xBG$	32 (11%)	17 (9%)	68 (14%)	59 (20%)	1031 (34%)	81 (27%)	870 (35%)
	RSE of $T_n > 20\%$	36 (12%)	27 (14%)	55 (11%)	59 (20%)	715 (24%)	61 (20%)	673 (27%)
2. Recuperation	>5%	8 (3%)	7 (4%)	26 (5%)	2 (1%)	15 (1%)	2 (1%)	56 (2%)
3. Poor DRC	FOM > 10%	89 (30%)	26 (13%)	117 (23%)	42 (14%)	374 (12%)	29 (10%)	426 (17%)
	RCS > 5	51 (17%)	45 (23%)	114 (23%)	12 (4%)	520 (17%)	2 (1%)	115 (5%)
Total rejected		216 (%)	122 (61%)	380 (76%)	174 (58%)	2655 (89%)	175 (58%)	2140 (86%)
Total accepted		84 (%)	78 (39%)	120 (24%)	126 (42%)	345 (12%)	125 (42%)	268 (11%)

To investigate whether the OSL signals from the rejected and accepted grains had distinctly different features, we compared the sensitivity (T_n) and OSL decay curve shapes (using the FR as proxy) of grains to their recuperation value and recycling ratio. For recuperation, we compared all grains with recuperation values <5% (accepted) and >5% (rejected). For the recycling ratio, we compared the grains that had ratios consistent with unity at 2σ (accepted) and those that were inconsistent with unity (rejected). Figure 3a,b shows the T_n and FR values for all accepted and rejected grains based on either the recuperation or recycling ratio from different samples. No distinctive differences were observed. Figure 4 shows all test-dose normalized sensitivity-corrected signals (L_x/T_x*D_t) from a total of 1146 grains that passed the rejection criteria. It shows large between-grain variability in DRCs from the same site, and a similar range of DRCs between grains from different sites. We also measured the DRCs for sample THD-OSL4 using different preheating temperatures ranging from 180 to 280 °C (Figure 4b), and no distinctive difference was observed in the variability in DRCs for different preheating temperatures. The results suggest that the variation in the DRCs is an intrinsic physical behavior rather than an artefact caused by the preheating conditions.

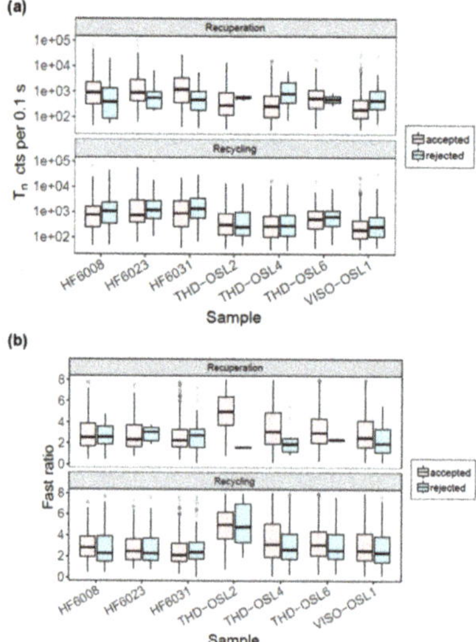

Figure 3. Boxplots showing the (**a**) sensitivity (T_n) and (**b**) fast ratio of grains with different recuperation percentages and recycling ratios. Grains that have recuperation values <5% or recycling ratios consistent with unity (at 2σ) are shown as 'accepted' (light blue bars), and those with recuperation values >5% or recycling ratios inconsistent with unity are shown as 'rejected' (pink bars). The center lines in each of the boxes show the data median. Boxes show the first and third quartiles (the 25th and 75th percentiles), and the whiskers extend from the upper and lower hinge to the largest and smallest values no further than 1.5 times the interquartile range from the hinge. Data beyond the end of the whiskers are outliers and are plotted individually.

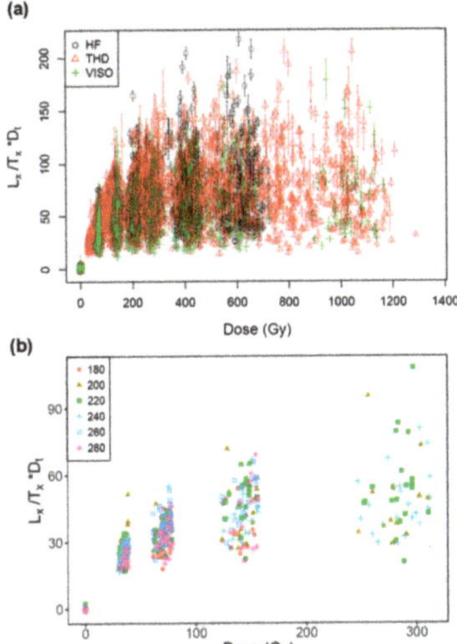

Figure 4. (**a**) Test-dose normalized sensitivity-corrected signal (L_x/T_x*D_t) plotted as a function of regenerative doses for a total of 1146 grains from the study sites. (**b**) Comparison of DRCs obtained for the sample THD-OSL4 using different preheating temperatures (shown in different symbols).

3.3. Sensitivity Change

Sensitivity change in quartz OSL are commonly observed during SAR measurements [37,38] and may be caused by irradiation, preheating, and/or OSL stimulation [39]. A reliable D_e determination relies on a successful correction of the sensitivity change, which is usually monitored through the use of a recycling ratio as part of the SAR procedure. Grains with poor recycling ratios are usually rejected for D_e analysis [1]. In this section, we investigated the extent of the sensitivity change for different grains during the SAR measurement procedure and its relationship to different OSL characteristics, such as the brightness, decay curve shape, recuperation, and reproducibility (e.g., recycling ratio). We used the ratio between T_x and T_n as an indicator of the extent of the sensitivity change through SAR cycles. Figure 5 plots the T_x/T_n values obtained for all grains from each site. The test dose signals measured during the first five SAR cycles (including those corresponding to the natural dose, T_n, and four regenerative doses, T_1–T_4) were normalized to T_n. Most of the grains show some sensitivity change from cycle to cycle, and were sensitized or de-sensitized by <50% during the first two measurement cycles (T_1/T_n). Most of them, however, gradually sensitized during subsequent measurement cycles, indicated by the systematic increase in T_x/T_n values. There are a small proportion of grains that produced a >100% change in sensitivity during two successive measurement cycles; this is most prominent for samples from HF and THD. Furthermore, the extent of sensitivity change is not significantly correlated to the sensitivity of the grains (Figure 5) or rate of decay of the OSL signals (Figure S1).

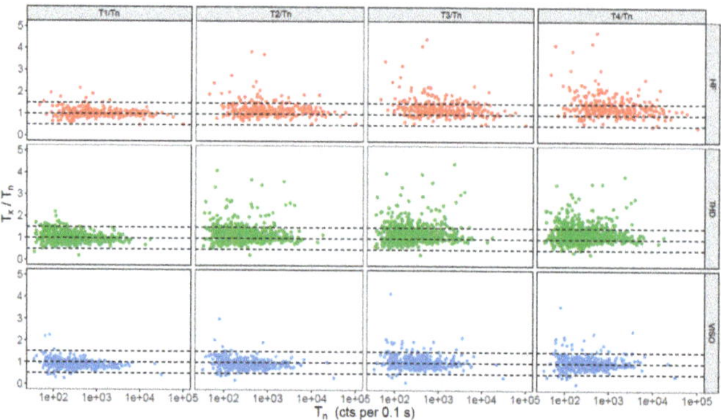

Figure 5. Ratios between test-dose signals (T_x) of the second to fifth SAR cycles (T_1–T_4) and the first cycle (T_n) plotted against inherent sensitivity (T_n) for grains from different sites (shown in different rows). The dashed horizontal lines represent values at 0.5, 1.0, and 1.5, respectively.

4. Grouping of Grains According to the Shape of Their DRCs

Li et al. [13] found that small-aliquot DRCs for samples from HF could be divided into three groups, which saturate at small, intermediate, and high doses, respectively. They proposed a method of grouping individual DRCs, by analyzing the ratio of L_x/T_x values between two regenerative dose points, and used the least-square normalization (LS-normalization) procedure to establish a standardized growth curve (SGC) for each of the groups. To test the variability of DRCs of samples from different regions, we grouped and applied the LS-normalization procedure to all grains measured for all the samples investigated in this study.

One way of comparing the shapes of DRCs is to fit the individual curve data from different grains with a single saturating exponential function, $f(x) = A [1 - \exp(-x/D_0)]$, where x is the dose, D_0 is the characteristic saturation dose, and A is a constant, and then use the D_0 estimate for comparison. However, there are several drawbacks to this method. Firstly, D_0 estimates are usually imprecise when only a few regenerative data (e.g., 5–7) and a relatively narrow dose range (e.g., <$2D_0$) are measured. Li et al. [21] demonstrated that the D_0 estimates are influenced significantly by measurement uncertainties and also the measurement strategy (such as the number and range of regenerative doses applied). Another issue with D_0 is that it only works when all the DRCs follow a single saturating exponential function, and it becomes problematic when there are many grains follow different growth patterns, such as double exponential or exponential plus linear. For this reason, we followed the method of Li et al. [13] by calculating the L_x/T_x ratios to quantify the saturation characteristic (e.g., shape of DRC) of different grains. We chose the L_x/T_x values of two regenerative doses, ~300 and ~70 Gy, respectively (see Li et al. [13] for a full discussion about how to choose the two regenerative doses). Since different grains have different regenerative doses due to the spatial variation in the dose rates of the beta sources, it is impossible to find exactly the same regenerative doses for all grains. To deal with this, we first fitted the measured L_x/T_x data for individual grains using a GOK function, and then estimated the L_x/T_x values at 300 and 70 Gy based on the best-fit DRCs for individual grains. In order to estimate the uncertainties of individual L_x/T_x ratios, we applied a Monte Carlo method. This involved generating random L_x/T_x values based on the experimental L_x/T_x values and their uncertainties according to Gaussian distributions (by taking the uncertainty of each L_x/T_x value as the standard deviation of the distribution). After that, the generated data were fitted using a GOK function, and an L_x/T_x ratio was calculated for each simulation. This process was repeated 500 times, so that a total of 500 L_x/T_x ratios were obtained. The final L_x/T_x ratio estimate and its uncertainty were then derived directly from the mean and the standard deviation of the sampling distribution of the 500 ratios.

The ratios for all the grains are shown in Figure 6a. A large range of ratios from ~1 to ~3 was observed, indicating that the grains have a wide range of saturation doses. For example, the grains with L_x/T_x ratios close to 1 correspond to early saturated grains (i.e., there was a negligible increase in the OSL signal beyond 70 Gy). In contrast, grains with higher L_x/T_x ratios have a larger saturation dose level.

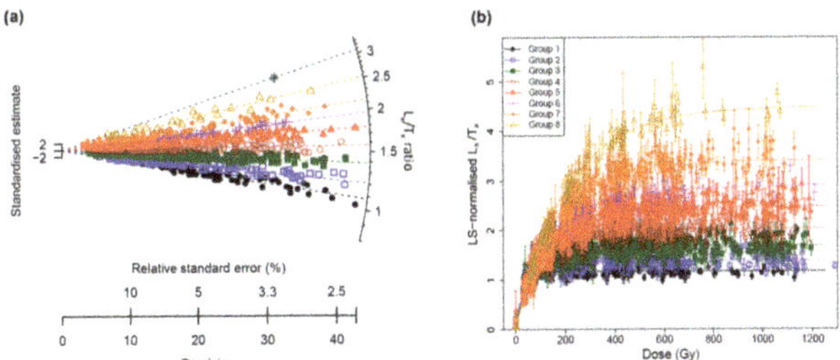

Figure 6. (a) Radial plot showing the distribution of the ratios of L_x/T_x values between two regenerative doses of 300 and 70 Gy for all accepted grains. The different colors and symbols represent different groups of grains identified using the FMM. (b) The LS-normalized L_x/T_x values plotted against regenerative doses for different groups. The data set for each group was fitted using a GOK function (full lines) and then normalized to unity at 50 Gy.

To find grains that share a similar saturation level, we applied the finite mixture model (FMM) [40–42] to analyze the ratios for all accepted grains. We ran the FMM using the build-in function from the R package 'Luminescence' [43] by setting the expected overdispersion (so-called sigma-b value) as zero (assuming that the sources of error associated with the signal intensity have been adequately taken into account) and increasing the number of groups from 3 to 10. The optimal number of groups was then estimated as the one associated with the lowest Bayesian information criterion. We found that nine groups is an optimum number needed to account for the observed spread in ratios for our samples (Figure 6a). As the ninth group (with the largest ratio of ~3) only contained one grain, we treated this grain as an outlier and ignored this group in all subsequent analyses.

We noted that the HF samples contained grains from all the eight groups identified in our study, which is different from the three groups identified by Li et al. [13] using the samples from the same sites. This is, however, expected as (1) this study is a true single-grain result while Li et al. [13] used a small aliquot (e.g., single-grain discs with smaller grains (90–125 um), where each hole contains about ~ eight grains); (2) the present study analyzed a much larger number of grains, which would reveal more detailed information; and (3) Li et al. [13] assumed an extra overdispersion value of 3.5% when applying FMM, but we assumed zero overdispersion in this study.

Since the L_x/T_x ratios from two regenerative doses do not necessarily tell us the shape of DRC, it is still possible that two grains with the same L_x/T_x ratios have different shaped DRC (e.g., one grain could follow a single-saturation exponential function, and the other could follow a single-saturation exponential plus linear function). Whether different grains share the same shape of DRCs can only be proved by establishing SGCs and investigating the goodness-of-fit of SGC to the data. In order to test whether the eight groups identified in Figure 6 actually did represent their difference in DRCs, the LS-normalization procedure was used to analyze the DRCs from each group. The GOK function was used to construct the DRCs for data from the same groups. Figure 6b shows the LS-normalized regenerative-dose data and their corresponding SGC curves for the eight groups. It can be seen that different groups have considerably different saturation dose levels. When the data for each group were fitted using a single saturating exponential function, D_0 values of 32 ± 2, 39 ± 2, 57 ± 1, 76 ± 2, 97 ± 2, 120 ± 5, 144 ± 6, and 197 ± 12 Gy were calculated for groups 1 to 8. To test the validity of the groupings and establishment of the SGCs, the ratios between the measured L_x/T_x values and the expected values based on the best-fitted SGCs were calculated and are shown in Figure S2. The results show that most of the measured-to-expected signal ratios (~90% or more) are statistically consistent with unity at 2σ for all the groups, indicating that most of the grains from the same groups classified based on the L_x/T_x ratios follow the same DRC shape.

We then investigated the relative proportions of grains making up the different groups for each of the samples (Figure 7a). Several patterns can be observed. Firstly, there are between-sample differences for the samples from the same site. For example, group 8 was identified in sample HF6008 but absent from samples HF6023 and HF6031. The proportion of grains in groups 1 and 2 vary significantly between samples from HF. Secondly, different sites appear to have different relative proportions in groups. For example, samples from HF generally contain more group 1 and 2 grains compared to the other two sites. In contrast, samples from THD generally have more group 7 grains. Third, all the samples have <10% of grains in groups 6 and 8, and more than half of the grains fall into groups 2 to 5.

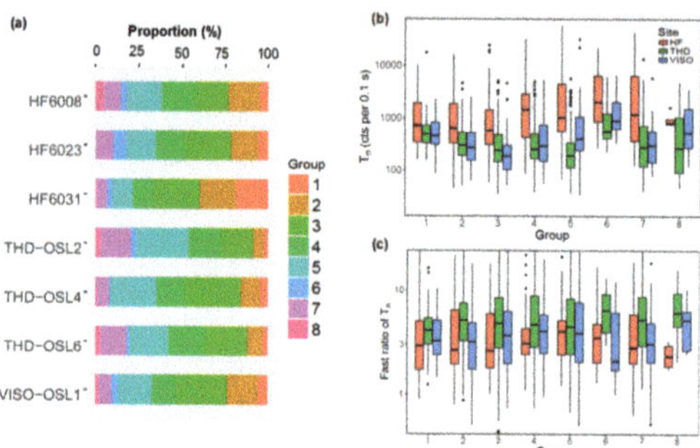

Figure 7. (**a**) Proportional distribution of accepted grains that make up the different DRC groups for each sample. (**b**) Boxplots showing the distribution of T_n for all grains from different DRC groups and different sites (shown as different colors). (**c**) Boxplots showing the fast ratio of T_n for all grains from different DRC groups and different sites (shown as different colors). Centre lines in each of the boxes show the data median. Boxes show the first and third quartiles (the 25th and 75th percentiles), and the whiskers extend from the upper and lower hinge to the largest and smallest values no further than 1.5 times the interquartile range from the hinge. Data beyond the end of the whiskers are outliers and are plotted individually.

5. Comparison of OSL Characteristics between Different DRC Groups

To investigate whether different DRC groups had distinctive OSL behaviors (other than their differences in saturation dose) that could be used to distinguish them from each other, we compared the OSL sensitivities (T_n) and decay curve shapes (FR) of the grains from different groups. The T_n of individual grains of different groups for the samples from different sites were compared and are shown in boxplots in Figure 7b. It shows that all groups contain grains with a wide range of sensitivities. There is no distinctive difference in the range of the sensitivities for different groups from the same site. However, it appears that the HF and VISO samples tend to have a larger number of brighter grains in the high-number groups associated with higher saturation doses than THD.

We also investigated the OSL decay shape of grains from different groups. Figure 7c shows the distribution of fast ratios for grains for different groups of different sites in boxplots. No distinctive difference in the range of fast ratios was observed between different groups and sites. However, the fast ratios for the grains from site THD are generally higher than those from the other two sites. The results of Figure 7b and c suggest that there is no discernible correlation between the sensitivity and decay curve shapes of the OSL signals and the shape of DRCs.

To check if different groups of grains had different extents of sensitivity change during SAR measurements, we compared the T_x/T_n values for the first five SAR cycles (similar to those shown in Figure 5) for the grains from different groups for each site (Figure 8). Again, no discernible difference in the extent of sensitivity changes among different DRC groups was observed.

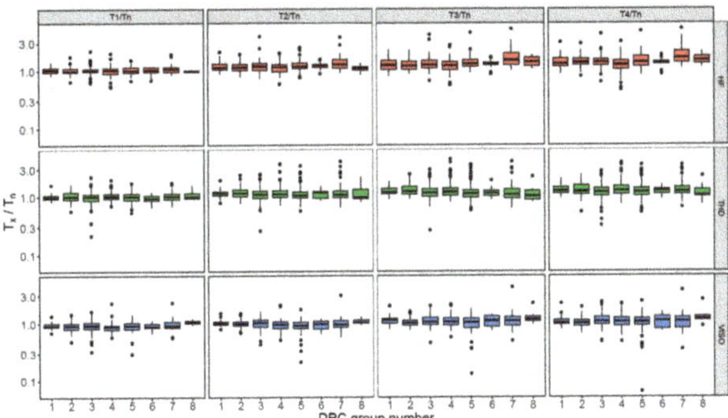

Figure 8. Boxplots showing the ratios between test-dose signals (T_x) of the second to fifth SAR cycles (T_1–T_4) and the first cycle (T_n) for grains from different groups and different sites (shown in different rows and colors). The center lines in each of the boxes show the data median. Boxes show the first and third quartiles (the 25th and 75th percentiles), and the whiskers extend from the upper and lower hinge to the largest and smallest values no further than 1.5 times the interquartile range from the hinge. Data beyond the end of the whiskers are outliers and are plotted individually.

6. Discussions and Conclusions

Based on single-grain measurements of seven samples from three sites on different continents, we demonstrated that the OSL signal sensitivity and decay shape are highly variable from grain to grain and site to site (Figure 1). Comparisons between the OSL signal sensitivity and decay curve shape suggest that the OSL signals of brighter grains may have faster decay rates (Figure 2), probably indicating that these grains are more dominated by the fast component. However, the brighter and fast component-dominant grains appear to have no advantage over dimmer grains in terms of SAR performance (i.e., recuperation and reproducibility) (Figure 3), and our results suggest that both bright and dim grains could behave poorly during SAR measurements. One of the explanations for this is that the observed variability in the OSL shape does not entirely reflect the physical behavior of the grains. It has been suggested that variation in the OSL decay rate could be caused by the differences in the effective stimulation power when the laser hits the surfaces of different grains [19]. However, previous studies have reported that the grains with fast-decaying OSL signals yield improved results in D_e estimation [10,11,44], indicating that the decay shape of laser-based OSL curves does reflect the physical behavior of the grains to some extent. Furthermore, if the stimulation mode is the primary cause, then we should see a similar correlation between the FR and T_n for different samples, which is not the case, as shown in Figure 2.

Large variation is observed in the DRCs for different grains, and the extent of variation appears to be similar for different sites and independent of measurement conditions (e.g., preheating temperature) (Figure 4). Although the grains from different sites show different extents of sensitivity change through SAR measurement cycles (Figure 5), the test dose used during the SAR procedure appears to be able to successfully monitor and correct for sensitivity changes for many grains, even though the sensitivity of some may change by >100% from cycle to cycle. Furthermore, there appears to be no correlation between the extent of the sensitivity change and the inherent brightness of the OSL signals or their decay curve shape (Figure 5 and Figure S1).

We analyzed 4500 quartz grains extracted from these sites, and about eight groups were identified and each group shared a similar DRC (Figure 6). Based on the establishment of an SGC for each group, we found that many of the grains (~50%) fall into groups 1 to 3, which are associated with low saturation doses (D_0 < 50 Gy). Groups 4 to 6, making up 40% to 50% of the accepted grains, have higher saturation doses (D_0 from ~70 to 110 Gy), and <5% of the grains fall into groups 7 and 8 that have D_0 values of up to ~200 Gy. Among the analyzed grains, only one 'super-grain' with a D_0 value as high as ~600 Gy was identified. Our results suggest that more than 99% of the grains have D_0 value

<200 Gy, restricting the dating samples with natural doses higher than ~400 Gy ($2D_0$) when using the conventional SAR method [45], in which D_e is determined for individual grains based on their corresponding DRCs.

Furthermore, since nearly half of the quartz grains have very low D_0 values (~30–50 Gy), it may result in "truncated" D_e distributions (and, hence, D_e underestimation) for samples with natural doses >100 Gy [21]. As a result, caution should be taken when dating samples with a large proportion of early saturation grains (e.g., [9–11,46]). For this case, it would be important to identify only the grains that have relatively larger D_0 values to estimate D_e values. To overcome this problem, previous studies suggested the selection of grains based on a range of different D_0 thresholds (e.g., [10–12,47]). However, this method has limitations if the D_0 values are obtained by fitting a few regenerative dose points for each grain, because they strongly depend on measurement uncertainties (e.g., especially for the signal from the largest regenerative dose) and strategy (e.g., the number and range of regenerative doses applied) [21]. This method is also not straightforward when a large number of grains are analyzed and different grains follow different growth patterns. For example, some grains may require a single-saturating exponential plus a linear component to fit their data (in this case, the D_0 value does not reflect the saturation dose), and some grains may require a double-saturating exponential function (with two different D_0 values, and only the larger one reflects the saturation dose), which makes the comparison of D_0 values not straightforward. In contrast, we demonstrated that it is possible to establish SGCs for different groups of quartz [48], which means that one can use the scaling factors of individual grains obtained from SGC analysis to normalize their corresponding natural signals and then apply the new method of Li et al. [21], which involves analyzing the re-normalized L_n/T_n values for grains sharing the same SGC (i.e., from the same group) and projecting the overall estimate (e.g., weighted mean) of L_n/T_n values onto the corresponding SGC to estimate the final D_e. Unlike the conventional method, in which individual L_n/T_n values are projected onto the corresponding DRCs to estimate D_e for individual grains, this method does not reject 'saturated' grains, so one can avoid the truncation problem and obtain the full distributions of L_n/T_n for individual DRC groups. This allows a reliable D_e estimation beyond the conventional limit of ~$2D_0$ using the standard SAR procedure. For the cases where the early saturation groups are saturated (i.e., their mean L_n/T_n values are consistent with the saturation levels of corresponding SGCs), only the later saturation groups can yield finite D_e values. This means that all the grains from the early saturation groups are rejected from the final D_e estimation. This method has been successfully applied to date several archaeological sites from southwest China [48] and Russia [49,50].

Supplementary Materials: The following are available online at http://www.mdpi.com/2409-9279/3/1/2/s1. Figure S1, Ratios between test-dose signals (T_x) of the 2nd–5th SAR cycles (T_1–T_4) and the first cycle (T_n) plotted against fast ratio for grains from different sites (shown in different rows); Figure S2, Radial plots showing the ratios between the LS-normalised L_x/T_x values and the expected values based on the best-fit SGC shown in Figure 6a; Table S1, The single-aliquot regenerative-dose (SAR) procedure for single grain;

Author Contributions: Conceptualization, B.L.; methodology, B.L. & Y.H.; validation, B.L. and Z.J.; formal analysis, Y.H.; investigation, Y.H.; resources, B.L. and Z.J.; writing—original draft preparation, Y.H.; writing—review and editing, B.L. and Z.J.; supervision, B.L. All authors have read and agreed to the published version of the manuscript.

Funding: This research was funded by the Australian Research Council through Future Fellowships to Li (FT140100384) and Jacobs (FT150100138).

Acknowledgments: Terry Lachlan and Yasaman Jafari are appreciated for assistance in sample preparation and laboratory analysis. We thank two anonymous reviewers for their constructive comments.

Conflicts of Interest: The authors declare no conflict of interest.

References

1. Jacobs, Z.; Duller, G.A.T.; Wintle, A.G. Interpretation of single grain De distributions and calculation of De. *Radiat. Meas.* **2006**, *41*, 264–277. [CrossRef]
2. Roberts, R.G.; Jacobs, Z.; Li, B.; Jankowski, N.R.; Cunningham, A.C.; Rosenfeld, A.B. Optical dating in archaeology: Thirty years in retrospect and grand challenges for the future. *J. Archaeol. Sci.* **2015**, *56*, 41–60. [CrossRef]

3. Arnold, L.J.; Demuro, M.; Pares, J.M.; Arsuage, J.L.; Aranburu, A.; Bermudez de Castro, J.M.; Carbonell, E. Luminescence dating and palaeomagnetic age constraint on hominins from Sima de los Huesos, Atapuerca, Spain. *J. Hum. Evolut.* **2014**, *67*, 85–107. [CrossRef] [PubMed]

4. Duller, G.A.T. Single grain optical dating of glacigenic deposits. *Quat. Geochronol.* **2006**, *1*, 296–304. [CrossRef]

5. Feathers, J.K. Single-grain OSL dating of sediments from the Southern High Plains, USA. *Quat. Sci. Rev.* **2003**, *22*, 1035–1042. [CrossRef]

6. Adamiec, G. Variations in luminescence properties of single quartz grains and their consequences for equivalent dose estimation. *Radiat. Meas.* **2000**, *32*, 427–432. [CrossRef]

7. Duller, G.A.T.; Bøtter-Jensen, L.; Murray, A.S. Optical dating of single sand-sized grains of quartz: Sources of variability. *Radiat. Meas.* **2000**, *32*, 453–457. [CrossRef]

8. Fan, C.A.; Li, S.H.; Li, B. Observation of unstable fast component in OSL of quartz. *Radiat. Meas.* **2011**, *46*, 21–28. [CrossRef]

9. Gliganic, A.L.; Jacobs, Z.; Roberts, R.G. Luminescence characteristics and dose distributions for quartz and feldspar grains from Mumba rockshelter, Tanzania. *Archaeol. Anthropol. Sci.* **2012**, *4*, 115–135. [CrossRef]

10. Duller, G.A.T. Improving the accuracy and precision of equivalent doses determined using the optically stimulated luminescence signal from single grains of quartz. *Radiat. Meas.* **2012**, *47*, 770–777. [CrossRef]

11. Thomsen, K.J.; Murray, A.S.; Buylaert, J.P.; Jain, M.; Hansen, J.H.; Aubry, T. Testing single-grain quartz OSL methods using sediment samples with independent age control from the Bordes-Fitte rockshelter (Roches d'Abilly site, Central France). *Quat. Geochronol.* **2016**, *31*, 77–96. [CrossRef]

12. Guo, Y.-J.; Li, B.; Zhang, J.-F.; Yuan, B.-Y.; Xie, F.; Roberts, R.G. New ages for the Upper Palaeolithic site of Xibaimaying in the Nihewan Basin, northern China: Implications for small-tool and microblade industries in north-east Asia during Marine Isotope Stages 2 and 3. *J. Quat. Sci.* **2017**, *32*, 540–552. [CrossRef]

13. Li, B.; Jacobs, Z.; Roberts, R.G. Investigation of the applicability of standardised growth curves for OSL dating of quartz from Haua Fteah cave, Libya. *Quat. Geochronol.* **2016**, *35*, 1–15. [CrossRef]

14. Jacobs, Z.; Li, B.; Farr, L.; Hill, E.; Hunt, C.; Jones, S.; Rabett, R.; Reynolds, T.; Roberts, R.G.; Simpson, D.; et al. The chronostratigraphy of the Haua Fteah cave (Cyrenaica, northeast Libya)—Optical dating of early human occupation during Marine Isotope Stages 4, 5 and 6. *J. Hum. Evol.* **2017**, *105*, 69–88. [CrossRef]

15. Jacobs, Z.; Duller, G.A.T.; Wintle, A.G. Optical dating of dune sand from Blombos Cave, South Africa: II—Single grain data. *J. Hum. Evol.* **2003**, *44*, 613–625. [CrossRef]

16. Anechitei-Deacu, V.; Timar-Gabor, A.; Thomsen, K.J.; Buylaert, J.-P.; Jain, M.; Bailey, M.; Murray, A.S. Single and multi-grain OSL investigations in the high dose range using coarse quartz. *Radiat. Meas.* **2018**, *117*, 1–86. [CrossRef]

17. Jain, M.; Choi, J.H.; Thomas, J.P. The ultrafast OSL component in quartz: Origins and implications. *Radiat. Meas.* **2008**, *43*, 709–714. [CrossRef]

18. Arnold, L.J.; Roberts, R.G.; MacPhee, R.D.E.; Willerslev, E.; Tikhonov, A.A.; Brock, F. Optical dating of perennially frozen deposits associated with preserved ancient plant and animal DNA in north-central Siberia. *Quat. Geochronol.* **2008**, *3*, 114–136. [CrossRef]

19. Thomsen, K.J.; Kook, M.; Murrary, A.S.; Jain, M.; Lapp, T. Single-grain results from an EMCCD-based imaging system. *Radiat. Meas.* **2015**, *81*, 185–191. [CrossRef]

20. Roberts, R.G.; Galbraith, R.; Olley, J.M.; Yoshida, H. Optical dating of single and multiple grains of quartz from jinmium rock shelter, northern Australia, part 2, Results and implications. *Archaeometry* **1999**, *41*, 365–395. [CrossRef]

21. Li, B.; Jacobs, Z.; Roberts, R.G.; Galbraith, R.; Peng, J. Variability in quartz OSL signals caused by measurement uncertainties: Problems and solutions. *Quat. Geochronol.* **2017**, *41*, 11–25. [CrossRef]

22. Ruan, Q.J.; Liu, J.H.; Hu, Y.; Li, B.; Yang, C.C.; Luo, X.R. A study of stone artifacts found in the Tianhuadong Paleolithic site, Heqing, Yunnan. *Acta Anthropol. Sin.* **2017**, *36*, 1–16.

23. Falguères, C.; Bahain, J.-J.; Tozzi, C.; Boschian, G.; Dolo, J.-M.; Mercier, N.; Vallabdas, H.; Yokoyama, Y. ESR/U-series chronology of the Lower Palaeolithic palaeoanthropological site of Visogliano, Trieste, Italy. *Quat. Geochronol.* **2008**, *3*, 390–398. [CrossRef]

24. Douka, K.; Jacobs, Z.; Lane, C.; Grun, R.; Farr, L.; Hunt, C.; Inglis, R.H.; Reynolds, T.; Albert, P.; Aubert, M.; et al. The chronostratigraphy of the Haua Fteah cave (Cyrenaica, northeast Libya). *J. Hum. Evol.* **2014**, *66*, 39–63. [CrossRef]

25. Hu, Y.; Ruan, Q.; Liu, J.; Marwick, B.; Li, B. Luminescence chronology and lithic technology of Tianhuadong Cave, an early Upper Pleistocene Paleolithic site in southwest China. *Quat. Res.* **2019**, 1–16. [CrossRef]
26. Wintle, A.G. Luminescence dating: Laboratory procedures and protocols. *Radiat. Meas.* **1997**, *27*, 769–817. [CrossRef]
27. Aitken, M.J. *Thermoluminescence Datin*; Academic Press: London, UK, 1985.
28. Bøtter-Jensen, L.; Andersen, C.E.; Duller, G.A.T.; Murray, A.S. Developments in radiation, stimulation and observation facilities in luminescence measurements. *Radiat. Meas.* **2003**, *37*, 535–541. [CrossRef]
29. Bøtter-Jensen, L.; Bulur, E.; Duller, G.A.T.; Murray, A.S. Advances in luminescence instrument systems. *Radiat. Meas.* **2000**, *32*, 523–528. [CrossRef]
30. Galbraith, R.F.; Roberts, R.G.; Laslett, G.M.; Yoshida, H.; Olley, J.M. Optical dating of single and multiple grains of quartz from Jinmium rock shelter, northern Australia, part 1, Experimental design and statistical models. *Archaeometry* **1999**, *41*, 339–364. [CrossRef]
31. Murray, A.S.; Wintle, A.G. Luminescence dating of quartz using an improved single-aliquot regenerative-dose protocol. *Radiat. Meas.* **2000**, *32*, 57–73. [CrossRef]
32. Duller, G.A.T. Distinguishing quartz and feldspar in single grain luminescence measurements. *Radiat. Meas.* **2003**, *37*, 161–165. [CrossRef]
33. Madsen, A.T.; Duller, G.A.T.; Donnelly, J.P.; Roberts, H.M.; Wintle, A.G. A chronology of hurricane landfalls at Little Sippewissett Marsh, Massachusetts, USA, using optical dating. *Geomorphology* **2009**, *109*, 36–45. [CrossRef]
34. Durcan, J.A.; Duller, G.A.T. The fast ratio: A rapid measure for testing the dominance of the fast component in the initial OSL signal from quartz. *Radiat. Meas.* **2011**, *46*, 1065–1072. [CrossRef]
35. Guralnik, B.; Li, B.; Jain, M.; Chen, R.; Paris, R.B.; Murray, A.S.; Li, S.-H.; Pagonis, V.; Valla, P.G.; Herman, F. Radiation-induced growth and isothermal decay of infrared-stimulated luminescence from feldspar. *Radiat. Meas.* **2015**, *81*, 224–231. [CrossRef]
36. Peng, J.; Li, B. Single-aliquot regenerative-dose (SAR) and standardised growth curve (SGC) equivalent dose determination in a batch model using the R Package 'numOSL'. *Ancient TL* **2017**, *35*, 32–53.
37. Murray, A.S.; Wintle, A.G. Sensitisation and stability of quartz OSL: Implications for interpretation of dose-response curves. *Radiat. Prot. Dosim.* **1999**, *84*, 427–432. [CrossRef]
38. Singhvi, A.K.; Stokes, S.C.; Chauhan, N.; Nagar, Y.C.; Jaiswal, M.K. Changes in natural OSL sensitivity during single aliquot regeneration procedure and their implications for equivalent dose determination. *Geochronometria* **2011**, *38*, 231–241. [CrossRef]
39. Bailey, R.M. Towards a general kinetic model for optically and thermally stimulated luminescence of quartz. *Radiat. Meas.* **2001**, *33*, 17–45. [CrossRef]
40. Galbraith, R.F.; Green, P.F. Estimating the component ages in a finite mixture. *Nucl. Tracks Radiat. Meas.* **1990**, *17*, 197–206. [CrossRef]
41. Galbraith, R.F.; Roberts, R.G. Statistical aspects of equivalent dose and error calculation and display in OSL dating: An overview and some recommendations. *Quat. Geochronol.* **2012**, *11*, 1–27. [CrossRef]
42. Roberts, R.G.; Galbraith, R.F.; Yoshida, H.; Laslett, G.M.; Olley, J.M. Distinguishing dose populations in sediment mixtures: A test of single-grain optical dating procedures using mixtures of laboratory-dosed quartz. *Radiat. Meas.* **2000**, *32*, 459–465. [CrossRef]
43. Kreutzer, S.; Schmidt, C.; Fuchs, M.C.; Dietze, M. Introducing an R package for luminescence dating analysis. *Ancient TL* **2012**, *30*, 1–8.
44. Feathers, J.K.; Pagonis, V. Dating quartz near saturation—Simulations and application at archaeological sites in South Africa and South Carolina. *Quat. Geochronol.* **2015**, *30*, 416–421. [CrossRef]
45. Wintle, A.G.; Murray, A.S. A review of quartz optically stimulated luminescence characteristics and their relevance in single-aliquot regeneration dating protocols. *Radiat. Meas.* **2006**, *41*, 369–391. [CrossRef]
46. Singh, A.; Singh, A.; Thomsen, K.J.; Sinha, R.; Buylaert, J.-P.; Carter, A.; Mark, D.F.; Mason, P.J.; Densmore, A.L.; Murrary, A.S.; et al. Counter-intuitive influence of Himalayan river morphodynamics on Indus Civilisation urban settlements. *Nat. Commun.* **2017**, *8*, 1617. [CrossRef]
47. Gliganic, L.A.; Jacobs, Z.; Roberts, R.G.; Domínguez, R.M.; Mabulla, A.Z. New ages for Middle and Later Stone Age deposits at Mumba rockshelter, Tanzania: Optically stimulated luminescence dating of quartz and feldspar grains. *J. Hum. Evol.* **2012**, *62*, 533–547. [CrossRef]

48. Hu, Y.; Li, B.; Marwick, B.; Zhang, J.-F.; Hou, Y.-M.; Huang, W.-W. Late Middle Pleistocene Levallois stone-tool technology in southwest China. *Nature* **2019**, *565*, 82–85. [CrossRef]
49. Jacobs, Z.; Li, B.; Shunkov, M.V.; Kozlikin, M.B.; Bolikhovskaya, N.S.; Agadjanian, A.K.; Uliyanov, V.A.; Vasiliev, S.K.; O'Gorman, K.; Derevianko, A.P.; et al. Timing of archaic hominin occupation of Denisova Cave in southern Siberia. *Nature* **2019**, *565*, 594–599. [CrossRef]
50. Derevianko, A.P.; Markin, S.V.; Rudaya, N.A.; Viola, B.; Zykin, V.S.; Zykina, V.S.; Chabay, V.P.; Kolobova, K.A.; Vasiliev, S.K.; Roberts, R.G.; et al. (Eds.) *Interdisciplinary Studies of Chagyrskaya Cave—Middle Paleolithic Site of Altai*; Institute of Archaeology and Ethnography, Siberian Branch of the Russian Academy of Sciences: Novosibirsk, Russia, 2018.

Article

Optically Stimulated Luminescence Sensitivity of Quartz for Provenance Analysis

André Oliveira Sawakuchi [1,*], Fernanda Costa Gonçalves Rodrigues [1], Thays Desiree Mineli [1], Vinícius Ribau Mendes [2], Dayane Batista Melo [2], Cristiano Mazur Chiessi [3] and Paulo César Fonseca Giannini [1]

[1] Luminescence and Gamma Spectrometry Laboratory (LEGaL), Instituto de Geociências, Universidade de São Paulo, São Paulo 05508-080, Brazil; cgr.fernanda@gmail.com (F.C.G.R.); thaysdesiree@gmail.com (T.D.M.); pcgianni@usp.br (P.C.F.G.)
[2] Institute of Marine Sciences, Federal University of São Paulo, Santos 11015-020, Brazil; vrm.unifesp@gmail.com (V.R.M.); d.batista.melo@hotmail.com (D.B.M.)
[3] School of Arts, Sciences and Humanities, University of São Paulo, São Paulo 03828-000, Brazil; chiessi@usp.br
* Correspondence: andreos@usp.br; Tel.: +55-11-3091-0496

Received: 1 October 2019; Accepted: 25 December 2019; Published: 13 January 2020

Abstract: Finding the source or provenance of quartz grains occurring in a specific location allows us to constrain their transport pathway, which is crucial information to solve diverse problems in geosciences and related fields. The optically stimulated luminescence (OSL) sensitivity (light intensity per unit mass per unit radiation dose) has a high capacity for discrimination of quartz sediment grains and represents a promising technique for provenance analysis. In this study, we tested the use of quartz OSL sensitivity (ultraviolet emission) measured under different preheating temperatures and with blue light stimulation at room temperature (~20 °C) for sediment provenance analysis. Quartz OSL sensitivity measured at 20 °C is positively correlated with the sensitivity of an OSL signal measured using procedures (preheat at 190 °C for 10 s, blue stimulation at 125 °C and initial 1 s of light emission) to increase the contribution of the fast OSL component, which has been successfully applied for sediment provenance analysis. The higher OSL signal intensity measured without preheating and with light stimulation at room temperature allows the use of lower given doses, thus reducing measurement time. Additionally, the OSL sensitivity measured at 20 °C in polymineral silt samples of a marine sediment core is also suitable for provenance analysis, as demonstrated by comparison with other independent proxies. OSL signals obtained through light stimulation at room temperature have thus the potential to considerably expand measurement possibilities, including in situ measurements using portable OSL readers.

Keywords: sediment provenance; quartz fingerprint; luminescence; source-to-sink system

1. Introduction

Mineral grains resting on or near Earth's surface can be dispersed through multiple ways, such as by landslides induced by earthquakes [1] or extreme rainfall events [2], winds [3], rivers [4], coastal waves, or tidal currents [5], soil-living insects [6], and diverse human activities [7]. Tracking the primary or previous source of mineral grains allows for the reconstruction of varied events, ranging from past climate changes [8] to a pathway taken by a crime suspect [9]. Thus, methods to find the source or the provenance of mineral grains have several applications in geosciences and related fields. Quartz (SiO_2) is the most frequent mineral in sediments [10], and it is widespread over the Earth's surface, especially in the sand (63–2000 µm) and silt (4–63 µm) fractions. Thus, quartz is a valuable material in provenance studies due to its ubiquitous occurrence and high stability in most Earth surface settings. However, quartz grains are poorly discriminated by methods commonly used for provenance analysis of sediments, which are based on the chemical [5] or mineralogical composition of sediment mixtures or on chemical variations of specific minerals [11]. The optically stimulated luminescence (OSL) is the light arising from charges trapped in crystal lattice defects when quartz is exposed to ionizing radiation and subsequently released for recombination when quartz is exposed to a stimulation light [12]. The OSL sensitivity is defined as the intensity of light emitted per unit mass of the material and unit radiation dose (photon counts Gy^{-1} mg^{-1}). The quartz OSL sensitivity ranges over five orders of magnitude [13,14] and can be measured in different types of sediment or sedimentary rock samples. Luminescence readers allow performing automated OSL measurements on aliquots (amount of grains) of silt, very fine to fine sand (62–250 µm), single grains of fine sand, or slices of sedimentary rocks [15–17] in a fast and low-cost way. Quartz has relatively low OSL sensitivity when crystallized in its parent igneous, hydrothermal, or metamorphic rocks [13,18], but it can be subsequently sensitized by artificial or natural heating [19] or illumination-irradiation cycles [20]. In nature, the OSL sensitization occurs after quartz is released from its parent rocks and is transported as a grain in sedimentary systems [13]. The OSL sensitization during the transport of quartz as sediment grain is attributed to repeated cycles of bleaching and irradiation [20–22]. Despite the fact that the increase of quartz OSL sensitivity due to bleaching–irradiation cycles is well demonstrated by laboratory experiments, e.g., [19,22], it is not conclusive if the sensitization is related with changes in the concentration of charge traps, recombination processes, or both [14,20]. However, this condition does not hinder the use of quartz OSL sensitivity for sediment provenance analysis, e.g., [21]. Previous studies to apply quartz OSL sensitivity in provenance analysis, e.g., [22–24] focused on the initial 1 s of the OSL decay curve obtained by constant energy blue light stimulation at 125 °C after heating the sample at temperatures from 160 to 260 °C, to eliminate unstable signals [25]. The OSL signal measured under this condition is presumably dominated by the fast OSL component [26], but quartz with low OSL sensitivity occurring in sediments around young mountain ranges can show the initial 1 s of light emission with significant contribution of medium and slow components [14]. Despite the successful use of the unstable 110 °C thermoluminescence (TL) peak of quartz in provenance analysis of coastal and marine sediments [27,28], little is known about the suitability of using unstable OSL signals for sediment-fingerprinting purposes. In this study, we performed experiments to improve measurements of quartz OSL sensitivity for provenance analysis. Experiments aimed to simplify protocols for OSL sensitivity measurements, including evaluation of heavy minerals contamination and whether thermal treatments to eliminate unstable OSL signals are necessary. Our experiments show that quartz OSL sensitivity measured without preheating and with optical stimulation at room temperature (~20 °C) can be used for provenance analysis in the same way as the OSL measured with preheating and light stimulation at 125 °C [21–23,28]. This expands the measurements conditions of quartz OSL sensitivity to in situ measurements in sediment cores and outcrop sedimentary successions, as well as in thermally sensitive materials hosting quartz.

2. Materials and Methods

OSL sensitivity measurements were carried out on quartz crystals from hydrothermal veins and metamorphic rocks (blue schist) and on quartz sand grains extracted from sediments from different geological settings (Table 1). Sediment samples were purposely selected to cover a wide range of quartz OSL sensitivity, which was appraised in a previous study [14]. Additionally, 16 samples from marine core GeoB16206-1 [29] were analyzed to further the application of OSL sensitivity provenance analysis to fine-grained (silt) sediments.

Rock samples (IP22B and BS223) were crushed using a hammer, and quartz-crystal fragments (1–5 mm) were handpicked for milling, manually, using a pestle and mortar. Afterward, these samples were submitted to sieving and chemical treatments, in the same way as sand samples, to isolate quartz grains. The procedures to obtain pure quartz concentrates for luminescence measurements consisted of: (1) wet sieving for isolation of 180–250 or 63–180 µm grain fraction; (2) treatment with H_2O_2 35% (not applied to quartz from rocks) and HCl 10%, to eliminate organic matter and carbonate minerals;

(3) density separation with lithium metatungstate (LMT) solutions at densities of 2.75 and 2.62 g/cm^3, to isolate quartz grains from heavy minerals and feldspar grains; (4) treatment with HF 40% for 40 min, to etch the outer rind of the quartz grains and to remove remaining feldspar; and (5) washing with distilled water and another sieving for isolation of the 180–250 or 63–180 μm grain fractions. The purity of quartz concentrates was tested by using infrared stimulation. Additional HF 5% (24 h) treatments were applied if a significant infrared stimulated luminescence signal was observed. It is also recommended to apply an additional HCl 10% wash, to eliminate possible fluorides formed as product of HF treatments. This step is especially important for samples rich in feldspar. For the fine-grained sediments of core GeoB16206-1, sample preparation [28] included the following: (1) sample drying at 60 °C; (2) weighing of 0.5 g of each sample; (3) treatment with H_2O_2 27% and HCl 10%, to remove organic matter and carbonate minerals; and (4) addition of acetone to the content until 5 mL.

Table 1. Summary of characteristics of quartz samples used in this study.

Code	Grain Size	* OSL Sensitivity	Source Material	Geological Setting
IP22B	180–250 μm	Low	Hydrothermal vein	Paraná Basin (Brazil)
BS223	180–250 μm	Low	Blue schist	Andes (Chile)
L0674	180–250 μm	Low	Alluvial sand	Atacama (Chile)
L0017	180–250 μm	Medium	Fluvial sand	Central Amazon
L0698	63–180 μm	Medium	Fluvial sand	Western Amazon
L0001	180–250 μm	High	Coastal sand	Santa Catarina coast
L0229	180–250 μm	High	Fluvial sand	Pantanal Wetland
L0231	180–250 μm	High	Fluvial sand	Pantanal Wetland
L0688	180–250 μm	High	Fluvial sand	Paraná River basin
** Cal. Qz	180–250 μm	Very high	Eolian sand	Jutland, Denmark

* OSL sensitivity appraised in a previous study [14]. ** "Cal. Qz" is quartz used for calibration of beta radiation sources of luminescence readers. It was artificially sensitized by heating at 700 °C for 1 h, 2 kGy gamma dose, and another heating at 450 °C for 1 h [30].

Two samples of Holocene coastal sands (L0387 and L0393) from Southern Brazil (Santa Catarina) were used to obtain heavy mineral concentrates, using the following procedures: (1) dry sieving to isolate the 63–125 μm grain-size fraction; (2) heavy liquid separation with bromoform, at a density of 2.85 g/cm^3, to isolate heavy and light minerals; (3) separation of magnetic and nonmagnetic heavy minerals, using a hand-magnet; and (4) weighing of heavy minerals. Aliquots of magnetic and nonmagnetic heavy minerals were used for luminescence measurements. Transparent nonmagnetic heavy mineral types were also identified and quantified under polarized light optical microscopy. The 63–125 μm fraction was chosen for heavy mineral analysis, to facilitate mineral identification under the optical microscope. However, the heavy mineral types from this grain-size fraction are representative of heavy minerals occurring in the grain-size fraction used in this study for the luminescence analysis of quartz sand grains (63–180 and 180–250 μm).

Luminescence measurements were carried out in the Luminescence and Gamma Spectrometry Laboratory at the Institute of Geosciences of the University of São Paulo. Measurements of quartz concentrates in the sand fraction were performed in a Risø TL/OSL DA-20 reader, equipped with a built-in beta source (^{90}Sr/^{90}Y; dose rate of 0.0981 Gy s^{-1} for 11.65 mm diameter stainless-steel cups), blue (470 nm, maximum power of 80 mW/cm^2) and infrared (870 nm, maximum power of 145 mW/cm^2) LEDs for stimulation, and Hoya U-340 filter for light detection in the ultraviolet band. Fine-grained polymineral sediments from core GeoB16206-1 were measured in a Lexsyg Smart TL/OSL reader, equipped with blue (458 nm, maximum power of 100 mW/cm^2) and infrared (850 nm, maximum power of 300 mW/cm^2) LEDs, filters for light detection in the ultraviolet band (365 nm, U340 + BP 365/50), and beta radiation source (^{90}Sr/^{90}Y), with a dose rate of 0.116 Gy s^{-1} for stainless steel discs.

In all experiments, the OSL sensitivity (blue stimulation, BOSL) was calculated by using the integral of the first 1 s of light emission, minus the last 10 s as background. In experiments carried out with quartz in the sand grain size, the OSL sensitivity calculation applied normalization by dose and aliquot mass (photon cts Gy^{-1} mg^{-1}). In the first experiment, OSL sensitivity was also calculated as a percentage of the initial 1 s of light emission in relation to the total OSL emission (%BOSL). This approach has been used to minimize the effect of aliquot mass variation on the OSL sensitivity [23]. However, this approach limits the range of sensitivity variation (0–100%) and reduces the capacity of sediment discrimination. The OSL sensitivity of fine-grained aliquots from core GeoB16206-1 was only calculated as a percentage of the first second of light emission in relation to the total emission (%BOSL). OSL decay curves for sensitivity calculation were analyzed by using the software Luminescence Analyst v4.31.9 [31].

For each experiment, three aliquots per sample were prepared by mounting quartz or heavy mineral grains in stainless-steel cups for measurements in the Risø TL/OSL DA-20 reader. An acrylic plate with a circular microhole of 1470 μm diameter, with 1860 μm depth, was used to prepare aliquots with similar volume [23], corresponding to 150–200 grains (180–250 μm grain size) and ensuring a monolayer of grains onto cups or discs. Each aliquot was weighed for mass normalization of luminescence signals. Aliquots of fine-grained sediments of the marine sediment core were mounted by using four drops of acetone–sediment solution over stainless-steel discs, for measurements in the Lexsyg Smart TL/OSL reader. Mounting of silt aliquots followed standard procedures used for luminescence dating [32].

The first experiment comprised the measurement of OSL sensitivity of quartz and heavy mineral concentrates, using the same protocol described in previous work [23] (Table 2). The first OSL stimulation (step 1) bleaches natural signals. A relatively high given dose of 50 Gy was used to guarantee measurable signals in low-sensitivity quartz and heavy mineral aliquots. After the given dose (step 2), the infrared stimulation (step 4) aims to identify quartz aliquots contaminated by feldspar or to bleach feldspar grains and measure quartz in polymineral aliquots [24]. The final blue stimulation (step 5) is used to determine the quartz OSL sensitivity (pure quartz aliquots) or a quartz-dominated OSL sensitivity (polymineral aliquots). Pure quartz aliquots with significant infrared-stimulated luminescence signal were not used for sensitivity calculation. The OSL sensitivity measured using step 5 of the protocol in Table 2, subsequent to a preheating at 190 °C for 10 s (in step 3) and using blue stimulation at 125 °C, is termed "$BOSL_F$"; this signal presumably has a higher contribution of the fast OSL component [23].

Table 2. Sequence of procedures used to measure OSL sensitivity of quartz and heavy mineral aliquots. $BOSL_F$ represents the signal used for sensitivity calculation, which was measured under varied preheat conditions (50–400 °C) and using blue stimulation at 125 °C.

Step	Treatment	Observed
1	Illumination with blue LED at 125 °C for 100 s	
2	Give dose, 50 Gy	
3	(1) Preheat at 190 °C for 10 s, (2) 50, 100, 150, 200, 300, or 400 °C for 10 s	
4	Infrared LED stimulation at 60 °C for 300 s	
5	Blue LED stimulation at 125 °C for 100 s	$BOSL_F$

The second experiment evaluated the influence of the preheat temperature on the OSL sensitivity. In this experiment, the same protocol described in Table 2 was applied, but using preheat temperatures of 50, 100, 150, 200, 300, and 400 °C, to measure the OSL sensitivity in samples IP22B, L0017 and L0229, which have low, medium, and high $BOSL_F$ sensitivities, respectively (Table 1). The OSL sensitivity from this experiment is also named $BOSL_F$. However, it is important to note that, despite the fact that higher preheating temperatures increase the fast-to-slow components ratio, the dominance of a fast OSL component in the $BOSL_F$ signal is sample-dependent [26]. The OSL sensitivity from other measurement conditions applied in this study is specified accordingly.

The third experiment was designed to evaluate the measurement of quartz OSL sensitivity, without preheating and with stimulation at room temperature. Six quartz samples with low (IP22B and BS223), medium (L0017 and L0698), and high (L0001 and L0229) $BOSL_F$ sensitivity were used for this test. The sequence of procedures (Table 3) comprises a longer bleaching with blue-light illumination for 500 s (step 1), a given beta dose of 50 Gy (step 2), no pause or pause of 16 h to eliminate unstable signals (step 3), OSL stimulation using blue light at 20 °C (step 4), and a second OSL stimulation using blue light at 125 °C (step 5). The longer bleaching (500 s) was applied because illumination was performed at room temperature, while the protocol from Table 2 used a bleaching at 125 °C. The infrared stimulation step was not included, in order to reduce protocol time for measurement of pure quartz aliquots. The pause before blue stimulation aimed to evaluate any effect of the 110 °C TL trap. The effect of a 16 h pause is similar to heating the sample at 125 °C for 10 s [20]. Thus, the pause was applied to eliminate the influence of the 110 °C TL trap on the further OSL measurement at 20 °C ($BOSL_{20°C}$) (step 4). Measurements conduced with a 16 h pause were used to validate measurements carried out with no pause. The blue stimulation at 125 °C (step 5) was performed to evaluate if the OSL sensitivity ($BOSL_{125°C}$) can be measured twice without signal reset and dosing.

Table 3. Sequence of procedures used to measure OSL sensitivity of pure quartz aliquots without preheating and with light stimulation at room temperature (20 °C). $BOSL_{20°C}$ and $BOSL_{125°C}$ represent the signals used for sensitivity calculation.

Step	Treatment	Observed
1	Illumination with blue LED at 20 °C for 500 s	
2	Give dose, 50 Gy	
3	Pause for 0 s or 16 h	
4	Blue LED stimulation at 20 °C for 100 s	$BOSL_{20°C}$
5	Blue LED stimulation at 125 °C for 100 s	$BOSL_{125°C}$

The sediment samples from marine core GeoB16206-1 were measured following the protocol described in Table 4. The $BOSL_{20°C}$ sensitivity for these samples were calculated as the percentage of initial 1 s of light emission in relation to the total OSL emission (0–100 s). Then, the $BOSL_{20°C}$ was normalized by subtracting the average sensitivity and dividing by the standard deviation of all samples. This normalization was applied in order to compare the $BOSL_{20°C}$ with $BOSL_F$ values [28].

Table 4. Sequence of procedures to measure $BOSL_{20°C}$ sensitivity in aliquots of fine-grained sediments from core GeoB16206-1.

Step	Treatment	Observed
1	Illumination with blue LED at 125 °C for 100 s	
2	Give dose, 10 Gy	
3	Blue LED stimulation at 20 °C for 100 s	$BOSL_{20°C}$

3. Results

$BOSL_F$ sensitivities of the studied quartz samples range in five orders of magnitude, from 0.42 ± 0.25 to 8565.00 ± 497.76 cts Gy^{-1} mg^{-1} (Figure 1 and Table 5). The lowest $BOSL_F$ sensitivity occurs in quartz from a hydrothermal vein (IP22B), and the highest sensitivity is observed in artificially sensitized quartz (CalQz). Magnetic heavy minerals, mainly represented by magnetite, have OSL sensitivity between 0.63 ± 0.29 and 1.19 ± 0.65 cts Gy^{-1} mg^{-1}. The OSL sensitivity of nonmagnetic heavy minerals varied between 20.80 ± 6.00 and 94.77 ± 103.03 cts Gy^{-1} mg^{-1}. Heavy minerals correspond to 18% and 12% (weight percent) of the 63–125 μm grain-size fraction in samples L0387 and L0393, respectively. The nonmagnetic and transparent heavy minerals in the studied samples include zircon (75–76%), rutile (6–7%), tourmaline (5–6%), kyanite (4%), staurolite (4%), epidote (1–3%), monazite (<1%), sillimanite (<1%), hornblende (<1%), and garnet (<1%). Figure 1 shows OSL decay curves of quartz, magnetic heavy minerals, and nonmagnetic heavy minerals for comparison.

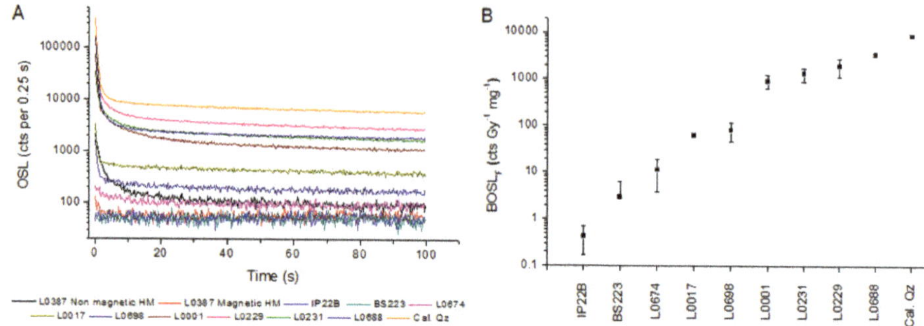

Figure 1. (**A**) OSL decay curves (step 5 of the protocol in Table 2) of quartz with different sensitivities. OSL decay curves of nonmagnetic and magnetic heavy minerals are also shown for comparison. Aliquots irradiated with 50 Gy. (**B**) Range of $BOSL_F$ sensitivity comprised in the studied quartz samples. Each sample is represented by the average of three aliquots. Error bars correspond to the standard deviation. Sample type is described in Table 1.

Table 5. Average ($n = 3$ aliquots) OSL sensitivity ($\pm$ standard deviation) of studied quartz samples. The OSL sensitivity of heavy minerals concentrates from coastal sands (samples L0387 and L0393) are also shown for comparison. The OSL sensitivities of magnetic (M) and nonmagnetic (NM) heavy minerals were calculated in the same way as the $BOSL_F$. OSL signal was not detected for the magnetic heavy mineral fraction of sample L0387. A given dose of 50 Gy was used for all aliquots.

Sample	%$BOSL_F$	$BOSL_F$ (cts Gy^{-1} mg^{-1})	$BOSL_{125°C}/0$ s (cts Gy^{-1} mg^{-1})	$BOSL_{125°C}/16$ h (cts Gy^{-1} mg^{-1})	$BOSL_{20°C}/0$ s (cts Gy^{-1} mg^{-1})	$BOSL_{20°C}/16$ h (cts Gy^{-1} mg^{-1})
IP22B	<1.00 ± 4.70	0.42 ± 0.25	<0.10 ± 0.17	<0.10 ± 0.15	1.31 ± 0.49	0.64 ± 0.16
BS223	8.43 ± 5.03	2.89 ± 3.34	1.47 ± 0.52	0.87 ± 0.32	11.74 ± 2.77	3.11 ± 1.65
L0674	12.33 ± 4.83	11.20 ± 7.36	-	-	-	-
L0017	19.36 ± 0.17	62.60 ± 4.98	55.68 ± 11.95	46.29 ± 8.60	814.58 ± 317.15	189.21 ± 101.99
L0698	19.40 ± 4.46	80.41 ± 35.04	52.61 ± 5.84	51.81 ± 3.52	917.67 ± 113.41	341.72 ± 58.18
L0001	50.28 ± 6.74	925.57 ± 295.55	94.91 ± 21.50	118.38 ± 25.81	4501.14 ± 807.25	1468.82 ± 566.53
L0229	44.78 ± 2.16	1963.84 ± 835.53	230.30 ± 43.89	254.61 ± 44.12	11,376.18 ± 3089.01	3794.16 ± 1489.61
L0231	36.31 ± 11.03	1311.64 ± 401.13	-	-	-	-
L0688	54.52 ± 6.48	3370.86 ± 205.27	-	-	-	-
CalQz	56.08 ± 2.94	8565.00 ± 497.76	-	-	-	-
L0387$_{NM}$	25.75 ± 6.28	20.80 ± 6.00	-	-	-	-
L0393$_{M}$	57.79 ± 49.30	1.19 ± 0.65	-	-	-	-
L0393$_{NM}$	33.31 ± 25.88	94.77 ± 103.03	-	-	-	-

The $BOSL_F$ sensitivity normalized by mass and dose is correlated with the $BOSL_F$, represented as a percentage of the total OSL emission (%$BOSL_F$). The advantage of $BOSL_F$ over %$BOSL_F$ is its larger amplitude of variation and, then, the higher capacity of quartz discrimination, especially among higher OSL sensitivity quartz (Figure 2).

The $BOSL_F$ sensitivity decreases with the increase of preheat temperature, but sensitivity differences among samples are not affected by the preheat temperature (Figure 3 and Table 6).

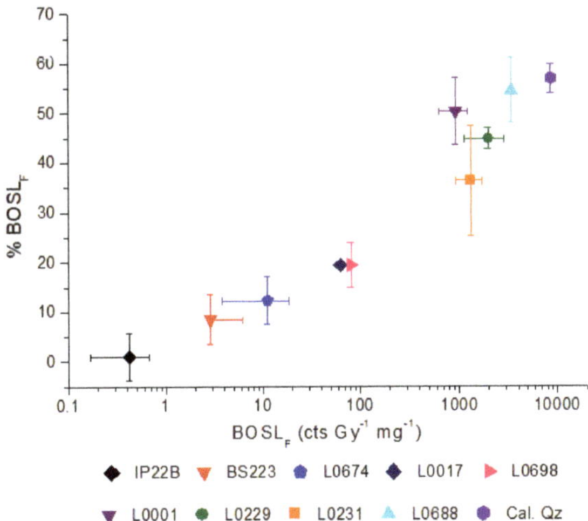

Figure 2. Quartz OSL sensitivity represented by the percentage of the initial 1 s of light emission (%BOSL$_F$) compared to the absolute quartz BOSL$_F$ sensitivity (normalized by dose and aliquot mass). Each sample is represented by the average of three aliquots. Error bars correspond to the standard deviation.

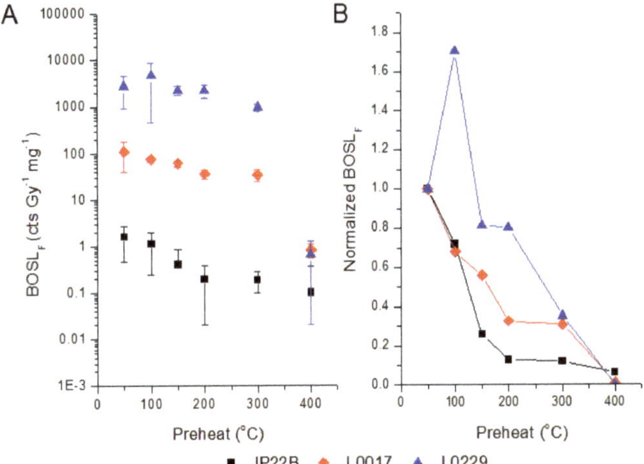

Figure 3. (A) Variation of BOSL$_F$ sensitivity (step 5 of Table 2), measured using preheat temperatures from 50 to 400 °C. (B) Variation of BOSL$_F$ sensitivity (preheat from 50 to 400 °C) normalized by the BOSL$_F$ sensitivity measured with preheat at 50 °C. Sensitivity values correspond to the average of three aliquots. Error bars are the standard deviation. Sample type is described in Table 1. BOSL$_F$ sensitivity values are shown in Table 6.

The $BOSL_{20°C}$ sensitivity measured at room temperature has positive correlation with the $BOSL_F$ (preheat at 190 °C). The correlation is observed both for the sensitivity measured with no pause or with pause of 16 h before blue stimulation (Figure 4), which confirms that the OSL sensitivity obtained without use of thermal treatments to eliminate unstable components can be used for quartz fingerprinting. The $BOSL_{20°C}$ sensitivity obtained for measurements after 16 h pause is two- to three-times lower than measured without pause (Table 5), confirming the significant contribution of unstable OSL components.

Polymineral fine-grained aliquots of core GeoB16206-1 show $BOSL_{20°C}$ sensitivity measured at room temperature have a significant correlation ($r^2 = 0.56$) with the $BOSL_F$ sensitivity measured with stimulation at 125 °C (Table 7). The $BOSL_{20°C}$ sensitivity at room temperature also reproduces the pattern that links OSL sensitivity and precipitation changes (Figure 5).

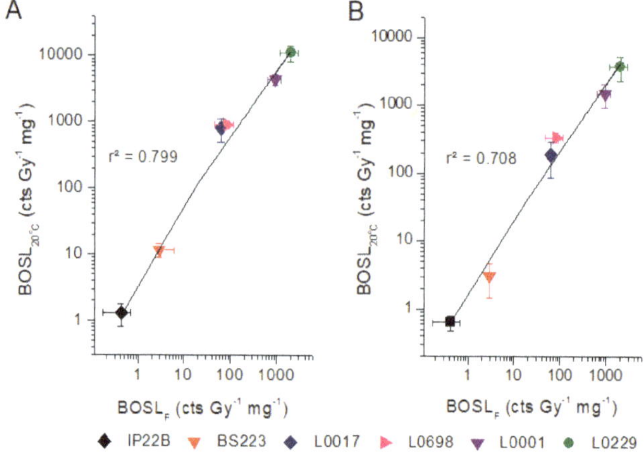

Figure 4. Comparison between $BOSL_F$ sensitivity and sensitivity measured at room temperature ($BOSL_{20°C}$). (**A**) $BOSL_{20°C}$ sensitivity measured without pause between dose and blue stimulation. (**B**) $BOSL_{20°C}$ sensitivity measured with 16 h pause between dose and blue stimulation. Each sample is represented by the average of three aliquots. Error bars correspond to the standard deviation.

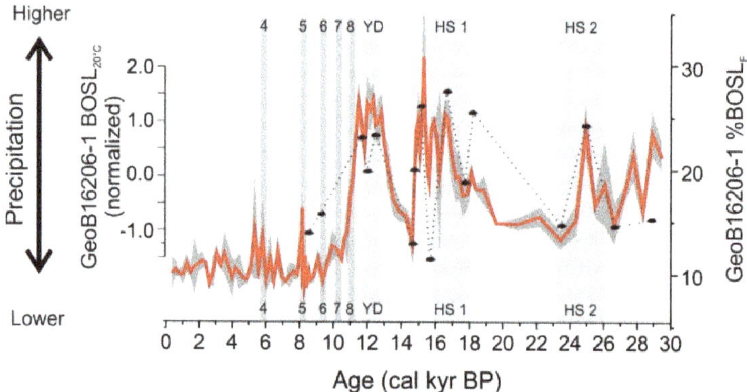

Figure 5. OSL sensitivity for GeoB16206-1 sediment core, with standard error (gray shading) [30] and normalized $BOSL_{20°C}$ (dashed black line, this study). Bond events 4, 5, 6, 7, and 8, Younger Dryas (YD), and the Heinrich Stadials (HS) 1 and 2 represent periods of higher precipitation in Northeast Brazil [33,34].

Table 6. Quartz BOSL$_F$ sensitivity measured using different preheat temperatures (step 5 of the protocol in Table 2). Sensitivity values correspond to the average of three aliquots (± standard deviation). Sample type is described in Table 1.

Preheat	BOSL$_F$ Sensitivity (cts Gy^{-1} mg^{-1})		
	IP22B	L0017	L0229
50 °C	1.60 ± 1.12	111.24 ± 71.03	2773.75 ± 1858.26
100 °C	1.15 ± 0.90	75.58 ± 10.31	4723.34 ± 4264.40
150 °C	0.41 ± 0.45	62.11 ± 9.11	2252.60 ± 512.44
200 °C	0.20 ± 0.18	36.30 ± 7.98	2227.47 ± 694.52
300 °C	0.19 ± 0.09	34.09 ± 9.75	981.85 ± 161.90
400 °C	<0.10 ± 0.27	0.84 ± 028	0.66 ± 064

Table 7. Comparison between BOSL$_F$ sensitivity [30] and sensitivity measured at room temperature (BOSL$_{20°C}$) for silt aliquots from GeoB16206-1 sediment core. Sensitivity values correspond to the average of three aliquots (± standard deviation). Data normalization included subtraction of mean and division by the standard deviation of the data set (%BOSL$_{20°C}$ or %BOSL$_F$).

Sample Depth (cm)	%BOSL$_{20°C}$	BOSL$_{20°C}$ (Normalized)	%BOSL$_F$	BOSL$_F$ (Normalized)
96	14 ± 2	−1.01	9 ± 1	−1.78
150	22 ± 9	−0.66	12 ± 1	−1.34
260	53 ± 5	0.75	27 ± 2	0.99
280	39 ± 5	0.13	22 ± 1	0.31
310	54 ± 7	0.79	27 ± 1	1.04
400	10 ± 2	−1.20	13 ± 2	−1.11
430	40 ± 5	0.16	25 ± 3	0.72
460	65 ± 2	1.31	31 ± 8	1.68
490	4 ± 1	−1.48	19 ± 2	−0.17
550	71 ± 1	1.58	26 ± 1	0.83
610	35 ± 5	−0.08	18 ± 2	−0.40
630	63 ± 3	1.20	20 ± 1	−0.05
700	18 ± 3	−0.85	13 ± 1	−1.05
720	58 ± 9	0.97	24 ± 2	0.60
750	17 ± 4	−0.88	15 ± 2	−0.82
790	20 ± 1	−0.75	24 ± 2	0.54

4. Discussion

When quartz OSL is measured by using preheat (160–260 °C), blue stimulation at 125 °C, and light detection in the ultraviolet band, the initial 1 s of light emission is presumably dominated by the fast OSL component [26], which was intensely characterized for dating purposes [25,35]. The extensive research to develop dating protocols based on the fast OSL component of quartz has also supported applications in sediment provenance studies, e.g., [21–24,27,28]. However, thermal treatments to isolate the fast OSL component usually require mounting samples on discs or cups for measurements in OSL readers built for dosimetry purposes. The decrease of the preheat temperature from 400 to 50 °C allows us to increase the intensity of the OSL signal (initial 1 s) and still preserve the pattern of sensitivity variation observed in the studied quartz samples using protocols [23] designed to measure a signal dominated by the fast OSL component. Higher signal intensity acquired by using lower preheat temperature is advantageous to measure low-sensitivity samples because it permits the use of lower given doses and reduce measurement time. In the same way, the OSL sensitivity measured without preheating and with stimulation at room temperature (~20 °C) is positively correlated with the sensitivity of the OSL signal with higher contribution of the fast component. The OSL sensitivities at room temperature (BOSL$_{20°C}$) recorded after a 16 h pause or without pause between dosing and stimulation have similar patterns of variation. Accordingly, the preheat to eliminate unstable OSL components and light stimulation at 125 °C to avoid any effect of the 110 °C TL on OSL signal measurement can be neglected in OSL sensitivity assessment for provenance analysis of sediments. The $BOSL_{20°C}$ signal measured without preheating has higher contribution of medium and slow components, which are partially removed if heating treatments are applied before light stimulation [26]. Its correlation with the $BOSL_F$ sensitivity agrees with the observation that low and medium OSL components are also sensitized in nature (bleaching–irradiation cycles), similarly to the fast OSL component [13,20]. The main advantages of the $BOSL_{20°C}$ measured without preheating are the reduction of measurement time and the possibility to perform in situ measurements on sediment cores or sedimentary successions exposed in outcrops, using portable OSL readers with an attached X-ray source, e.g., [36].

The assessment of quartz OSL sensitivity in polymineral aliquots of sand requires bleaching other minerals presenting BOSL signals, but preserving the quartz OSL signal. Besides quartz, main mineral components of sands are feldspar and heavy minerals. Feldspar OSL signals can be bleached by using infrared stimulation, allowing for the measurement of quartz OSL signals in the presence of feldspar [37]. However, little is known about the effect of heavy minerals for measurement of quartz OSL sensitivity in polymineral aliquots. Magnetic heavy mineral concentrates have OSL sensitivities similar to low-sensitivity quartz, while nonmagnetic heavy minerals can show OSL sensitivities similar to medium-sensitivity quartz (Figure 1 and Table 5). Heavy minerals are minor components of sands, with higher concentrations (>10%) occurring only in sediments from specific geological settings, such as in areas around mafic parent rocks or environments where hydraulic conditions promoted the selective transport of lighter minerals like quartz and feldspar [38]. Thus, OSL signals arising from heavy minerals have minor influence on the record of quartz OSL sensitivity in bulk samples. However, the tracking of sediment layers enriched in heavy minerals (placers) is recommended to validate that the bulk OSL signal is dominated by quartz. This can be performed by using elemental composition characterized through portable X-ray fluorescence spectrometers [39].

Quartz OSL sensitivity ($BOSL_F$) measured in polymineral fine-grained sediments (silt) can also be applied in provenance analysis, e.g., [23]. A recent example of a successful application of OSL sensitivity for provenance analysis comprised measurements in silt of a marine sediment core [30]. Marine sediment cores are one of the most commonly used archives for the reconstruction of past climatic and oceanographic changes, e.g., [40–44]. The OSL sensitivity represented by the percentage of the initial (1 s) light emission in relation to the total OSL emission ($\%BOSL_F$) was used to track changes in precipitation over the Parnaíba River watershed in Northeastern Brazil [28], where higher values of sensitivity resemble with moments of increased precipitation (i.e., Heinrich Stadials 2 and 1, and the Younger Dryas) [36]. Our data show that BOSL sensitivity measured at room temperature ($BOSL_{20°C}$) in silt samples of core GeoB16206-1 can be used in the same way as the $BOSL_F$ signal with stimulation at 125 °C (Figures 3 and 5). This allows us to increase data resolution through in situ measurements and to couple the provenance analysis of different grain-size fractions (silt and sand), using a single mineral component (quartz). Provenance analysis based on a single and chemically resistant mineral type avoids spurious effects associated with hydraulic sorting during sediment transport or with compositional changes due to weathering or diagenesis.

Therefore, the quartz OSL sensitivity measured without preheating and using blue light stimulation at room temperature after a given dose is recommended for provenance analysis of quartz. Future research can evaluate the use of other ionizing radiation sources, such as portable X-ray sources, and test in situ OSL-sensitivity measurements, using a similar approach as employed, for example, in bulk sediment geochemistry, using X-ray fluorescence. In this case, it is also necessary to improve protocols to measure quartz OSL signals in polymineral material, especially in the presence of feldspar [37], but without the use of thermal treatments.

Author Contributions: A.O.S. conceptualized the research, prepared the manuscript draft, and raised funding. F.C.G.R. carried out experiments and data analysis with the help of T.D.M., V.R.M., and D.B.M. performed preparation and measurements of marine core sediment samples and discussed the study case. C.M.C. contributed with applications to marine sediment cores. P.C.F.G. supervised heavy mineral analysis and raised funding. All authors contributed with revision and editing of the final version of the manuscript.

Funding: A.O.S., P.C.F.G., and C.M.C. are funded by Conselho Nacional de Desenvolvimento Científico e Tecnológico (CNPq grants 304727/2017-2, 308772/2018-0 and 422255/2016-5). F.C.G.R. and V.R.M. are funded by Fundação de Amparo à Pesquisa do Estado de São Paulo (FAPESP grants 2018/12472-8 and 2013/21942-4).

Acknowledgments: We thank Fabiano Pupim and Ian del Río for providing sediment samples of Pantanal and Chile, respectively. We also acknowledge the GeoB Core Repository at MARUM-University of Bremen for supplying samples (core GeoB16206-1) used in this study.

Conflicts of Interest: The funders had no role in the design of the study; in the collection, analyses, or interpretation of data; in the writing of the manuscript; or in the decision to publish the results.

References

1. Meunier, P.; Hovius, N.; Haines, J. Regional patterns of earthquake-triggered landslides and their relation to ground motion. *Geophys. Res. Lett.* **2007**, *34*, L20408. [CrossRef]
2. Chen, H.; Dadson, S.; Chi, Y.-G. Recent rainfall-induced landslides and debris flow in northern Taiwan. *Geomorphology* **2006**, *77*, 112–125. [CrossRef]
3. Stuut, J.-B.; Zabel, M.; Ratmeyer, V.; Helmke, P.; Schefuß, E. Provenance of present-day eolian dust collected off NW Africa. *J. Geoph. Res.* **2005**, *110*, D04202. [CrossRef]
4. Goodbred, S.L., Jr. Response of the Ganges dispersal system to climate change: A source-to-sink view since the last interstade. *Sed. Geol.* **2003**, *162*, 83–104. [CrossRef]
5. Yang, S.Y.; Jung, H.S.; Lim, D.I.; Li, C.X. A review on the provenance discrimination of sediments in the Yellow Sea. *Earth Sci. Rev.* **2003**, *63*, 93–120. [CrossRef]
6. Martin, S.J.; Funch, R.R.; Hanson, P.R.; Yoo, E.-H. A vast 4000-year-old spatial pattern of termite mounds. *Curr. Biol.* **2018**, *28*, R1292–R1293. [CrossRef]
7. Allan James, L. Legacy sediment: Definitions and processes of episodically produced anthropogenic sediment. *Anthropocene* **2013**, *2*, 16–26. [CrossRef]
8. Bond, G.; Broecker, W.; Johnsen, S.; McManus, J.; Labeyrie, L.; Jouzel, J.; Bonani, G. Correlations between climate records from North Atlantic sediments and Greeland ice. *Nature* **1993**, *365*, 143–147. [CrossRef]
9. Concheri, G.; Bertoldi, D.; Polone, E.; Otto, S.; Larcher, R.; Squartini, A. Chemical elemental distribution and soil DNA fingerprints provide the critical evidence in murder case investigation. *PLoS ONE* **2011**. [CrossRef]
10. Dickinson, W.R. Interpreting provenance relations from detrital modes of sandstones. In *Provenance of Arenites*; Zuffa, G.G., Ed.; Reidel Publishing: Dordrecht, The Netherlands, 1985; pp. 333–361.
11. Morton, A. Geochemical studies of detrital heavy minerals and their application to provenance research. In *Developments in Sedimentary Provenance Studies*; Morton, A.C., Todd, S.P., Haughton, P.D.W., Eds.; Geological Society London, Special Publications: London, UK, 1991; Volume 57, pp. 31–45.
12. Gray, H.J.; Jain, M.; Sawakuchi, A.O.; Mahan, S.A.; Tucker, G.E. Luminescence as a sediment tracer and provenance tool. *Rev. Geophys.* **2019**. [CrossRef]
13. Sawakuchi, A.O.; Blair, M.W.; De Witt, R.; Faleiros, F.M.; Hyppolito, T.; Guedes, C.C.F. Thermal history versus sedimentary history: OSL sensitivity of quartz grains extracted from rocks and sediments. *Quat. Geochronol.* **2011**, *6*, 261–272. [CrossRef]
14. Mineli, T.D.; Sawakuchi, A.O.; Guralnik, B.; Lambert, R.; Jain, M.; Pupim, F.N.; del Río, I.; Guedes, C.C.F.; Nogueira, L. Variation of luminescence sensitivity, characteristic dose and trap parameters of quartz. *Radiat. Meas.* **2020**. under review.
15. Bøtter-Jensen, L.; Thomsen, K.J.; Jain, M. Review of optically stimulated luminescence (OSL) instrumental developments for retrospective dosimetry. *Radiat. Meas.* **2010**, *45*, 253–257. [CrossRef]
16. Richter, D.; Richter, A.; Dornich, K. Lexsyg—A new system for luminescence research. *Geochronometria* **2013**, *40*, 220–228. [CrossRef]
17. Kook, M.; Lapp, T.; Murray, A.S.; Thomsen, K.J.; Jain, M. A luminescence imaging system for the routine measurement of single-grain OSL dose distributions. *Radiat. Meas.* **2015**, *81*, 171–177. [CrossRef]

18. Guralnik, B.; Ankjærgaard, C.; Jain, M.; Murray, A.S.; Müller, A.; Wällea, M.; Lowick, S.E.; Preusser, F.; Rhodes, E.J.; Wu, T.-S.; et al. OSL-thermochronometry using bedrock quartz: A note of caution. *Quat. Geochronol.* **2015**, *25*, 37–48. [CrossRef]

19. Bøtter-Jensen, L.; Agersnap Larsen, N.; Mejdahl, V.; Poolton, N.R.J.; Morris, M.F.; McKeever, S.W.S. Luminescence sensitivity changes in quartz as a result of annealing. *Rad. Meas.* **1995**, *24*, 535–541. [CrossRef]

20. Moska, P.; Murray, A.S. Stability of the quartz fast-component in insensitive samples. *Radiat. Meas.* **2006**, *41*, 878–885. [CrossRef]

21. Tsukamoto, S.; Nagashima, K.; Murray, A.S.; Tada, R. Variations in OSL components from quartz from Japan sea sediments and the possibility of reconstructing provenance. *Quat. Int.* **2011**, *234*, 182–189. [CrossRef]

22. Pietsch, T.J.; Olley, J.M.; Nanson, G.C. Fluvial transport as a natural luminescence sensitiser of quartz. *Quat. Geochronol.* **2008**, *3*, 365–376. [CrossRef]

23. Sawakuchi, A.O.; Jain, M.; Mineli, T.D.; Nogueira, L.; Bertassoli, D.J., Jr.; Häggi, C.; Sawakuchi, H.O.; Pupim, F.N.; Grohmann, C.H.; Chiessi, C.M.; et al. Luminescence of quartz and feldspar fingerprints provenance and correlates with the source area denudation in the Amazon River basin. *Earth Planet. Sci. Lett.* **2018**, *492*, 152–162. [CrossRef]

24. Lü, T.; Sun, J.; Li, S.-H.; Gong, Z.; Xu, L. Vertical variations of luminescence sensitivity of quartz grains from loess/paleosol of Luochuan section in the central Chinese Loess Plateau since the last interglacial. *Quat. Geochronol.* **2014**, *22*, 107–115. [CrossRef]

25. Murray, A.S.; Wintle, A.G. The single aliquot regenerative dose protocol: Potential for improvements in reliability. *Radiat. Meas.* **2003**, *37*, 377–381. [CrossRef]

26. Jain, M.; Murray, A.S.; Bøtter-Jensen, L. Characterization of blue-light stimulated luminescence components in different quartz samples: Implications for dose measurement. *Radiat. Meas.* **2003**, *37*, 441–449. [CrossRef]

27. Zular, A.; Sawakuchi, A.O.; Guedes, C.C.; Giannini, P.C.F. Attaining provenance proxies from OSL and TL sensitivities: Coupling with grain size and heavy minerals data from southern Brazilian coastal sediments. *Radiat. Meas.* **2015**, *81*, 39–45. [CrossRef]

28. Mendes, V.R.; Sawakuchi, A.O.; Chiessi, C.M.; Giannini, P.C.F.; Rehfeld, K.; Mulitza, S. Thermoluminescence and optically stimulated luminescence measured in marine sediments indicate precipitation changes over northeastern Brazil. *Paleoceanography* **2019**, *34*, 1476–1486. [CrossRef]

29. Mulitza, S.; Chiessi, C.M.; Cruz, A.P.S.; Frederichs, T.W.; Gomes, J.G.; Gurgel, M.H.C.; Haberkern, J.; Huang, E.; Jovane, L.; Kuhnert, H.; et al. *Response of Amazon Sedimentation to Deforestation, Land Use and Climate Variability—Cruise No. MSM20/3—Recife, Brazil–Bridgetown, Barbados, 19 February–11 March 2012. MARIA S. MERIAN-Berichte, MSM20/3*; DFG Senatskommission für Ozeanographie: Bonn, Germany, 2012; p. 86. [CrossRef]

30. Hansen, V.; Murray, A.S.; Buylaert, J.P.; Yeo, E.Y.; Thomsen, K. A new irradiated quartz for beta source calibration. *Radiat. Meas.* **2015**, *81*, 123–127. [CrossRef]

31. Duller, G.A.T. The Analyst software package for luminescence data: Overview and recent improvements. *Anc. TL* **2015**, *33*, 35–42.

32. Timar, A.; Vandenberghe, D.; Panaiotu, E.C.; Panaiotu, C.G.; Necula, C.; Cosma, C.; van den Haute, P. Optical dating of Romanian loess using fine-grained quartz. *Quat. Geochronol.* **2010**, *5*, 143–148. [CrossRef]

33. Stríkis, N.M.; Cruz, F.W.; Cheng, H.; Karmann, I.; Edwards, R.L.; Vuille, M.; Wang, X.; de Paula, M.S.; Novello, V.F.; Auler, A.S. Abrupt variations in South American monsoon rainfall during the Holocene based on a speleothem record from central-eastern Brazil. *Geology* **2011**, *39*, 1075–1078. [CrossRef]

34. Mulitza, S.; Chiessi, C.M.; Schefuß, E.; Lippold, J.; Wichmann, D.; Antz, B.; Mackensen, A.; Paul, A.; Prange, M.; Rehfeld, K.; et al. Synchronous and proportional deglacial changes in Atlantic meridional overturning and northeast Brazilian precipitation. *Paleoceanography* **2017**, *32*, 622–633. [CrossRef]

35. Wintle, A.G.; Murray, A.S. A review of quartz optically stimulated luminescence characteristics and their relevance in single-aliquot regeneration dating protocols. *Radiat. Meas.* **2006**, *41*, 369–391. [CrossRef]

36. Blair, M.W.; Yukihara, E.G.; McKeever, S.W.S. A system to irradiate and measure luminescence at low temperatures. *Radiat. Prot. Dosim.* **2006**, *119*, 454–457. [CrossRef] [PubMed]

37. Wallinga, J.; Murray, A.S.; Bøtter-Jensen, L. Measurement of the dose in quartz in the presence of feldspar contamination. *Radiat. Prot. Dosim.* **2002**, *101*, 367–370. [CrossRef] [PubMed]

38. Garzanti, E.; Andò, S. Heavy mineral concentration in modern sands: Implications for provenance interpretation. *Develop. Sedim.* **2007**, *58*, 517–545. [CrossRef]

39. Ryan, J.G.; Shervais, J.W.; Li, Y.; Reagan, M.K.; Li, H.Y.; Heaton, D.; Godard, M.; Kirchenbaur, M.; Whattam, S.A.; Pearce, J.A.; et al. Application of a handheld X-ray fluorescence spectrometer for real-time, high density quantitative analysis of drilled igneous rocks and sediments during IODP Expedition 352. *Chem. Geol.* **2017**, *451*, 55–66. [CrossRef]

40. Molfino, B.; McIntyre, A. Precessional forcing of nutricline dynamics in the equatorial Atlantic. *Science* **1990**, *249*, 766–769. [CrossRef]

41. Duplessy, J.C.; Labeyrie, L.; Juillet-Leclerc, A.; Maitre, F.; Duprat, J.; Sarnthein, M. Surface salinity reconstruction of the North Atlantic Ocean during the last glacial maximum. *Oceanol. Acta* **1992**, *14*, 311–324.

42. Nürnberg, D.; Bijma, J.; Hemleben, C. Assessing the reliability of magnesium in foraminiferal calcite as a proxy for water mass temperatures. *Geochim. Cosmochim. Acta* **1996**, *60*, 803–814. [CrossRef]

43. Mulitza, S.; Dürkoop, A.; Hale, W.; Wefer, G.; Stefan Niebler, H. Planktonic foraminifera as recorders of past surface-water stratification. *Geology* **1997**, *25*, 335–338. [CrossRef]

44. Govin, A.; Holzwarth, U.; Heslop, D.; Ford Keeling, L.; Zabel, M.; Mulitza, S.; Collins, J.A.; Chiessi, C.M. Distribution of major elements in Atlantic surface sediments (36 N–49 S): Imprint of terrigenous input and continental weathering. *Geochem. Geophy. Geosy.* **2012**, *13*. [CrossRef]

Letter

ESR and Radiocarbon Dating of Gut Strings from Early Plucked Instruments

Sumiko Tsukamoto [1,*], Taro Takeuchi [2], Atsushi Tani [3], Yosuke Miyairi [4] and Yusuke Yokoyama [4]

[1] Leibniz Institute for Applied Geophysics, Stilleweg 2, 30655 Hannover, Germany
[2] The consortium for Guitar Research, Sidney Sussex College, Cambridge CB2 3HU, UK; guitarouk@aol.com
[3] Graduate School of Human Development and Environment, Kobe University, Kobe 657-8501, Japan; tani@carp.kobe-u.ac.jp
[4] Atmosphere and Ocean Research Institute, The University of Tokyo, Chiba 277-8564, Japan; miyairi@aori.u-tokyo.ac.jp (Y.M.); yokoyama@aori.u-tokyo.ac.jp (Y.Y.)
*   Correspondence: sumiko.tsukamoto@leibniz-liag.de

Received: 14 October 2019; Accepted: 17 January 2020; Published: 28 January 2020

Abstract: Early European plucked instruments have recently experienced a great revival, but a few aspects remain unknown (e.g., the gauge of gut strings). Here we report, for the first time, that the electron spin resonance (ESR) signal intensity of oxidized iron, Fe(III), from gut strings at $g = 2$ increases linearly with age within a few hundred years. The signal increase in the remaining old strings on early instruments can be used to judge if they are as old as or younger than the instrument. Obtaining the authenticity information of gut strings contributes to the revival of the old instruments and the music.

Keywords: ESR; gut strings; Fe(III); early plucked instruments; radiocarbon

1. Introduction

Electron spin resonance (ESR) has been utilized as a geochronometer, based upon the increase in the number of trapped electrons and holes in crystal lattices induced by natural radiation with time [1–3]. ESR also detects unpaired electrons in organic radicals and transition metals in organic substances; the intensity of such ESR signals also increases with time, mainly by thermal activation processes. The possibility of dating organic materials using organic radicals was first tested using potato crisps [4]. The day-by-day increase in the organic radical intensity was used to estimate the production date of the potato crisps. ESR signals of organic radicals and Fe(III) in organic matters, e.g., animal skins, papers, silks and mummies were also investigated, and the intensity showed a positive correlation with age [5–8]. Furthermore, Fe(II) in heme-proteins in human blood starts to oxidize to Fe(III) after exposure to air, and therefore the increase in Fe(III) in bloodstains detected by ESR has been suggested for use in forensic investigations [9].

In this study, we test the potential for establishing a relative chronology of gut strings from early plucked instruments using the ESR signals of Fe(III) and organic radicals. Gut strings are made from sheep guts containing heme-proteins with Fe(II), which is hardly detected with ESR. Once Fe(II) changes to Fe(III) by the catalytic cycle reaction of the heme-proteins or oxidation reaction, it becomes detectable with ESR. The g-values of the Fe(III) signals in oxidized heme-proteins are affected by the ligand fields [10–12]; for instance, the Fe(III) signal at $g = 6$ (high spin state, $S = 5/2$) is originated from methemoglobin coordinated with H_2O, whereas the signal at $g = 2$ (low spin state, $S = 1/2$) is that coordinated to CN^- or OH^- [12,13].

Early European plucked string instruments (e.g., lutes, early guitars and harp-lutes) had fallen out of use by the middle of the 19th century, but have been revived with the increased interest in early music in the 20th century. Old gut stings are found occasionally on the instruments or in the cases. We examine five old strings from guitars and harp-lutes, which were made in the 19th century, as well as four strings with known ages. Radiocarbon dating of the strings is also conducted for six strings for a comparison. Although radiocarbon dating has been frequently applied to archaeological music instruments [14,15], little work has been done on recent plucked instruments [16]. To our knowledge, this is the first study of radiocarbon dating on modern (19th–20th centuries) gut strings.

2. Samples and Results

The nine gut strings used for the ESR investigation are listed in Table 1. The approximate ages of four gut string samples are known. Two of them (LHE-14 and -27) were obtained shortly before the first measurement of the samples in April 2015. Another string (LHE-15) was acquired around the year 2000. LHE-16 was provided with a memorandum of the supplier, G. Butler and Sons in London, in 1913 (Figure 1a). Five string samples were collected from four original instruments. Two harp-lutes, which are both assumed to have been made ca. 1815, had remaining old gut strings at the bridge. Two string samples (LHE-17A, -17B) were collected from the 14-string harp-lutes (Figure 1b), and a sample LHE-28 was taken from the 12-string instrument (Figure 1c). A separate string from the 14-string harp-lute (LHE-17C) was used for radiocarbon dating. Sample LHE-19 was found in the original case of a guitar, which was made approximately in ca. 1840 by D&A Roundhloff in London, and LHE-44 was obtained from a guitar, estimated to have been made between 1850 and 1860. All these original instruments were produced and preserved in the United Kingdom.

Table 1. List of sample strings for ESR.

Sample ID	Expected Production Year	Weight (mg)	Instrument
LHE-14	ca. 2013	13.6	
LHE-15	ca. 2000	15.3	
LHE-16	ca. 1913	1.5	
LHE-17A	ca. 1815	4.9	14-string harp-lute
LHE-17B	ca. 1815	5.6	14-string harp-lute (same as above)
LHE-19	ca. 1840	1.6	a case of a D&A Roundfloff guitar
LHE-27	ca. 2013	24.8	
LHE-28	ca. 1815	26.2	12-string harp-lute
LHE-44	ca. 1850–1860	20.1	from a guitar

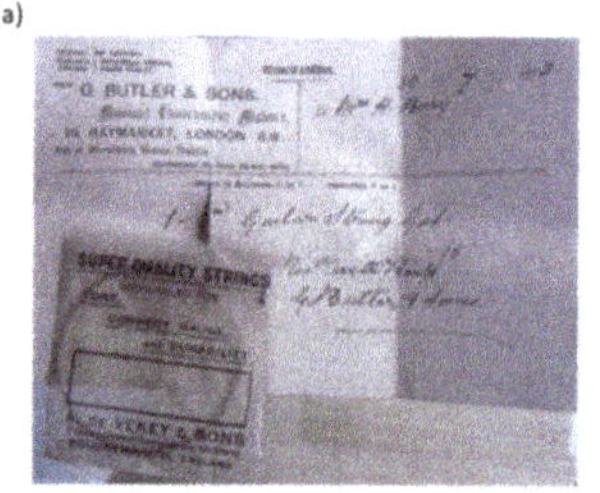

Figure 1. Photos of (**a**) gut string and its receipt (LHE-16), (**b**) a 14-string harp-lute (LHE-17A and -17B) and (**c**) a 12-string harp-lute (LHE-28). Gut string samples were collected from them.

ESR spectra of a recent (LHE-15) and an old string (LHE-17A) are shown for wider (250 ± 250 mT; Figure 2a) and narrower (324 ± 5 mT; Figure 2b) scanned magnetic fields. All gut string samples showed ESR signals from Fe(III) signal at $g = 2$ (low spin states, $S = 1/2$) (Figure 2a) and organic radicals at around $g = 2.005$ (Figure 2b). The Fe(III) signal at $g = 6$ (high spin states, $S = 5/2$) (Figure 2a) was observed in all samples except LHE-28.

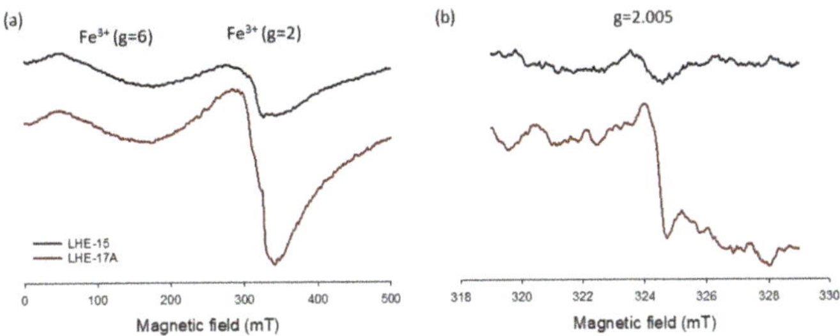

Figure 2. ESR spectra of LHE-15 and -17A for the magnetic field of (**a**) 250 ± 250 mT and (**b**) 324 ± 5 mT. Note that the spectra in (**b**) were recorded with eight times larger receiver gain than (**a**) due to the weak signal intensity.

All these strings are assumed to have been produced within ~2 years before the time of acquisition. Crude but supporting evidence was obtained by the increasing intensity of the Fe(III) signal at $g = 2$ within the 11 months for the two modern strings, LHE-14 and LHE-27 (Figure S1). When the intensity is plotted against the time after the first measurement, April 2015, the x-intercept of the fitted straight line points to 15 months and 19 months before the first measurements for LHE-14 and LHE-27, respectively.

The correlation of the ESR signal intensity with age was first examined using four string samples with known ages (LHE-14, -15, -16, and -27). The results are plotted in Figure 3a. Although the number of known age samples is limited, it is clear that the best correlation between the ESR signal intensity and age is obtained from the Fe(III) signal at $g = 2$ ($r = 0.95$, Spearman's correlation coefficient). The other two signals showed much less correlation with age, with correlation coefficient of 0.63 for both organic radicals and the Fe(III) at $g = 6$. This indicates that the Fe(III) signal at $g = 2$ increases linearly with time, for at least 100 years, and therefore can be used for the relative age estimation of older gut strings.

In Figure 3b–d, the three ESR intensities of all samples are plotted against the known string age (filled symbols) or the expected instrument age (open symbols). The results of Fe(III) at $g = 6$ and the organic radicals are highly scattered, suggesting these two signals are not suitable as chronometers. For the Fe(III) at $g = 2$, the intensity of three of the five old string samples is consistent with the extrapolated fitted line of the known age strings within the 1-σ uncertainty (LHE- 17A, -17B, and -28) and one is consistent within the 2- σ uncertainty (LHE-19). This suggests that these strings are as old as the instruments. However, the Fe(III) signal ($g = 2$) of LHE-44 has a much lower intensity when compared to the expected intensity from the age of the instrument.

The results of the radiocarbon dating are summarized in Table 2. The method successfully distinguished the strings before and after 1950. The three recent samples (LHE-14, -15, -27) yielded > 100% modern carbon (pMC), and are therefore judged as modern. The ^{14}C ages of the older three string samples (LHE-16, -17, and -44) ranged from 167 ± 26 y BP to 309 ± 29 y BP. These were calibrated using OxCal v.3.10 based on Intcal13. Detailed results of the calibrated ^{14}C ages are given in Figure S2.

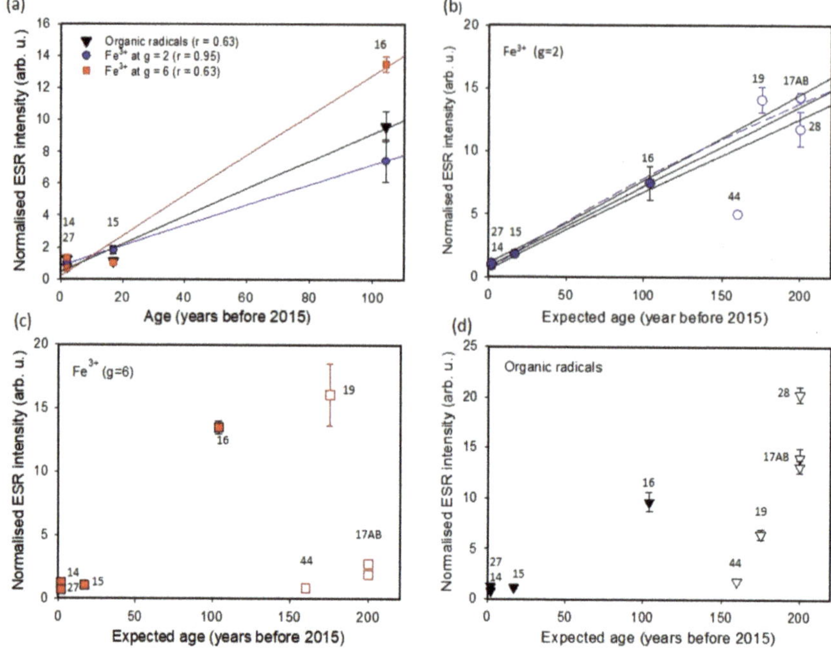

Figure 3. Plot of ESR intensity against the known (filled symbols) or the expected instrument age (open symbols) for (**a**) all ESR signals of the known age samples, (**b**) Fe(III) at $g = 2$, (**c**) Fe(III) signal at $g = 6$, and (**d**) organic radicals. The numbers besides the symbols are sample IDs. All error bars in vertical axis are 1-σ standard error.

Table 2. Results of radiocarbon dating of the strings.

Lab. Code	Sample ID	$\delta^{13}C$ (‰)		pMC (%)		^{14}C Age (Year BP)	
YAUT-021827	LHE-14	−25.2	± 1.7	104.57	± 0.38	modern	
YAUT-021828	LHE-15	−25.4	± 1.8	114.34	± 0.42	modern	
YAUT-021830	LHE-16	−30.1	± 0.5	97.73	± 0.27	185	± 22
YAUT-021832	LHE-17C	−24.0	± 1.7	96.23	± 0.35	309	± 29
YAUT-021833	LHE-27	−16.5	± 1.3	118.13	± 0.37	modern	
YAUT-021834	LHE-44	−20.4	± 1.3	97.95	± 0.32	167	± 26

3. Discussion

The oxidation of Fe(II) to Fe(III) should be sensitive to various environmental factors, e.g., temperature and moisture. As shown in Figure 3a, the very good positive correlation of the Fe(III) signal at $g = 2$ is probably due to the fact that all the string samples used in this study were preserved in the United Kingdom, and therefore the environmental factors were similar. A possible explanation of contrasting correlation with age for these two types of Fe(III) is that the low spin heme-proteins are contained relatively uniformly in all gut strings, but the concentration of the high spin heme-proteins is different from sample to sample.

The observed increase in the Fe(III) signal at $g = 2$ does not show clear tendency toward saturation, although it is natural to assume that the signal intensity reaches saturation over a longer time period, since a limited number of Fe(II) in heme-proteins should be available in the gut strings. The blue dashed line in Figure 3b shows a regression line, fitted to a single saturation exponential function when all Fe(III) data at $g = 2$ except LHE-44 are used. The fitted line is not significantly deviated from the linear regression line (middle of black solid lines in Figure 3b), suggesting that the signal increase is still very close to linear for ~200 years. This result is in contrast with the study of the Fe(III) signal in coagulated human blood; in which the intensity of high spin state Fe(III) reached close to saturation in a few hundred hours at room temperature after coagulation [9], presumably because the Fe(III) in blood was directly exposed to air and the oxidization process was faster.

To investigate how the Fe(III) intensity at $g = 2$ in gut strings changes over a longer timescale, aging experiments were conducted using LHE-27 by heating at 60 and 70 °C in air. The result is shown in Figure S3. At 60 °C, the signal increased only about 14% from the modern string intensity and reached saturation, whereas at 70 °C the signal intensity once increased slightly then decreased. In nature, the gut strings of original instruments of ~200 years old (LHE-17A, -17B, -19, and -28) showed 12–14 times larger signal intensity than the modern strings. We conclude that the heating at higher temperature cannot reproduce the natural signal growth.

Using the linear increase in the Fe(III) signal ($g = 2$) of gut strings, it is possible to estimate the age ranges of strings with unknown ages. For the five string samples obtained from early plucked string instruments, the consistent ESR signal intensity with the linear regression line of the known age strings for the four samples (LHE-17A, -17B, -19, and -28) indicate that the strings are as old as the instruments. This indicates that these instruments (early guitars and harp lutes) were not used for a long time after they were made. This also suggests that remaining strings can be used as a record of the original gauges. One string sample, LHE-44 yielded much a lower Fe(III) intensity than expected considering the instrument age. By comparing the intensity with the regression line, it is assumed that this string was made 60–70 years before 2015.

Currently, it is still unknown how variable the signal increase is in different storage conditions. More data of string samples with known ages and from different environments should be accumulated to establish the method to be a robust dating technique.

AMS radiocarbon dating can accurately judge whether a gut string was made before or after 1950 (Table 2). However, calibration of a ^{14}C age into a calendar age generates a large uncertainty (Figure S2). Sample LHE-16, which was sold in 1913, yielded a calibrated age of AD1660–1690 (18.2%), AD1730–1810 (53.4%) and later than AD1920 (23.8%) in 95.4% probability. As mentioned above, the ESR of LHE-44 gave an assumption that the string was much younger than the instrument (AD1850–1860). However, the calibrated ^{14}C age of this sample yielded AD1660–1700 (16.7%), AD1720–1820 (51.9%), AD1830–1880 (6.5%) and later than AD1910 (20.5 %), which cover both the instrument age and ESR age of this string. For LHE-17, the calibrated ^{14}C age yielded a much older age than the instrument (ca. 1815)—AD1480–1650—which might be due to the contamination of old carbon in the production process.

4. Methods

All ESR measurements were conducted using a JEOL-FA-100 X-band spectrometer at the Leibniz Institute for Applied Geophysics, Hannover, Germany. The string samples, which were cut into ~1 cm length, were inserted into quartz glass tubes of 4 mm outer diameter (3 mm inner diameter), and used for the ESR measurements. The measurements were made at room temperature, three times in April 2015, January 2016 and February 2016. The measurement parameters for the Fe(III) signals were; 250 ± 250 mT magnetic field, 1 mW microwave power, 1 min scan time for 3 times, 0.5 mT modulation amplitude and 0.1 s time constant. The organic radical signal ($g = 2.005$) was measured with the following conditions; 324 ± 5 mT magnetic field, 2 mW microwave power, 30 s scan time up to 50 times, 0.3 mT modulation width and 0.1 s time constant. The mean peak to peak ESR intensity of the three measurements and its 1-σ standard error was calculated. For one sample, which we obtained in December 2015 (LHE-44) the mean signal intensity of two measurements was used. Since the available sample mount varied significantly (1.5 and 26 mg; Table 1), all ESR signal intensities were normalized to the weight of each string sample.

The radiocarbon dating of the six string samples (LHE-14, -15, -16, -17C, -19 and -44) was conducted with an NEC 250 kV single-stage accelerator mass spectrometer (AMS) at the Atmosphere and Ocean

Research Institute, The University of Tokyo [17]. The string samples were washed with ultra-pure water and dried. These samples were oxidized using a Vario Micro Cube elemental analyzer manufactured by Elementar Analysensysteme GmbH., and an AMS analytical target was prepared using an automated graphitization process device manufactured by Koshin Rikagaku Seisakusho Co., Ltd (Tokyo, Japan).

5. Conclusions

A very good correlation between the Fe(III) signal and the known ages of gut strings was observed. This probably indicates that Fe(II) in the gutstrings has been oxidized to Fe(III) with age. We conclude that it is possible to assume whether a gut string is as old as the instrument or much younger, using the correlation of the Fe(III) signal ($g = 2$) with age. Using ^{14}C dating it is also possible to judge a string is a recent one or older, but a calibrated age has a large uncertainty.

Supplementary Materials: The following are available online at http://www.mdpi.com/2409-9279/3/1/13/s1, Figure S1: The increase of the Fe(III) ($g = 2$) signal after the first ESR measurement, April 2015 with time. The production time before the first measurement was roughly estimated by linearly extrapolating the fitted line to zero intensity. Data were normalised to the same factor as was used for Figure 3. Unfortunately further measurements were not achieved due to a problem with the ESR spectrometer, Figure S2: Results of calibration for the radiocarbon ages, Figure S3: Result of the aging experiment of a string sample LHE-27 heated at 60 °C and 70 °C. The intensity of Fe(III) at $g = 2$ after different durations of heating is shown.

Author Contributions: S.T. and T.T. designed the study. S.T. conducted all ESR measurements and wrote the paper. T.T. collected gut string samples and contributed in the sample descriptions. A.T. contributed to the discussion of the physics of Fe^{3+}. Y.M. and Y.Y. conducted radiocarbon dating and contributed to the corresponding discussion. All authors have read and agreed to the published version of the manuscript.

Acknowledgments: We are very grateful to James Westbrook for providing string samples and photos. Gwynyn Buchanan is thanked for the language edits.

Conflicts of Interest: The authors declare no conflict of interest.

References

1. Ikeya, M. *New Applications of Electron Spin Resonance Dating, Dosimetry and Microspcopy*; World Scientific: Singapore, 1993; p. 520.
2. Rink, W.J. Electron spin resonance (ESR) dating and ESR applications in quaternary science and archaeometry. *Radiat. Meas.* **1997**, *27*, 975–1025. [CrossRef]
3. Schellmann, G.; Beerten, K.; Radtke, U. Electron spin resonance (ESR) dating of Quaternary materials. *Quat. Sci. Jour.* **2008**, *57*, 150–178.
4. Ikeya, M.; Miki, T. A new dating using digital ESR method. *Naturwissenschaften* **1980**, *67*, 191–192. [CrossRef]
5. Ikeya, M.; Miki, T. ESR dating of organic materials: from potato-chips to a dead body. *Nucl. Tracks. Radiat. Meas.* **1985**, *10*, 909–912. [CrossRef]
6. Miki, T.; Yahagi, T.; Ikeya, M.; Sugawara, N.; Furuno, J. ESR dating of organic substances: corpse for forensic medicine. In *ESR Dating and Dosimetry*; Ikeya, M., Miki, T., Eds.; Ionics: Tokyo, Japan, 1985; pp. 447–451.
7. Miki, T.; Kai, A.; Ikeya, M. ESR dating utilizing in valency or ligand of transition metals. *Jpn. J. Appl. Phys.* **1987**, 972–973. [CrossRef]
8. Miki, T.; Kai, A.; Ikeya, M. ESR dating of organic substances utilizing paramagnetic degradation products. *Nucl. Tracks. Radiat. Meas.* **1988**, *14*, 253–258. [CrossRef]
9. Miki, T.; Kai, A.; Ikeya, M. Electron spin resonance of bloodstains and its application to time after bleeding. *Forensic Sci. Intern.* **1987**, *35*, 149–158. [CrossRef]
10. Banerjee, R.; Stetzkowski, F. Heme transfer from haemoglobin and ferrihemoglobin to some new ligands and its implication in the mechanism of oxidation of ferrohemoglobin by air. *Biochem. Biophys. Acta.* **1970**, *221*, 636–639. [PubMed]
11. Peisach, J.; Blumberg, W.E.; Wittenberg, B.A.; Wittenberg, J.B.; Kampa, L. Hemoglobin a: an electron paramagnetic resonance study of the effects of interchain contacts on the heme symmetry of high-spin and low-spin derivatives of ferric alpha chains. *Proc. Nat. Acad. Sci. USA* **1969**, *63*, 934–939. [CrossRef] [PubMed]
12. Hori, H. Analysis of the principal g-tensors in single crystal of ferrimyoglobin cpmplexes. *Biochem. Biophys. Acta* **1971**, *251*, 227–235. [PubMed]

13. Masuda, H.; Tagam, T. The crystal and molecular structures of metalloporphyrin complexes with anomalous electronic states. *Jour. Crystal. Soc. Jpn.* **1984**, *25*, 289–298. [CrossRef]

14. Zhang, J.; Harbottle, G.; Wang, C.; Kong, Z. Oldest playable musical instruments found at Jiahu early Neolithic site in China. *Nature* **1999**, *401*, 366–368. [CrossRef] [PubMed]

15. Conrad, N.J.; Malina, M.; Münzel, S.C. New flutes document the earliest musical tradition in southwestern Germany. *Nature* **2009**, *460*, 737–740. [CrossRef] [PubMed]

16. Durier, M.G.; Bruguière, P.; Hatté, C.; Vaiedelich, S. Radiocarbon dating of legacy music instrument collections: Example of traditional indian Vina from the Musée De La Musique, Paris. *Radiocarbon* **2018**, *61*, 17–22. [CrossRef]

17. Yokoyama, Y.; Miyairi, Y.; Aze, T.; Yamane, M.; Sawada, C.; Ando, Y.; de Natris, M.; Hirabayashi, S.; Ishiwa, T.; Sato, N.; et al. A single stage Accelerator Mass Spectrometry at the Atmosphere and Ocean Research Institute, The University of Tokyo. *Nucl. Inst. Meth. Phys. Res. B* **2019**, *455*, 311–316. [CrossRef]

Protocol

Determining the Age of Terrace Formation Using Luminescence Dating—A Case of the Yellow River Terraces in the Baode Area, China

Jia-Fu Zhang [1],*, Wei-Li Qiu [2], Gang Hu [1,3] and Li-Ping Zhou [1]

[1] MOE Laboratory for Earth Surface Processes, College of Urban and Environmental Sciences, Peking University, Beijing 100871, China; hugang@ies.ac.cn (G.H.); lpzhou@pku.edu.cn (L.-P.Z.)
[2] Faculty of Geographical Science, Beijing Normal University, Beijing 100875, China; qiuweili@bnu.edu.cn
[3] Institute of Geology, China Earthquake Administration, Beijing 100029, China
* Correspondence: jfzhang@pku.edu.cn

Received: 12 November 2019; Accepted: 17 February 2020; Published: 20 February 2020

Abstract: Dating fluvial terraces has long been a challenge for geologists and geomorphologists, because terrace straths and treads are not usually directly dated. In this study, the formation ages of the Yellow River terraces in the Baode area in China were determined by dating fluvial deposits overlying bedrock straths using optically stimulated luminescence (OSL) dating techniques. Seven terraces (from the lowest terrace T1 to the highest terrace T7) in the study area were recognized, and they are characterized by thick fluvial terrace deposits overlaid by loess sediments. Twenty-five samples from nine terrace sections were dated to about 2–200 ka. The OSL ages (120–190 ka) of the fluvial samples from higher terraces (T3–T6) seem to be reliable based on their luminescence properties and stratigraphic consistency, but the geomorphologic and stratigraphic evidence show that these ages should be underestimated, because they are generally similar to those of the samples from the lower terrace (T2). The formation ages of the terrace straths and treads for the T1 terrace were deduced to be about 44 ka and 36 ka, respectively, based on the deposition rates of the fluvial terrace deposits, and the T2 terrace has the same strath and tread formation age of about 135 ka. The incision rate was calculated to be about 0.35 mm/ka for the past 135 ka, and the uplift rate pattern suggests that the Ordos Plateau behaves as a rigid block. Based on our previous investigations on the Yellow River terraces and the results in this study, we consider that the formation ages of terrace straths and treads calculated using deposition rates of terrace fluvial sediments can overcome problems associated with age underestimation or overestimation of strath or fill terraces based on the single age of one fluvial terrace sample. The implication is that, for accurate dating of terrace formation, terrace sections should be systematically sampled and dated.

Keywords: terrace strath and tread ages; luminescence dating; deposition rate; age-depth model; Yellow River terrace

1. Introduction

Fluvial terrace deposits and landforms can provide important information about river incision, tectonic activity, climate change and archaeological traces of hominid activity [1–5], and terraces record different stages of fluvial evolution and sedimentation. Yet, understanding the formation of terraces is a perennial problem in geomorphology and has been hampered by the exact timing of terrace formation. Determining the formation age of terraces is a challenge for geomorphologists and geologists. Recently, the successful application of optically stimulated luminescence (OSL) dating techniques [6–8] to fluvial sediments [9,10] has made it possible to establish the chronology of terraces by dating fluvial sediments atop terrace straths [11–15]. Conventionally, only a single age is assigned to a terrace, but this may not reflect actual fluvial processes [16,17]. The development of OSL techniques allows dating a series of samples from terrace deposits by systematic sampling, which helps to establish the chronology of deposition and then deduce the formation ages of terrace straths and treads [18–20].

The investigations of Yellow River fluvial terraces have become one of the most important topics in Chinese geomorphology and are also of interest to researchers in sedimentology, engineering, hydrology and archaeology. The fluvial terraces in the Jinshaan Canyon in the middle reaches of the Yellow River have been widely investigated and dated mainly using luminescence and paleomagnetic dating methods [18,19,21–26]. However, the dates of the terraces obtained are not enough to construct the precise chronology of a series of terraces in the region. This is further complicated by the fact that some terraces with the same elevations above the modern river have different formation ages, because their formations are controlled by knickpoint migration [19,25]. Therefore, it is necessary to determine the ages of more Yellow River terraces in different localities of the region. On the other hand, one sample for OSL dating is often collected from terrace deposits in fluvial terrace investigations [21], and its age is usually assumed to be the terrace formation age. This assumption may be incorrect, and its validity needs to be tested. In this study, we first identified terraces in the two banks of the Yellow River in the Baode area in the middle reaches of the river, and terrace deposits were systematically sampled for OSL dating, i.e., a series of samples were collected from channel facies and floodplain facies and overlying loess deposits on terraces. The reliability of their ages was evaluated based on the internal consistency of the OSL ages and the lithostratigraphic and geomorphological consistency in the OSL ages, and the formation ages of terrace straths and treads were then inferred.

2. Geological and Geomorphologic Setting

The Ordos Plateau, with an area of 320,000 km^2 located in the western part of the North China block, is a relatively stable rigid block (Figure 1a) [27,28]. It is bounded by a series of marginal faults and orogenic belts [29–34]. The basement of the plateau is composed of Archean and Proterozoic metamorphic rocks that are overlain by Cambrian and Ordovician marine carbonates with the lack of Silurian and Devonian strata. Carboniferous marine limestone and fluvial-deltaic sandstone are overlain by fluvial Permian strata. Triassic and Jurassic strata are composed of fluvial and lacustrine deposits, which are overlain by Cretaceous fluvial and eolian redbeds. The Ordos block evolved from a basin to a plateau during the late Miocene–Pliocene, which influences the environmental effect on the topography of the area [35]. Geomorphologically, the Ordos Plateau is surrounded by the Yin Mountains to the north, Luliang Mountains to the east, Qinling Mountains to the south, Liupan and Helan Mountains to the west (Figure 1). The arc-shaped Yinchuan–Hetao graben system along the northwestern and northern margins of the block and the S-shaped Weihe-Shanxi graben system along the southeastern margins bound the plateau [29,36]. The plateau is covered by Quaternary eolian sediments. The northern part, the Mu Us Desert, is mantled by dune sands, and the southern part, the Chinese Loess Plateau, is covered by loess/paleosol deposits.

The Yellow River (called Huanghe in Chinese), well-known for its tremendous sediment load, originates in the northeast of the Tibetan Plateau. In its middle reaches, the river flows along the northeastern and northern margins of the Ordos Plateau and cuts through the eastern plateau at an average elevation ranging from 1000 to 1500 m above sea level (asl), from Lamawan in the north to Yumenkou in the south (Figure 1a). The river then turns east and finally overflows into the Bohai Sea. The river downcutting between Lamawan and Yumenkou leads to the formation of a deep and narrow gorge called Jinshaan Canyon, which connects the Yinchuan-Hetao graben on the north and the Weihe graben on the south. A series of strath terraces along the banks of the canyon have been found based on the presence of fluvial deposits and bedrock strath. The terrace treads are covered by loess with various thicknesses.

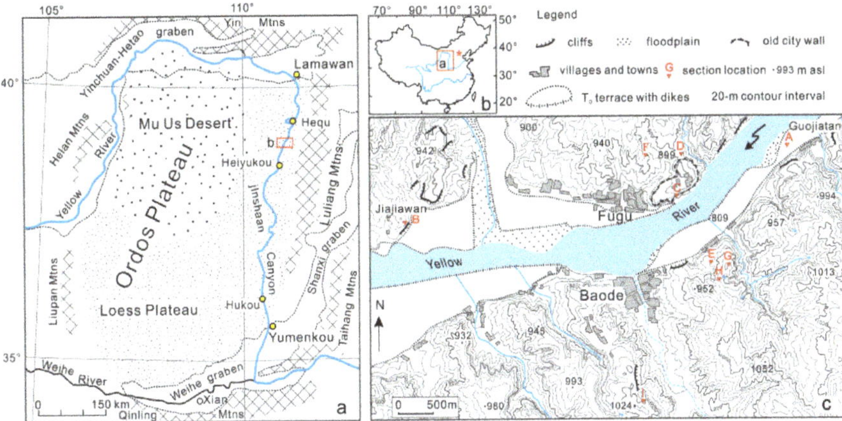

Figure 1. (**a,b**) Map showing the Yellow River course along the margins of the Ordos Plateau and location of the study area (Baode); (**c**) topographic map showing the localities of the sections (A–I) shown in Figure 2. The topographic map with contour intervals of 20 m was constructed based on a Chinese 1:50,000 topographic map (unpublished).

3. Methodology

3.1. Field Work

As shown in Figure 1a, the two banks of the Yellow River in the study area are covered by loess/paleosol deposits. This implies that the original stair-stepped topography of the fluvial terraces in the area is not directly observed, and the situation is further complicated by the modification of the loess landscape due to soil erosion. The fluvial terraces cannot be mapped from topographic maps, satellite images or air photos or even identified in the field based on surface landform. In this case, the identification of remnants of the terraces was only conducted by observing fluvial sediments exposed in natural outcrops or road-cuts in the field. This means that the terraces were found as isolated remnants. The heights of terrace strath (bedrock surface) and tread (here the "tread" is referred to as the near-horizontal top surface of fluvial deposits, including channel or overbank facies resting on terrace straths) above the modern river level and the thickness of the terrace deposits were measured using a total station and a tape measure. The precise locations and altitude of the exposures were recorded using a hand-held global positioning system (GPS) receiver with a barometric altimeter (Garmin GPSmap 60CSx) with an accuracy of ±~3 m in altitude. Stratigraphy and lithology of the exposures were described in detail.

A systematic sampling strategy was adopted for dating the sediments on terrace straths using OSL techniques. Systematic sampling means that a series of samples are taken from fluvial and loess deposits on a terrace, and their OSL ages are used to construct an age-depth model for the terrace deposits. The model helps us to evaluate the reliability of OSL ages obtained for the terraces and to explain the history of river incision. The formation ages of the strath and tread for a terrace can be inferred from the age-depth relationship. OSL samples were taken from each unit of terrace deposits and overlying loess/paleosol (see Figure 2). The samples were collected by hammering 3.5-cm-diameter and 30-cm-long stainless-steel tubes horizontally into freshly cleaned exposures. The tubes were wrapped with aluminum foil and adhesive tape in order to prevent further exposure to light and moisture loss.

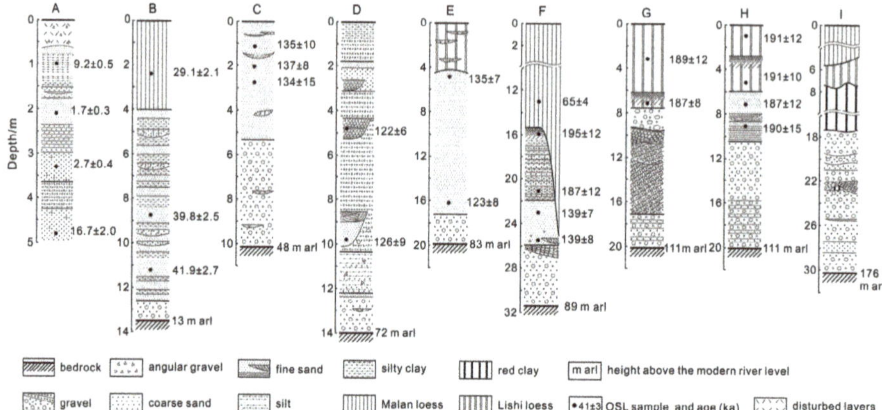

Figure 2. Stratigraphic columns of the terrace deposits in outcrops (A–I) marked in Figure 1c, and the positions of optically stimulated luminescence (OSL) samples and ages in ka. The elevations of bedrock surfaces (strath surface) above the modern river are displayed.

3.2. Optical Dating

3.2.1. Equivalent Dose Measurements

The samples for OSL measurements were prepared in our dark room with a dim red light at the Peking University, Beijing, China. Coarse- and fine-grained quartz were extracted from sand- and silt-sized samples using the procedures in our laboratory, respectively [37,38]. In the laboratory, the light-exposed ends (about 2–3 cm of sediment) of a sample tube were first removed, and this removed material was used for dose-rate analysis. The remaining material from the middle of the tube was treated with 10% hydrochloric acid to dissolve carbonates and then 30% hydrogen peroxide to remove organic material, respectively. For sand-sized samples, the material was then washed with water to eliminate finer grains, followed by drying and sieving to select coarse grains (grain sizes for each sample are listed in Table 1) for luminescence measurements. The coarse-grained quartz was obtained by immersing the sieved sample in 40% HF for 40 or 80 min and then 10% HCl to remove feldspar contaminants. For fine-grained samples, the samples were then deflocculated using a dilute sodium oxalate solution, and polymineral fine grains (4–11 µm) were isolated by settling the polymineral fine silt fraction in the solution. Fine-grained quartz was obtained by treating the polymineral extracts with silica-saturated fluorosilicic acid (H_2SiF_6) at room temperature to dissolve feldspars, amorphous silica and other contaminant minerals, followed by a treatment with 10% HCl to remove any fluorides produced. The purity of the quartz extracts was checked by infrared stimulation. The results showed that the infrared-stimulated luminescence signals were negligible, indicating feldspar contaminants were almost entirely removed. The chemically purified quartz was prepared for luminescence measurements by settling the fine grains in acetone onto 0.97-cm-diameter aluminum discs or mounting the coarse grains as a monolayer on 0.97-cm-diameter aluminum discs with the grains covering the area with a diameter of ~5 mm (medium aliquots) using silicone oil as an adhesive.

Table 1. The results of optical dating of the samples from the Yellow River terraces in the Baode area.

Section No.	Lab Code	Field No.	Depth (m)	Sediment	Grain Size (µm)	K * (%)	K ** (%)	U ** (ppm)	Th ** (ppm)	Water Content # (%)	Dose rate (Gy/ka)	No. of Aliquots Measured	Arithmetic Mean De (Gy)	Central Age Modeling (CAM)		
														OD+ (%)	Mean De (Gy)	Age (ka)
Terrace T0																
A	L552	BD-OSL06	0.98	Overbank silt	4–11	1.8	1.87 ± 0.10	2.36 ± 0.10	9.81±0.24	10	3.30 ± 0.12	3	30.4 ± 1.1		30.5 ± 1.2	9.2 ± 0.5
	L551	BD-OSL05	2.05	Channel sand	90–125	1.72	1.63 ± 0.09	2.02 ± 0.09	10.10 ± 0.22	5	2.89 ± 0.08	21	6.7 ± 1.5	72	5.0 ± 0.8	1.7 ± 0.3
	L550	BD-OSL04	3.30	Channel sand	150–250	2.98	3.25 ± 0.11	0.68 ± 0.06	5.52 ± 0.17	5	3.69 ± 0.10	25	13.5 ± 2.4	69	10.0 ± 1.5	2.7 ± 0.4
	L549	BD-OSL03	4.77	Channel sand	150–250	2.35	2.32 ± 0.10	0.83 ± 0.07	5.63 ± 0.16	5	2.84 ± 0.09	21	53.7 ± 5.5	53	47.4 ± 5.6	16.7 ± 2.0
Terrace T1																
B	L784	BD06-OSL03	2.40	Loess	4–11	1.4		3.01 ± 0.37	11.41 ± 1.24	10	3.12 ± 0.15	4	91.3 ± 5.4		90.7 ± 4.8	29.1 ± 2.1
	L783	BD06-OSL02	8.80	Channel sand	150–250	1.85		2.48 ± 0.33	9.13 ± 1.10 *	5	2.95 ± 0.11	21	122.7 ± 7.2	22	117.6 ± 6.0	39.8 ± 2.5
	L782	BD06-OSL01	11.20	Channel sand	150–250	1.65		2.01 ± 0.37	12.24 ± 1.24 *	5	2.85 ± 0.11	21	125.2 ± 7.2	21	119.5 ± 6.3	41.9 ± 2.7
Terrace T2																
C	L567	BD-OSL21	1.05	Overbank silt	90–125	1.72	1.68 ± 0.10	1.59 ± 0.07	8.75 ± 0.22	10	2.64 ± 0.09	16(2)	378.9 ± 65.8	21	357 ± 24	135 ± 10
	L568	BD-OSL22	2.00	Overbank silt	90–125	1.72	1.64 ± 0.10	1.90 ± 0.09	8.02 ± 0.20	10	2.60 ± 0.09	13	369.4 ± 17.6	12	355 ± 18	137 ± 8
	L569	BD-OSL23	2.70	Overbank silt	90–125	1.72	1.72 ± 0.10	1.80 ± 0.08	7.32 ± 0.20	10	2.59 ± 0.09	8	366.0 ± 41.8	29	346 ± 38	134 ± 15
Terrace T3																
D	L571	BD-OSL25	4.8	Channel sand	150–250	2.2	2.35 ± 0.09	1.09 ± 0.08	6.49 ± 0.18	5	2.98 ± 0.08	20	366.9 ± 14.5	13	365 ± 13	122 ± 6
	L570	BD-OSL24	9.8	Channel sand	90–150	1.88	1.09	1.70 ± 0.10	8.62 ± 0.21	5	2.73 ± 0.09	22	361.8 ± 25.5	29	344 ± 23	126 ± 9
Terrace T4																
E	L557	BD-OSL11	4.80	Channel sand	125–150	1.8	1.80 ± 0.09	1.23 ± 0.07	5.83 ± 0.17	5	2.50 ± 0.08	23	357.3 ± 21.0	23	339 ± 15	135 ± 7
	L556	BD-OSL10	16.2	Channel sand	150–250	2.2	2.10 ± 0.09	0.80 ± 0.07	5.03 ± 0.16	5	2.51 ± 0.08	23	321.1 ± 20.3	26	310 ± 18	123 ± 8
Terrace T5																

Table 1. *Cont.*

Section No.	Lab Code	Field No.	Depth (m)	Sediment	Grain Size (μm)	K * (%)	K ** (%)	U ** (ppm)	Th ** (ppm)	Water Content # (%)	Dose rate (Gy/ka)	No. of Aliquots Measured	Arithmetic Mean De (Gy)	Central Age Modeling (CAM)		
														OD+ (%)	Mean De (Gy)	Age (ka)
F	L787	BD06-OSL06	12.9	Loess	4–11	1.95		2.62 ± 0.40	13.19 ± 1.33 *	10	3.55 ± 0.17	4	228.3 ± 3.2		228.9 ± 6.6	65 ± 4
	L786	BD06-OSL05	16.00	Overbank silty clay	4–11	1.65		2.01 ± 0.35	10.89 ± 1.18 *	10	2.93 ± 0.14	3	578.0 ± 20.4		570 ± 23	195 ± 12
	L785	BD06-OSL04	22.20	Overbank silty clay	4–11	1.65		2.37 ± 0.35	10.72 ± 1.18 *	10	2.99 ± 0.14	4	590.1 ± 53.6		559 ± 26	187 ± 12
	L573	BD-OSL27	23.00	Channel sand	90–125	2.67	2.69 ± 0.10	1.00 ± 0.08	6.42 ± 0.18	5	3.27 ± 0.09	52	481.9 ± 20.6	24	456 ± 19	139 ± 7
	L572	BD-OSL26	25.7.00	Channel sand	90–125	2.35	2.45 ± 0.09	1.18 ± 0.08	11.50 ± 0.25	5	3.43 ± 0.08	25	509.3 ± 35.3	23	476 ± 26	139 ± 8
Terrace T6																
G	L559	BD-OSL13	3.10	Loess	4–11	1.72	1.78 ± 0.09	2.57 ± 0.09	9.67 ± 0.22	10	3.21 ± 0.12	5	612.8 ± 25.0		607 ± 32	189 ± 12
	L558	BD-OSL12	7.00	Paleosol	4–11	1.96	2.07 ± 0.09	2.07 ± 0.09	11.50 ± 0.25	15	3.26 ± 0.12	5	642.9 ± 38.0		609 ± 17	187 ± 8
H	L564	BD-OSL18	1.00	Loess	4–11	1.88	1.73 ± 0.10	1.88 ± 0.09	9.15 ± 0.22	10	3.00 ± 0.12	6	593.1 ± 37.1		571 ± 29	191 ± 12
	L563	BD-OSL17	5.15	Paleosol	4–11	2.04	2.23 ± 0.10	2.00 ± 0.08	9.60 ± 0.22	15	3.27 ± 0.12	6	605.1 ± 20.3		623 ± 24	191 ± 10
	L562	BD-OSL16	7.10	Channel sand	150–250	2.2	2.32 ± 0.09	0.96 ± 0.07	5.64 ± 0.16	5	2.84 ± 0.08	21	547.0 ± 28.4	23	531 ± 29	187 ± 12
	L561	BD-OSL15	9.10	Channel sand	150–250	2.2	2.10 ± 0.09	0.80 ± 0.07	4.39 ± 0.15	5	2.50 ± 0.08	25	497.6 ± 32.2	31	474 ± 31	190 ± 15

*: Determined using flame photometry; **: Determined using neutron-activation-analysis (NAA), except for samples L782–787, for these six samples whose U and Th contents were determined using the alpha counting method; the K contents obtained using NAA were used for dose rate calculation, except for samples L782–787 (for these six samples, the K contents obtained using flame photometry were used, and a relative uncertainty of 5% was assumed); OSL: optically stimulated luminescence; #: Relative uncertainty of 20% is assumed for water contents for all samples; +: OD = overdispersion; for fine grains, the OD value is not calculated. All uncertainties are reported as one sigma.

The improved single-aliquot regenerative dose procedure (SAR) [39,40] was used to measure the single-aliquot equivalent dose (De) of the quartz extracts at Peking University. The regenerative beta doses used in the SAR procedure included a zero dose used for monitoring recuperation effects and a repeat of the first regeneration dose used to check the reproducibility of the sensitivity correction (i.e., recycling ratio). Based on the results of the preheat plateau and dose recovery tests (see below), the preheat temperature was set to 220 °C or 260 °C, and the cut-heat temperature to 160 °C. OSL signals were measured for 40 s at 125 °C. In addition, a 20-s IR stimulation at room temperature before each OSL measurement was carried out to remove the possible effect of feldspar contamination [38], although IRSL signals were negligible, and a 40-s blue light stimulation at 280 °C at the end of each cycle was also carried out for reducing recuperation. The signals were analyzed using late background subtraction (the intensity of the initial 0.16 s minus a background (normalized to 0.16 s) from the last 3.2 s), and the value of De was estimated by interpolating the sensitivity-corrected natural OSL onto the dose-response curve using the Analyst software [41]. The error on individual De values was calculated using the counting statistics and an instrumental uncertainty of 1.0%.

All luminescence measurements, beta irradiation and preheat treatments were carried out in automated Risø TL/OSL (DA-15 and DA-20) readers equipped with a $^{90}Sr/^{90}Y$ beta source (the Risoe National Laboratory, Denmark Technical University, Denmark) [42]. Blue light LED (470 ± 30 nm) stimulation was used for quartz OSL measurements and an IR laser diode (830 ± 10 nm) stimulation for scanning feldspar contamination and for feldspar IRSL measurements. Luminescence was detected by an EMI 9235QA photomultiplier tube with two Hoya U-340 filters (290–370 nm) in front of it.

3.2.2. Dose Rate Determination

The uranium and thorium contents of samples L782–787 were determined using thick-source alpha counting (a Littlemore low-level alpha counter 7286 with 42-mm-diameter ZnS screens) (Littlemore Scientific, UK), and other samples were analyzed using the neutron-activation-analysis (NAA) for U, Th and K contents. The K content of all samples was also measured using flame photometry. The present-day water contents (ratio of mass of water/dry-sample [6]) of all the samples were measured in the laboratory to be 1.0%–8.4%, with an average of 3.1±0.5%, by weight. These samples were taken from the subsurface position of the sections, and they have been partly dried due to exposure to air before sampling. These values are clearly not to be representative of the long-term water contents in the natural conditions during most of the burial history. In this case, the water contents used for dose rate calculation were assumed and taken as 5% for sand, 10% for loess and silt sediments (overbank deposits) and 15% for paleosol. A relative uncertainty is taken as 20% to the long-term water content values. This large uncertainty on the water content should cover the water content fluctuations during burial. An alpha efficiency factor (*a*-value) of 0.038 ± 0.003 for quartz [43] was used to calculate the alpha contribution to the total dose rate. Based on the above measurements, the effective dose rates and ages were calculated using the online dose rate and age calculator DRAC v1.2 [44], in which cosmic ray contribution and conversion factors [45] are involved, and the alpha [46] and beta [47] grain attenuation factors were used.

4. Results

4.1. Terraces and Deposits

A total of nine exposures of fluvial sediments in the study area were found and observed. Their locations (numbered A to I) are marked in Figure 1c, and their detailed lithology are shown in Figure 2. It can be seen that the sections (exposures) are mainly composed of fluvial sediments called terrace deposits and overlying eolian sediments. The fluvial sediments consist of channel gravels and sands capping the strath surfaces and overbank silt or silty clay overlying the channel deposits. The top eolian layers composed of loess, paleosol or red clay were deposited after paleo-floodplains were completely abandoned. The bedrock surfaces (strath) of Sections B–I were presented, and their elevations range from about 13 m to 176 m above the modern river level (arl) (Figure 2). Based on the elevation of the straths, Exposures B–I are considered to represent seven terraces, from the lowest T1 terrace with the elevation of about 13 m arl and the highest T7 terrace with the elevation of 176 m arl. Accordingly, a schematic composite across-section of the terraces was generated based on the elevations of the terraces and their spatial distribution and is shown in Figure 3. Note that T0 (Section A) is referred to the high floodplain of the river. Exposures (Sections) G and H have the same elevation of strath (111 m arl) and belong to the T6 terrace. It is noted that the effect of the variations in river gradients on terrace height in the study area is negligible.

The high floodplain (T0) is composed of sand interbedded with gravel and laminated silty (Figure 2A), and the thickness of the fluvial deposits is more than 4.5 m. The top layer of this section was disturbed by human activities. The thickness of the terrace deposits, including channel facies (gravels and sands) and overbank facies (silts), for the eight sections ranges from ~10 to ~16 m. The fluvial gravels have moderate-to-high sphericity and are well-rounded. The gravel layers have a thickness ranging from 1 to 12 m, within which, crude imbrication is present locally, such as in Section G. The overbank silts overlying gravel layers are characterized by thin horizontal bedding. Some silty layers are interbedded with sand or gravel or sand lenses. The thickness of the silt layers varies from 0 to 8 m. The top loess/paleosol deposits are characterized by massive structure and vertical joints. On the highest terrace (T7, Section I), the red clay deposits were found between the top loess and fluvial sandy gravels resting on the bedrock. It is noted that the pre-Quaternary red clay deposits [48,49] are far beyond the upper limit of luminescence dating. Therefore, the T7 terrace is not sampled for OSL dating. The bedrock mainly comprises Triassic sandstone with a horizontal bedding, and the surface of bedrock is loose because of weathering. As shown in Figure 2, a total of 25 samples for OSL dating were taken from sections A to H. The information about the samples are listed in Table 1, and their positions are also shown in Figure 2.

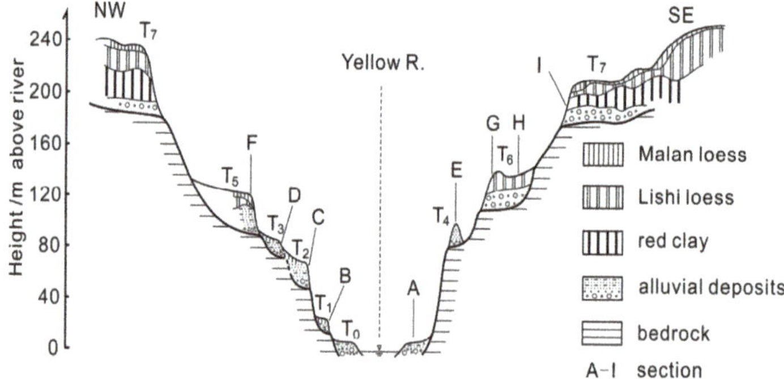

Figure 3. Schematic composite cross-section across the Jinshaan Canyon at the Baode area showing the Yellow River terrace sequence and fluvial and overlying loess/paleosol deposits, which are shown in details in Figure 2 (here, the letters refer to the number of sections in Figure 2).

4.2. OSL Ages

Two dose-response curves (DRC) and two typical OSL decay curves for a fine quartz aliquot from a loess sample from the T6 terrace and a coarse quartz aliquot from a fluvial sample from the T5 terrace are shown in Figure 4. The decay curves demonstrate that the quartz OSL signals are easily bleached and dominated by fast components. The dose response curves are well-fitted with a double-saturating exponential function or a saturating exponential plus linear function. The recycling ratios are close to unity, and the recuperation values are less than 0.5%. The DRCs for all the samples are very similar, and the natural signals are apparently not close to saturation.

Preheat plateau and dose recovery tests were carried out on two samples to find the most suitable preheat temperature in the SAR procedure for our samples. Preheat plateau tests were performed using the SAR procedure with different preheat temperatures ranging from 160 °C to 300 °C at an interval of 20 °C. The results shown in Figure 5 demonstrate that the De values are independent of preheat temperatures, at least between 220 °C and 280 °C, for both samples. Dose recovery tests were performed on the same samples to further confirm the results of the preheat plateau tests. After removing the natural OSL signals by exposing aliquots to blue light within the readers at room temperature for 40 s, the residual OSL signals were examined by a second 40-s OSL measurement ~10,000 s after bleaching, and no detectable OSL signals could be observed. The aliquots were then irradiated with a laboratory beta dose approximately equal to the natural dose (De) of the sample. This artificial dose (given dose) was then taken as unknown, and the aliquots were treated as "natural samples". After a storage of at least 10 h, the irradiated aliquots were then measured using the SAR procedure with the preheat temperatures of 160–280 °C with an interval of 20 °C for a relatively young sample (L782) from the T1 terrace and with the temperature of 260 °C for a relatively old sample (L557) from the T4 terrace. The dose recovery ratios (ratio of measured dose to given dose) were plotted as a function of the preheat temperature and shown in Figure 6. It can be seen that the average dose recovery ratios for each temperature are close to unity between the preheat temperatures of 180 °C and 280 °C for sample L782, and at 260 °C for sample L557, respectively. Based on the above results, the preheat of 220 °C for 10 s was adopted for the samples from the T1 terraces and 260 °C for 10 s for the samples from the higher terraces.

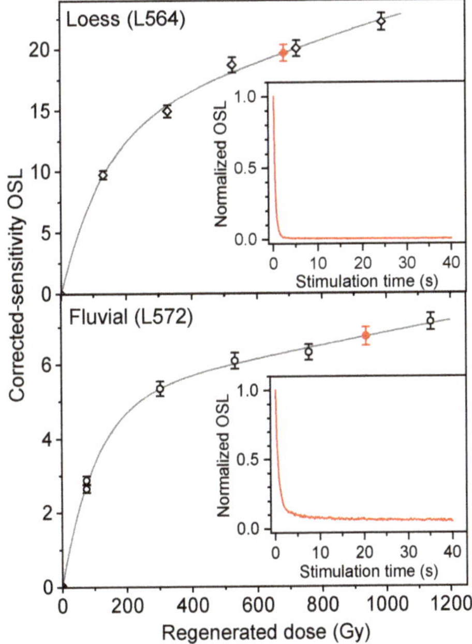

Figure 4. Dose response curves for fine quartz grains from loess sample (L564) and coarse quartz grains from fluvial sand sample (L572). The two curves are fitted by the functions of y = 13.8(1−exp(−(x + 0.139)/136) + 0.00869x and y = 3.17(1−exp(−x/83.3)) + 4.4(1−exp(−x/406)) + 0.0406, respectively. filled diamonds and circles represent the corrected sensitivity natural signals. The insets show the decay curves for the natural signals from the two quartz samples. Note that the signals were normalized to unity at the first point.

The dating results are summarized in Table 1, in which the arithmetic means and weighted means of individual De estimates are presented. The latter values and the overdispersion (OD) values are obtained using the "central age model" (CAM) of Galbraith et al. [50]. For most of the samples, their unweighted average De values are larger than their CAM values. It can be seen that the OD values for the sand samples from section A for the T0 floodplain vary from 53% to 72%, which are much larger than those (13%–31% with an average of 22.9% ± 1.5%) for the terrace samples.

Representative De distributions shown as abanico plots [51] are presented in Figure 7. Here, OSL ages were obtained by dividing the CAM De values by dose rates, and the age values are also shown in Figure 2

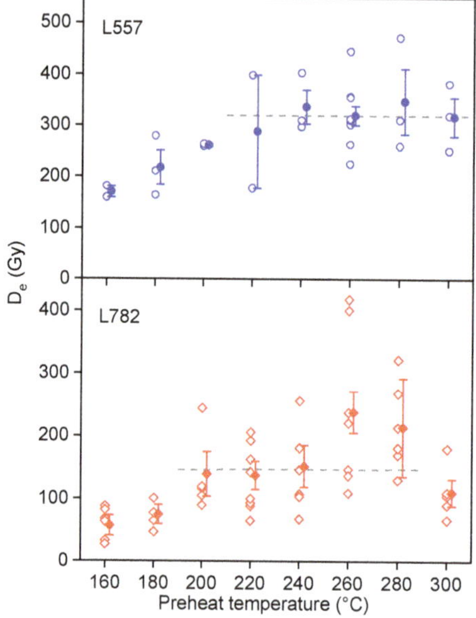

Figure 5. Dependence of De on preheat temperatures for samples L557 and L782. The open circles and diamonds represent the values of individual aliquots, and the filled circles and diamonds represent the average with their associated errors (one standard error).

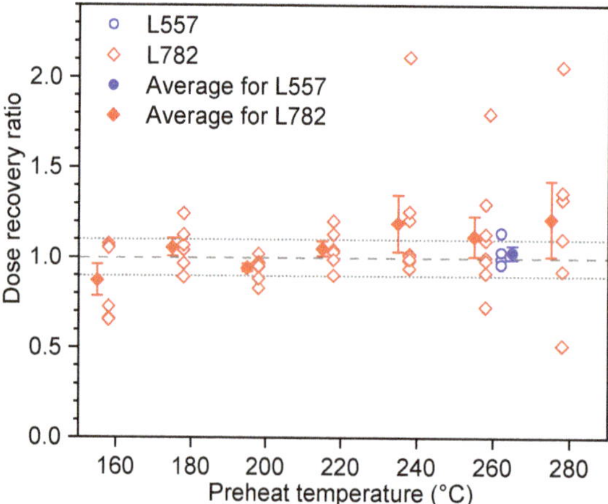

Figure 6. Plot of dose recovery ratios as a function of preheat temperatures for sample L557 and L782. The dose recovery ratios were obtained by dividing the recovery De values by the given doses (see text for details). The open circles and diamonds represent the values of individual aliquots, and the filled circles and diamonds represent the average with their associated errors (one standard error).

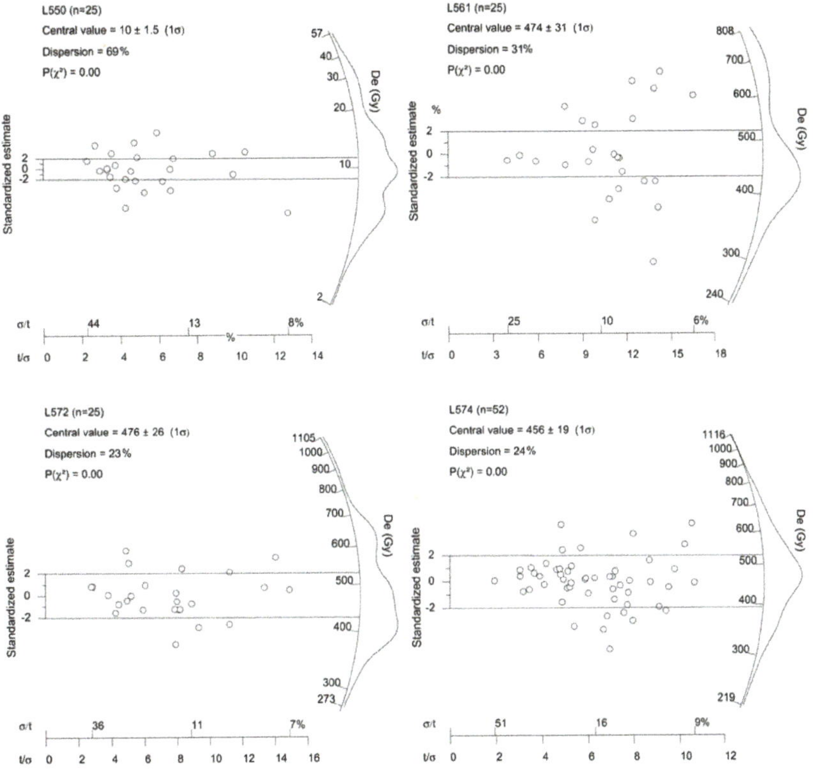

Figure 7. Abanico plots [51] showing the De distribution of four representative samples. The plots were generated using the RadialPlotter software (version 9.5) [52]. An abanico plot is the combination of both radial and kernel density estimate (KDE) plots [53]. It shows a visual correlation between De errors (radial plot) and De frequency distribution (KDE, on the z-axis of the radial plot). σ/t and t/σ refer to relative error (%) and precision, respectively.

5. Discussion

5.1. Reliability of OSL Ages

As mentioned above, the good luminescence characteristics of the studied samples for the SAR protocol indicate that the SAR protocol is suitable for our samples [40]. However, the bleaching of fluvial sediments prior to burial for young samples may be problematic and which are generally evaluated by the OD values of their single-grain or single-aliquot De distributions. The three sand samples (L549, 550 and 551) from section A have large OD values (53%, 69% and 72%, respectively), indicating a large De scatter. The comparison with the global average value of 9 ± 3% published for well-bleached large-sized aliquots [54] implies that these three samples were poorly bleached at the time of deposition. Their CAM De values are 47.4 ± 5.6, 10.0 ± 1.5 and 5.0 ± 0.8 Gy, respectively; corresponding to the OSL ages of 16.7 ± 2.0, 2.7 ± 0.4 and 1.7 ± 0.3 ka, they are in stratigraphical order. Even if we assume that the De values of 10.0 ± 15 and 5.0 ± 0.8 Gy are the residual doses of the two samples at the time of deposition, the corresponding ages of 2.7 ± 0.4 and 1.7 ± 0.3 are similar to the age errors (one sigma) of the two sand samples (L551 and 550) from the T1 terraces. Their OSL ages are 39.8 ± 2.5 and 41.9 ± 2.7 ka, respectively (Table 1). Actually, previous investigations have demonstrated that the residual dose of modern fluvial sand samples from the middle reaches of the Yellow River are only 0.1 to 2.4 Gy, and have a large scatter, with OD values up to 90% [18,20,55]. The small residual doses of the modern fluvial sand samples are also confirmed by those of modern analogues from different rivers in the world (e.g., [56–59]). These suggest that the effect of the residual dose on the old samples from the terrace deposits in this study are insignificant. It is noted that the silty clay sample (L552) from the top of section A was much overestimated based on the comparison with the OSL ages of the underlying sand samples. That some fine grains from the Yellow River were relatively poorly bleached at deposition time is supported by the residual dose of the modern samples [20,55]. We deduced that fine grains were derived from nearby loess deposits due to storm and gravitational erosion [60]. Some grains were transported as aggregates, and there is not enough time to expose to sunlight because of near-distance transport from the source areas and rapid deposition. Relative to the large OD values of the samples from the floodplain of section A, the terrace deposits have smaller OD values of 12–31%. For example, there are 52 aliquots of sample L573 measured, and the sample exhibits log-normal De distribution with the OD value of 24% (Figure 7). The relatively large OD values may be attributed to the variations in intrinsic brightness among the individual aliquots and the large De values of the samples [15,61–64] and beta microdosimetry [65]. The difference in luminescence properties between coarse grains may be attributed to the different sources of the sediments associated with the Yellow River [55,60,66,67], including some grains from the local weathered sandstone bedrock [68]. In summary, the effect of the residual dose on the terrace samples in this study can be neglected.

As shown in Table 1, the samples from the T2–T6 terraces have the De values (arithmetic mean) ranging from 228 to 643 Gy, and most of them are >320 Gy. The reliability of the De values obtained in the high-dose region of dose-response curves has been debated [69–74]. For practical purposes, the 2D0 value (characteristic saturation dose) of a dose response curve fitted with a single saturating exponential function is usually used as a criterion to evaluate the upper limit for precise age determination [40,74]. However, the D0 value has been found to be varied with the size of the maximum regeneration dose [74]. This is also the case for D01 and D02, when a double-saturating exponential function is applied [75]. In this study, the two typical dose response curves fitted with a double-saturating exponential function and a saturating exponential plus linear function, respectively, and shown in Figure 4 demonstrate that the two curves are not fully saturated at the dose of up to 1000 Gy, which are larger than the De values obtained for our samples. This means that the reliability of the OSL ages for our samples cannot be evaluated only on the basis of the shapes of the dose-response curves. Furthermore, there are no independent age controls in this study. In this case, the reliability of the OSL ages obtained for the terrace samples are assessed in terms of internal stratigraphical consistency of the OSL ages and/or their geomorphological consistency.

5.2. Terrace Ages and River Incision Rates

The formation ages of terrace treads and straths are constrained by dating the overlying loess and fluvial deposits between tread and straths and can be inferred from the OSL ages and/or the deposition rates of the loess and fluvial sediments.

T1: The two fluvial sand samples (L782 and 783) from section B for the T1 terrace were dated to 41.9 ± 2.7 and 39.8 ± 2.5 ka, respectively, and the overlying loess sample (L784) to 29.1 ± 2.1 ka (Figures 2B and 3). These ages are in stratigraphic order. The deposition rate of the silt sediments was calculated to be about 1.1 mm/a for the sediments between the two fluvial samples if errors are not included in the analysis. If this deposition rate is assumed to be constant for the whole overbank silt sediments, the formation age of the terrace tread is calculated to be ~36 ka using the deposition rate of 1.1 mm/a, and the age of the silt at the bottom is induced to be about 43 ka. If the deposition rate is also used for the gravel layer, the age of the terrace strath is inferred to be about 44 ka. The accumulation of the terrace deposits lasts about 8 ka. This is consistent with the period suggested by Weldon [76] and Pazzaglia and Brandon [77]. The tread age of ~36 ka is also constrained by the age (29.1 ± 2.1 ka) of the overlying loess sample (Figure 2B).

T2: The three sand samples (L567, 568 and 569) taken from section C for the T2 terrace (Figures 2C and 3) were OSL dated to 135 ± 10, 137 ± 8 and 134 ± 15 ka, respectively. They are consistent within errors, suggesting that the sand sediments were rapidly deposited. We then deduced that the formation age of the terrace tread is about 135 ka. If we assume that the deposition of the channel gravels was rapid or the deposition and strath carving were simultaneous, the strath age is also inferred to be about 135 ka.

T3: The two samples (L570 and 571) from sand lens within the channel gravel facies in section D for the T3 terrace were from the depths of 9.8 and 4.5 m, respectively, and their OSL ages are respectively 126 ± 9 and 122 ± 6 ka (Figures 2D and 3). The consistency in age between them indicates a higher deposition rate for this terrace. Even so, the deposition rate was calculated to be about 1.3 mm/a based on the ages and the difference in depth between the two samples, and we deduced that the ages of the tread and strath are about 119 and 129 ka, respectively. The deposition of the fluvial sediments occurred during about 10 ka.

T4: The two channel sand samples (L556 and 557) from section E for the T4 terrace (Figures 2F and 3) were determined to be 123 ± 8 and 135 ± 7 ka, respectively. They are not stratigraphically consistent if errors are excluded in the analysis but are in agreement within error limits. We thus deduce that the ages of the tread and strath may be both about 130 ka.

T5: The five samples from section F for the T5 terrace (Figures 2F and 3) were dated, and the two channel sand samples (L572 and 573) were determined to be 139 ± 8 and 139 ± 7 ka, respectively. The overlying overbank samples (L785 and 786) and the top loess sample (L787) were dated to 187 ± 12, 195 ± 12 and 65 ± 4 ka, respectively. The OSL ages of the overbank deposits are larger than those of the underlying channel sands, which can be explained by the age overestimation of the overbank deposits. This is because that the overbank deposits consist largely of reworked bedrock silty-clay pellets which were not well-bleached prior to burial. Therefore, we infer that the strath and the tread ages of this terrace are about 139 ka.

T6: A total of six OSL samples were taken from the two sections (sections G and H) for the T6 terrace (Figure 2G,H and Figure 3). In section G, paleosol immediately overlies the channel gravel layer. The paleosol sample (L558) was dated to 187 ± 8 ka and the overlying loess sample (L559) to 189 ± 12 ka. For section H at about 350 m distance from section G, the two channel sand samples (L561 and 562) were respectively dated to 190 ± 15 and 187 ± 12 ka and the overlying paleosol (L563) and loess (L564) samples to 187 ± 8 and 189 ± 12 ka, respectively. The OSL ages of the six samples are in agreement within errors. The actual difference in ages between them may be masked by their errors. In this case, we deduced that the ages of the tread and strath are about 190 ka.

As discussed above, the luminescence properties of the samples and stratigraphic consistency of the OSL ages obtained for the terraces appear that the OSL ages are reliable. In order to further evaluate the reliability of the ages, the strath ages obtained for the terraces and strath elevations above the modern river level are plotted in Figure 8. It is known that higher strath terraces are generally formed earlier than lower terraces for a river, implying that the OSL ages of samples from higher terraces should be older than the ages of lower terraces. From Figure 8, it can be seen that the formation ages of the straths for the T2, T3, T4 and T5 terraces are very similar, but their elevations increase from 48 m arl for the T2 terrace to 89 m arl for the T5 terrace. The only reasonable explanation for this situation is that the OSL ages of the fluvial samples from the T3 to T6 terraces were underestimated or regarded as the minimum ages of these terraces.

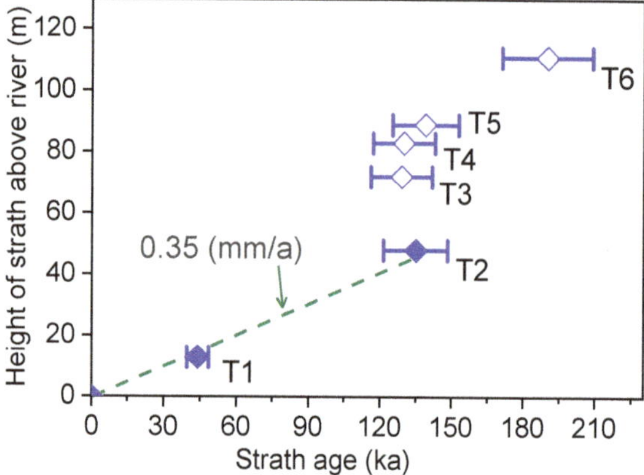

Figure 8. Plot of elevation of strath above the modern river level as a function of strath ages. The error of 10% was assumed for the inferred strath ages. The OSL ages of the T3 to T6 terraces marked as open diamonds are obviously underestimated (see text for details). The dash regression line represents a time-averaged incision rate 0.35 mm/a for the past 135 ka.

The elevation of terrace strath above a modern river level is often used to calculate the mean bedrock river incision rates [78,79], which in some cases can be used as a proxy for rock-uplift rates [1]. Although Figure 8 shows that the ages of the T3–T6 terraces were underestimated, the strath heights and ages of the T1 and T2 terraces can be used for calculating the average incision rate for the past 135 ka. The regression of the strath ages versus the elevations for the T1 and T2 terraces above the modern river level defines a mean incision rate of 0.35 mm/a for the past 135 ka (Figure 8). The incision rate for the past 44 ka was calculated to be about 0.35 mm/a by dividing the elevation by the age of the T1 terrace. The incision rate of 0.35 mm/a in this study is similar to the rates of 0.34 mm/a for the past 108 ka in the Heiyukou area [20] and 0.35 mm/ka for the past 70 ka in the Hukou area [19] (Figure 1a). This can be explained by the fact that the Ordos Plateau, an uplifted basin, behaves as a rigid block (Figure 1a). On the other hand, the incision rate also represents the uplift rates of the block.

5.3. Implication for Dating Fluvial Terraces

Fluvial terraces are morphostratigraphic units, and terrace deposits are channel sand and/or gravel facies overlain by fine overbank facies deposits bound by bottom strath and upper tread. Loess/paleosol deposits accumulate on the tread in our studied area (Figure 9). Fluvial terraces are geomorphologically classified into strath terraces and fill terraces, and the only difference between them is that there is only a thin layer of fluvial sands or gravels atop bedrock strath surfaces for strath terraces [80]. Practically, different thicknesses (h1 in Figure 9) of the "thin layer" have been used when a terrace is assigned as a strath terrace in the field. The thickness varies from about <3 m [4,77,81–83] or <10 m [2,84] or >10 m [85–87]. Pazzaglia and Brandon [77] proposed the criterion of the thickness of <3 m of coarse fluvial sediments for strath terraces and considered that this thickness represents the sediments in transport in a channel and the approximate scour depth during bankfull or larger discharges in a river the size of their studied river. This implies that the river downcutting and deposition of channel gravels for a strath terrace are simultaneous, and the thickness of channel gravels for strath terraces is associated with river size.

As shown in Figure 9, both strath age (t_s) and tread age (t_t) can be obtained for a terrace. t_s refers to the end of an interval of strath formation [88], and t_t represents the last time that overbank or channel deposition occurred on the strath. t_s is often used to calculate the river incision rate [4,77,78,89] and t_t to calculate the age of the deposit and, when displaced by tectonic, the slip rate of faulting [13,90,91] or strath incision rate [90]. As mentioned above, typical terrace deposits include channel gravels unconformably overlying bedrock strath, both generally covered by fine-grained overbank sediments. The overbank deposits may be in unconformable contact with the underlying channel gravels because of erosion; also, when gravels are exposed to air due to river downcutting. The strath ages are often inferred from the ages of the directly overlying fluvial sediments, assuming that the sediments are intimately associated with the beveling of the strath surface and considering that the sediments and the strath have correlative ages [16]. It is noted that the tread ages refer to the timing of abandonment of the tread, not the ages for the deposition of the terrace deposits. The ages can be inferred from the ages of overlying loess and overbank deposits (Figure 9) for some terraces.

Figure 9. Photograph of an exposure in the Heiyukou area showing the lithostratigraphic units of a fluvial terrace of the Yellow River (flow south (S)) and the boundaries between the units. It also shows the difference (Δ) between the formation ages (t_s and t_t) of the strath, and the tread varies from terrace to terrace. h, h1 and h2 represent, respectively, the thickness of the terrace deposits, channel gravels and overbank sediments, h = h1 + h2 (see text for details). The upper overbank facies may be in unconformable contact with underlying channel facies because of erosion or depositional hiatus. The gate (about 2-m-wide) of a cave house excavated in loess provides a scale.

Figure 2 shows that the terrace types and the relationship among the ages of strath, tread, fluvial terrace deposits and overlying loess are complicated. This is supported by those for the terrace sections at other localities along the banks of the Jinshan Canyon previously investigated [18–20]. Almost all of the terrace deposits for these Yellow River terraces are composed of channel gravels and overbank silts, and these fluvial deposits are mantled by loess/paleosol. Single layers of channel gravels for most of the terraces are >3-m-thick, even up to 20 m, and the thickness of the overbank silts varies from 0 to about 10 m. For the higher terraces (T2–T6 in Figure 3, and the corresponding sections in Figure 2C–F,H) in this study, the OSL ages of the fluvial samples from each section are older than 100 ka and consistent within errors. The consistency may be attributed to the fact that (1) the age differences between samples are less than their errors or (2) the upper age limits of OSL dating for the samples are reached. The above discussion implies that the formation ages of terrace strath and tread can be inferred based on deposition rate and thickness of terrace deposits, regardless of terrace types. Using deposition rates to infer strath ages can overcome the potential age underestimation caused by dating samples from thick-fill terraces [4,77]. For older terraces, the accumulation time of fluvial sediment atop terrace strath may be negligible relative to the formation ages of terrace strath, meaning that the ages of terrace deposits are approximately equal to the formation ages of terrace strath and tread.

All the above analyses suggest that terrace deposits should be systemically sampled for dating when dating fluvial terraces. Additionally, the ages of overlying loess deposits are often regarded as strath ages [92,93]. Sections B and F in this study show that the burial ages of loess are much younger than the underlying fluvial deposits, but other sections indicate that the ages of the two types of sediments are almost similar. This implies that whether the ages of terrace loess are approximately equal to the fluvial sediments varies from terrace to terrace. On the other hand, the age of terrace loess can at least be used to constrain the tread age. Here, it should be pointed out that the formation of terrace strath or tread should be specified when we report the formation ages of fluvial terraces.

6. Conclusions

In the Baode area in the middle reaches of the Yellow River, seven fluvial terraces (T1 to T7) were identified in the field based on the presence of fluvial sediments and the elevation of terrace straths observed on nine sediment exposures, and the thicknesses of fluvial sediments on these terraces are about 10–16 m. Twenty-five samples for OSL dating using the SAR protocol on quartz grains were collected from the terraces, except for the highest terrace (T7), whose age is beyond the upper limit of OSL dating, and they were dated to ~2–200 ka. The luminescence properties of the samples and stratigraphic consistency demonstrate that the OSL ages obtained appear to be reliable. However, the geomorphologic evidence shows that the ages (>120 ka) of the fluvial samples from the higher terraces (T3 to T6) should be underestimated, because these ages are generally similar to those of the samples from the lower terrace (T2). Based on the deposition rate of fluvial sediments atop the straths, we infer that the formation ages of both terrace strath and tread are, respectively, about 44 ka and 36 ka for the T1 terrace, and the strath and tread of the T2 terrace has the same formation age of about 135 ka. Based on our previous investigations on the Yellow River terrace and the results in this study, we argue that the formation ages of terrace straths and treads calculated using deposition rates of terrace fluvial sediments may be more accurate than those obtained from a single age of one sample. The incision rate was calculated to be about 0.35 mm/ka for the past 135 ka. This is similar to the rates for the past 108 ka in the Heiyukou area and for the past 70 ka in the Hukou area in the middle reaches of the river, implying the uplift rates of the Ordos Plateau as a rigid block.

Author Contributions: J.-F.Z., W.-L.Q and L.-P.Z. designed the research. J.-F.Z., W.-L.Q and G.H. conducted field work. J.-F. Z and G.H. performed all experiments. J.-F.Z analyzed OSL data. J.-F.Z. and W.-L.Q. wrote the first draft. J.-F.Z. revised the manuscript. All authors have read and agree to the published version of the manuscript.

Funding: The work was supported by the National Natural Science Foundation of China (No.: 41771003, 41471003 and 41171007).

Acknowledgments: We are grateful to the two anonymous reviewers and the guest editor (James K. Feathers); their valuable comments and constructive suggestions significantly improved the quality of the manuscript.

Conflicts of Interest: The authors declare no conflict of interest.

References

1. Merritts, D.J.; Vincent, K.R.; Wohl, E.E. Long river profiles, tectonism, and eustasy: A guide to interpreting fluvial terraces. *J. Geoph. Res.* **1994**, *99*, 14031–14050. [CrossRef]

2. Lavé, J.; Avouac, J.P. Fluvial incision and tectonic uplift across the Himalayas of central Nepal. *J. Geoph. Res. Solid Earth* **2001**, *106*, 26561–26591. [CrossRef]

3. Mishra, S.; White, M.J.; Beaumont, P.; Antoine, P.; Bridgland, D.R.; Limondin-Lozouet, N.; Santisteban, J.I.; Schreve, D.C.; Shaw, A.D.; Wenban-Smith, F.F.; et al. Fluvial deposits as an archive of early human activity. *Quat. Sci. Rev.* **2007**, *26*, 2996–3016. [CrossRef]

4. Wegmann, K.W.; Pazzaglia, F.J. Late Quaternary fluvial terraces of the Romagna and Marche Apennines, Italy: Climatic, lithologic, and tectonic controls on terrace genesis in an active orogen. *Quat. Sci. Rev.* **2009**, *28*, 137–165. [CrossRef]

5.   Chauhan, P.R.; Bridgland, D.R.; Moncel, M.-H.; Antoine, P.; Bahain, J.-J.; Briant, R.; Cunha, P.; Despriée, J.; Limondin-Lozouet, N.; Locht, J.-L.; et al. Fluvial deposits as an archive of early human activity: Progress during the 20 years of the Fluvial Archives Group. *Quat. Sci. Rev.* **2017**, *166*, 114–119. [CrossRef]

6.   Aitken, M.J. *An Introduction to Optical Dating: The Dating of Quaternary Sediments by the Use of Photon-stimulated Luminescence*; Oxford University Press: Oxford, UK, 1998.

7.   Rhodes, E.J. Optically stimulated luminescence dating of sediments over the past 200000 years. *Annu Rev. Earth Planet. Sci.* **2011**, *39*, 461–488. [CrossRef]

8.   Rittenour, T.M. Dates and rates of earth-surface processes revealed using Luminescence dating. *Elements* **2018**, *14*, 21–26. [CrossRef]

9.   Wallinga, J. Optically stimulated luminescence dating of fluvial deposits: A review. *Boreas* **2002**, *31*, 303–322. [CrossRef]

10. Traher, I.M.; Mauz, B.; Chiverrell, R.C.; Lang, A. Luminescence dating of glaciofluvial deposits: A review. *Earth Sci. Rev.* **2009**, *97*, 133–146.

11. Fuller, T.K.; Perg, L.A.; Willenbring, J.K.; Lepper, K. Field evidence for climate-driven changes in sediment supply leading to strath terrace formation. *Geology* **2009**, *37*, 467–470. [CrossRef]

12. Gong, Z.J.; Li, S.H.; Li, B. The evolution of a terrace sequence along the Manas River in the northern foreland basin of Tian Shan, China, as inferred from optical dating. *Geomorphology* **2014**, *213*, 201–212. [CrossRef]

13. Gong, Z.J.; Li, S.H.; Li, B. Late Quaternary faulting on the Manas and Hutubi reverse faults in the northern foreland basin of Tian Shan, China. *Earth Planet. Sci. Lett.* **2015**, *424*, 212–225. [CrossRef]

14. Fu, X.; Cohen, T.J.; Fryirs, K. Single-grain OSL dating of fluvial terraces in the upper Hunter catchment, southeastern Australia. *Quat. Geochronol.* **2019**, *49*, 115–122. [CrossRef]

15. Fu, X.; Li, S.-H.; Li, B.; Fu, B. A fluvial terrace record of late Quaternary folding rate of the Anjihai anticline in the northern piedmont of Tian Shan, China. *Geomorphology* **2017**, *278*, 91–104. [CrossRef]

16. Hancock, G.S.; Anderson, R.S. Numerical modeling of fluvial strath-terrace formations in response to oscillating climate. *Geol. Soc. Am. Bull.* **2002**, *114*, 1131–1142.

17. Foster, M.A.; Anderson, R.S.; Gray, H.J.; Mahan, S.A. Dating of river terraces along Lefthand Creek, western High Plains, Colorado, reveals punctuated incision. *Geomorphology* **2017**, *295*, 176–190. [CrossRef]

18. Zhang, J.F.; Qiu, W.L.; Li, R.Q.; Zhou, L.P. The evolution of a terrace sequence along the Yellow River (HuangHe) in Hequ, Shanxi, China, as inferred from optical dating. *Geomorphology* **2009**, *109*, 54–65. [CrossRef]

19. Guo, Y.J.; Zhang, J.F.; Qiu, W.L.; Hu, G.; Zhuang, M.G.; Zhou, L.P. Luminescence dating of the Yellow River terraces in the Hukou area, China. *Quat. Geochronol.* **2012**, *10*, 129–135. [CrossRef]

20. Yan, Y.Y.; Zhang, J.F.; Hu, G.; Zhou, L.P. Luminescence chronology of the Yellow River terraces in the Heiyukou area, China, and its implication for the uplift rate of the Ordos Plateau. *Geochronometria*. in press.

21. Cheng, S.P.; Deng, Q.D.; Zhou, S.W.; Yang, G.Z. Strath terraces of Jinshaan Canyon, Yellow River, and Quaternary tectonic movements of Ordos Plateau, North China. *Terra Nova* **2002**, *14*, 215–224. [CrossRef]

22. Pan, B.T.; Hu, Z.B.; Wang, J.P.; Vandenberghe, J.; Hu, X.F. A magnetostratigraphic record of landscape development in the eastern Ordos Plateau, China: Transition from Late Miocene and Early Pliocene stacked sedimentation to Late Pliocene and Quaternary uplift and incision by the Yellow River. *Geomorphology* **2011**, *125*, 225–238. [CrossRef]

23. Pan, B.T.; Hu, Z.B.; Wang, J.P.; Vandenberghe, J.; Hu, X.F.; Wen, Y.H.; Li, Q.; Cao, B. The approximate age of the planation surface and the incision of the Yellow River. *Palaeogeogr. Palaeoclimatol. Palaeoecol.* **2012**, *356*, 54–61. [CrossRef]

24. Hu, Z.; Pan, B.; Bridgland, D.R.; Vandenberghe, J.; Guo, L.; Fan, Y.; Westaway, R. The linking of the upper-middle and lower reaches of the Yellow River as a result of fluvial entrenchment. *Quat. Sci. Rev.* **2017**, *166*, 324–338. [CrossRef]

25. Qiu, W.L.; Zhang, J.F.; Wang, X.Y.; Guo, Y.J.; Zhuang, M.G.; Fu, X.; Zhou, L.P. The evolution of the Shiwanghe River valley in response to the Yellow River incision in the Hukou area, Shaanxi, China. *Geomorphology* **2014**, *215*, 34–44. [CrossRef]

26. Meng, Y.M.; Zhang, J.F.; Qiu, W.L.; Fu, X.; Guo, Y.J.; Zhou, L.P. Optical dating of the Yellow River terraces in the Mengjin area (China): First results. *Quat. Geochronol.* **2015**, *30*, 219–225. [CrossRef]

27. Chen, X.B.; Zang, S.X.; Wei, R.Q. Is the stable Ordos block migrating as an entire block? *Chin. J. Geophy* **2011**, *54*, 1750–1757.

28. Bao, X.W.; Xu, M.J.; Wang, L.S.; Mi, N.; Yu, D.Y.; Li, H. Lithospheric structure of the Ordos Block and its boundary areas inferred from Rayleigh wave dispersion. *Tectonophysics* **2011**, *499*, 132–141. [CrossRef]

29. Zhang, Y.Q.; Mercier, J.L.; Vergely, P. Extension in the graben systems around the Ordos (China), and its contribution to the extrusion tectonics of south China with respect to Gobi-Mongolia. *Tectonophysics* **1998**, *285*, 41–75. [CrossRef]

30. Zhang, Y.Q.; Liao, C.Z.; Shi, W.; Hu, B. Neotectonic evolution of the peripheral zones of the Ordos Basin and geodynamic setting. *Geol. J. China Univ.* **2006**, *12*, 285–297, (In Chinese with English abstract).

31. Deng, G.D.; Cheng, S.P.; Min, W.; Yang, G.Z.; Reng, D.W. Discussion on Cenozoic tectonics and dynamics of Ordos block. *J. Geom.* **1999**, *5*, 13–21. (In Chinese)

32. Lin, A.M.; Yang, Z.Y.; Sun, Z.M.; Yang, T.S. How and when did the Yellow River develop its square bend? *Geology* **2001**, *29*, 951–954. [CrossRef]

33. Darby, B.J.; Ritts, B.D. Mesozoic contractional deformation in the middle of the Asian tectonic collage: The intraplate Western Ordos fold–thrust belt, China. *Earth Planet. Sci. Lett.* **2002**, *205*, 13–24. [CrossRef]

34. Yang, M.H.; Liu, C.Y.; Zheng, M.L.; Lan, C.L.; Tang, X. Sequence framework of two different kinds of margins and their response to tectonic activity during the Middle-Late Triassic, Ordos Basin. *Sci. China Ser. D* **2007**, *50*, 203–216. [CrossRef]

35. Yue, L.; Li, J.; Zheng, G.; Li, Z. Evolution of the Ordos Plateau and environmental effects. *Sci. China Ser. D* **2007**, *50*, 19–26. [CrossRef]

36. Zhang, Y.Q.; Ma, Y.S.; Yang, N.; Shi, W.; Dong, S.W. Cenozoic extensional stress evolution in North China. *J. Geodyn.* **2003**, *36*, 591–613.

37. Zhang, J.F.; Zhou, L.P.; Yue, S.Y. Dating fluvial sediments by optically stimulated luminescence: Selection of equivalent doses for age calculation. *Quat. Sci. Rev.* **2003**, *22*, 1123–1129. [CrossRef]

38. Zhang, J.F.; Zhou, L.P. Optimization of the 'double SAR' procedure for polymineral fine grains. *Radiat. Meas.* **2007**, *42*, 1475–1482. [CrossRef]

39. Murray, A.S.; Wintle, A.G. Luminescence dating of quartz using an improved single-aliquot regenerative-dose protocol. *Radiat. Meas.* **2000**, *32*, 57–73. [CrossRef]

40. Wintle, A.G.; Murray, A.S. A review of quartz optically stimulated luminescence characteristics and their relevance in single-aliquot regeneration dating protocols. *Radiat. Meas.* **2006**, *41*, 369–391. [CrossRef]

41. Duller, G.A.T. Assessing the error on equivalent dose estimates derived from single aliquot regenerative dose measurements. *Anc. TL* **2007**, *25*, 15–24.

42. Bøtter-Jensen, L.; Bulur, E.; Duller, G.A.T.; Murray, A.S. Advances in luminescence instrument systems. *Radiat. Meas.* **2000**, *32*, 523–528. [CrossRef]

43. Rees-Jones, J. Optical dating of young sediments using fine-grain quartz. *Anc. TL* **1995**, *13*, 9–14.

44. Durcan, J.A.; King, G.E.; Duller, G.A.T. DRAC: Dose rate and age calculator for trapped charge dating. *Quat. Geochronol.* **2015**, *28*, 54–61. [CrossRef]

45. Guerin, G.; Mercier, N.; Adamiec, G. Dose-rate conversion factors: Update. *Anc. TL* **2011**, *29*, 5–8.

46. Brennan, B.J.; Lyons, R.G.; Phillips, S.W. Attenuation of alpha particle track dose for spherical grains. *Int. J. Radiat. Appl. Instrum. Part D* **1991**, *18*, 249–253. [CrossRef]

47. Guerin, G.; Mercier, N.; Nathan, R.; Adamiec, C.; Lefrais, Y. On the use of the infinite matrix assumption and associated concepts: A critical review. *Radiat. Meas.* **2012**, *47*, 778–785. [CrossRef]

48. Ding, Z.L.; Sun, J.M.; Liu, T.S.; Zhu, R.X.; Yang, S.L.; Guo, B. Wind-blown origin of the Pliocene red-clay formation in the central Loess Plateau, China. *Earth Planet. Sci. Lett.* **1998**, *161*, 135–143. [CrossRef]

49. Zhu, Y.M.; Zhou, L.P.; Mo, D.W.; Kaakinen, A.; Zhang, Z.Q.; Fortelius, M. A new magnetostratigraphic framework for late Neogene Hipparion Red Clay in the eastern Loess Plateau of China. *Palaeogeogr. Palaeocl. Palaeoecol.* **2008**, *268*, 47–57. [CrossRef]

50. Galbraith, R.F.; Roberts, R.G.; Laslett, G.M.; Yoshida, H.; Olley, J.M. Optical dating of single and multiple grains of quartz from Jinmium rock shelter, Northern Australia: Part I, experimental design and statistical models. *Archaeometry* **1999**, *41*, 339–364. [CrossRef]

51. Dietze, M.; Kreutzer, S.; Burow, C.; Fuchs, M.C.; Fischer, M.; Schmidt, C. The abanico plot: Visualising chronometric data with individual standard errors. *Quat. Geochronol.* **2016**, *31*, 12–18. [CrossRef]

52. Vermeesch, P. RadialPlotter: A Java application for fission track, luminescence and other radial plots. *Radiat. Meas.* **2009**, *44*, 409–410. [CrossRef]

53. Galraith, R.F.; Green, P.F. Estimating the component ages in a finite mixture. *Nucl. Tracks Radiat. Meas.* **1990**, *17*, 197–206. [CrossRef]
54. Arnold, L.J.; Bailey, R.M.; Tucker, G.E. Statistical treatment of fluvial dose distributions from southern Colorado arroyo deposits. *Quat. Geochronol.* **2007**, *2*, 162–167. [CrossRef]
55. Hu, G.; Zhang, J.F.; Qiu, W.L.; Zhou, L.P. Residual OSL signals in modern fluvial sediments from the Yellow River (HuangHe) and the implications for dating young sediments. *Quat. Geochronol.* **2010**, *5*, 187–193. [CrossRef]
56. Alexanderson, H. Residual OSL signals from modern Greenlandic river sediments. *Geochronometria* **2007**, *26*, 1–9. [CrossRef]
57. Jain, M.; Murray, A.S.; Bøtter-Jensen, L. Optically stimulated luminescence dating: How significant is incomplete light exposure in fluvial environments. *Quaternaire* **2004**, *15*, 143–157. [CrossRef]
58. Toth, O.; Sipos, G.; Kiss, T.; Bartyik, T. Variation of OSL residual doses in terms of coarse and fine grain modern sediments along the Hungarian section of the Danube. *Geochronometria* **2017**, *44*, 319–330. [CrossRef]
59. Chamberlain, E.L.; Walling, J. Seeking enlightenment of fluvial sediment pathways by optically stimulated luminescence signal bleaching of river sediments and deltaic deposits. *Earth Surf. Dynam.* **2019**, *7*, 723–736. [CrossRef]
60. Xu, J.X.; Yang, J.S.; Yan, Y.X. Erosion and sediment yields as influenced by coupled eolian and fluvial processes: The Yellow River, China. *Geomorphology* **2006**, *73*, 1–15. [CrossRef]
61. Duller, G.A. Single-grain optical dating of Quaternary sediments: Why aliquot size matters in luminescence dating. *Boreas* **2008**, *37*, 589–612. [CrossRef]
62. Thomsen, K.J.; Murray, A.S.; Jain, M. The dose dependency of the over- dispersion of quartz OSL single grain dose distributions. *Radiat. Meas.* **2012**, *47*, 732–739. [CrossRef]
63. Reimann, T.; Lindhorst, S.; Thomsen, K.J.; Murray, A.S.; Frechen, M. OSL dating of mixed coastal sediment (Sylt, German Bight, North Sea). *Quat. Geochronol.* **2012**, *11*, 52–67. [CrossRef]
64. Rui, X.; Li, B.; Guo, Y.J.; Zhang, J.F.; Yuan, B.Y.; Xie, F. Variability in the thermal stability of OSL signal of single-grain quartz from the Nihewan Basin, North China. *Quat. Geochronol.* **2019**, *49*, 25–30. [CrossRef]
65. Jankowski, N.R.; Jacobs, Z. Beta dose variability and its spatial contextualisation in samples used for optical dating: An empirical approach to examining beta microdosimetry. *Quat. Geochronol.* **2018**, *44*, 23–37. [CrossRef]
66. Zhang, J.F.; Qiu, W.L.; Wang, X.Q.; Hu, G.; Li, R.-Q.; Zhou, L.-P. Optical dating of a hyperconcentrated flow deposit on a Yellow River terrace in Hukou, Shaanxi, China. *Quat. Geochronol.* **2010**, *5*, 194–199. [CrossRef]
67. Zhang, J.F.; Wang, X.Q.; Qiu, W.L.; Shelach, G.; Hu, G.; Fu, X.; Zhuang, M.G.; Zhou, L.P. The paleolithic site of Longwangchan in the middle Yellow River, China: Chronology, paleoenvironment and implications. *J. Archaeol. Sci.* **2011**, *38*, 1537–1550. [CrossRef]
68. Sawakuchi, A.O.; Jain, M.; Mineli, T.D.; Nogueira, L.; Bertassoli, D.J.; Häggi, C.; Sawakuchi, H.O.; Pupim, F.N.; Grohmann, C.H.; Chiessi, C.M.; et al. Luminescence of quartz and feldspar fingerprints provenance and correlates with the source area denudation in the Amazon River basin. *Earth Planet. Sci. Lett.* **2018**, *492*, 152–162. [CrossRef]
69. Murray, A.; Buylaert, J.P.; Henriksen, M.; Svendsen, J.I.; Mangerud, J. Testing the reliability of quartz OSL ages beyond the Eemian. *Radiat. Meas.* **2008**, *43*, 776–780. [CrossRef]
70. Lai, Z.P. Chronology and the upper dating limit for loess samples from Luochuan section in the Chinese Loess Plateau using quartz OSL SAR protocol. *J. Asian Earth Sci.* **2010**, *37*, 176–185. [CrossRef]
71. Lowick, S.E.; Buechi, M.W.; Gaar, D.; Graf, H.R.; Preusser, F. Luminescence dating of Middle Pleistocene proglacial deposits from northern Switzerland: Methodological aspects and statigraphical conclusions conclusions. *Boreas* **2015**, *44*, 459–482. [CrossRef]
72. Zhang, J.F.; Huang, W.W.; Hu, Y.; Yang, S.X.; Zhou, L.P. Optical dating of flowstone and silty carbonate-rich sediments from Panxian Dadong Cave, Guizhou, southwestern China. *Quat. Geochronol.* **2015**, *30*, 479–486. [CrossRef]
73. Timar-Gabor, A.; Buylaertm, J.P.; Guralnik, B.; Trandafir-Antohi, O.; Constantin, D.; Anechitei-Deacu, V.; Jain, M.; Murray, A.S.; Porat, N.; Hao, Q.; et al. On the importance of grain size in luminescence dating using quartz. *Radiat. Meas.* **2017**, *106*, 464–471. [CrossRef]
74. Buechi, M.W.; Lowick, S.E.; Anselmetti, F.S. Luminescence dating of glaciolacustrine silt in overdeepened basin fills beyond the last interglacial. *Quat. Geochronol.* **2017**, *37*, 55–67. [CrossRef]

75. Anechitei-Deacu, V.; Timar-Gabor, A.; Constantinm, D.; Trandafir-Antohim, O.; Del Valle, L.; Fornos, J.J.; Gomez-Pujol, L.; Wintle, A.G. Assessing the maximum limit of SAR-OSL dating using quartz of different grain sizes. *Geochronometria* **2018**, *45*, 146–159. [CrossRef]

76. Weldon, R.J. Late Cenozoic Geology of Cajon Pass; Implications for Tectonics and Sedimentation Along the San Andreas Fault. Ph.D.'s Thesis, California Institute of Technology, Pasadena, CA, USA, 1986; p. 1986.

77. Pazzaglia, F.J.; Brandon, M.T. A fluvial record of long-term steady-state uplift and erosion across the Cascadia forearc high, western Washington State. *Am. J. Sci.* **2001**, *301*, 385–431. [CrossRef]

78. Maddy, D. Uplift-driven valley incision and river terrace formation in southern England. *J. Quat. Sci.* **1997**, *12*, 539–545. [CrossRef]

79. Mady, D.; Bridgland, D.R.; Green, C.P. Crustal uplift in southern England: Evidence from the river terrace records. *Geomorphology* **2000**, *33*, 167–181. [CrossRef]

80. Bull, W.B. *Geomorphic Responses to Climate Change*; Oxford Univ. Press: Oxford, UK, 1991.

81. Wegmann, K.W.; Pazzaglia, F.J. Holocene strath terraces, climate change, and active tectonics: The Clearwater River basin, Olympic Peninsula, Washington State. *Geol. Soc. Am. Bull.* **2002**, *114*, 731–744. [CrossRef]

82. Wilson, L.F.; Pazzaglia, F.J.; Anastasio, D.J. A fluvial record of active fault-propagation folding, Salsomaggiore anticline, northern Apennines, Italy. *J. Geoph. Res.* **2009**. [CrossRef]

83. Hedrick, K.; Owen, L.A.; Rockwell, T.K.; Meigs, A.; Costa, C.; Caffee, M.W.; Masana, E.; Ahumada, E. Timing and nature of alluvial fan and strath terrace formation in the Eastern Precordillera of Argentina. *Quat. Sci. Rev.* **2013**, *80*, 143–168. [CrossRef]

84. Yildirim, C.; Schildgen, T.F.; Echtler, H.; Melnick, D.; Bookhagen, B.; Çiner, A.; Niedermann, S.; Merchel, S.; Martschini, M.; Steier, P.; et al. Tectonic implications of fluvial incision and pediment deformation at the northern margin of the Central Anatolia Plateau based on multiple cosmogenic nuclides. *Tectonics* **2013**, *32*, 1107–1120. [CrossRef]

85. Gold, R.D.; Cowgill, E.; Wang, X.F.; Chen, X.H. Application of trishear fault-propagation folding to active reverse faults: Examples from the Dalong fault, Gansu Province, NW China. *J. Struct. Geol.* **2006**, *28*, 200–219. [CrossRef]

86. Harkins, N.; Kirby, E.; Heimsath, A.; Robinson, R.; Reiser, U. Transient fluvial incision in the headwaters of the Yellow River, northeastern Tibet, China. *J. Geoph. Res.* **2007**, *112*, F03S04. [CrossRef]

87. Craddock, W.H.; Kirby, E.; Harkins, N.W.; Zhang, H.P.; Shi, X.H.; Liu, J.H. Rapid fluvial incision along the Yellow River during headward basin integration. *Nat. Geosci.* **2010**, *3*, 209–213. [CrossRef]

88. Bull, W.B. *Tectonic Geomorphology of Mountains: A New Approach to Paleoseismology*; Wiley-Blackwell: Oxford, UK, 2007.

89. Burbank, D.W.; Leland, J.; Fielding, E.; Anderson, R.S.; Brozovic, N.; Reid, M.R.; Duncan, C. Bedrock incision, rock uplift and threshold hillslopes in the northwestern Himalayas. *Nature* **1996**, *379*, 505–510. [CrossRef]

90. Mason, D.P.M.; Little, T.A.; Van Dissen, R.J. Rates of active faulting during the late Quaternary fluvial terrace formation at Saxton River, Awatere fault, New Zealand. *Geol. Soc. Am. Bull.* **2006**, *118*, 1431–1446. [CrossRef]

91. Cowgill, E. Impact of riser reconstructions on estimation of secular variation in rates of strike-slip faulting: Revisiting the Cherchen River site along the Altyn Tagh Fault, NW China. *Earth Planet. Sci. Lett.* **2007**, *254*, 239–255. [CrossRef]

92. Pan, B.T.; Wang, J.P.; Gao, H.S.; Guan, Q.Y.; Wang, Y.; Su, H.; Li, B.Y.; Li, J.J. Paleomagnetic dating of the topmost terrace in Kouma, Henan and its indication to the Yellow River's running through Sanmen Gorges. *Chin. Sci. Bull.* **2005**, *50*, 657–664. [CrossRef]

93. Pan, B.; Su, H.; Hu, Z.; Hu, X.; Gao, H.; Li, J.; Kirby, E. Evaluating the role of climate and tectonics during non-steady incision of the Yellow River: Evidence from a 1.24 Ma terrace record near Lanzhou, China. *Quat. Sci. Rev.* **2009**, *28*, 3281–3290. [CrossRef]

MDPI

St. Alban-Anlage 66

4052 Basel

Switzerland

Tel. +41 61 683 77 34

Fax +41 61 302 89 18

www.mdpi.com

Methods and Protocols Editorial Office

E-mail: mps@mdpi.com

www.mdpi.com/journal/mps